Neritic Carbonate Sediments in a Temperate Realm

Noel P. James • Yvonne Bone

Neritic Carbonate Sediments in a Temperate Realm

Southern Australia

 Springer

Dr. Noel P. James
Department of Geological Sciences
and Geological Engineering,
Queen's University, Kingston K7L 3N6,
Ontario, Canada
james@geol.queensu.ca

Dr. Yvonne Bone
School of Earth & Environmental Sciences,
University of Adelaide, 5005 Adelaide,
South Australia, Australia
yvonne.bone@adelaide.edu.au

ISBN: 978-90-481-9288-5   e-ISBN: 978-90-481-9289-2
DOI: 10.1007/978-90-481-9289-2
Springer Dordrecht Heidelberg London New York

Library of Congress Control Number: 2010934295

Cover illustration: Force 6, the Southern Ocean, Great Australian Bight, October 1998 from RV *JOIDES
Resolution*. The seafloor beneath these cold, stormy waters is the site of prolific temperate water carbonate
sediment production and accumulation.

Cover design: deblik, Berlin

Printed on acid-free paper

Springer is part of Springer Science+Business Media (www.springer.com)

To
Vic Gostin & Chris von der Borch

The Pioneers

Preface

Carbonate sediments deposited on shelves and ramps across the globe and in the geological record have traditionally been viewed as tropical, warm-water deposits (Bathurst 1975; Wilson 1975; Tucker and Wright 1990; James and Kendall 1992). Although it has been recognized for more than 50 years that carbonate sediments do accumulate in cool-water temperate and cold, polar environments (Chave 1952), it is only in the last several decades that these sediments have been studied seriously in the modern ocean (Nelson 1988a; James 1997; Pedley and Carannante 2006). This relative neglect is largely because they occur in environments that are difficult to document. The mid latitudes are stormy and the waters are cool but above all the shelves are mostly deep and so not amenable to research using SCUBA. The system must, as a result, be studied by remote sensing, chiefly through shipborne sampling, acoustic profiling, towed imaging and tethered water characterization. Scientific appreciation of the temperate carbonate depositional realm has thus lagged behind our understanding of the warm-water tropical environment. A direct consequence of this knowledge gap is that actualistic cool-water depositional models are not being routinely considered when interpreting the older rock record.

The Australian continent with its old, topographically subdued landscape, has a continental shelf that is almost entirely covered with carbonate sediment. The southern part of the continental shelf is the largest area of temperate, cool-water carbonate deposition in the modern world. Sediments in this vast southern region, in environments ranging from paralic to deep sea, have been examined by a variety of workers but the resultant information is scattered throughout the scientific literature (von der Borch et al. 1970; Wass et al. 1970; Belperio et al. 1988; James et al. 1992; Boreen et al. 1993; James et al. 1994; James et al. 1997; James et al. 2001; James et al. 2008), or presented as short, general summaries in special publications (James and Clarke 1997) and textbooks (Tucker and Wright 1990).

The purpose of this volume is to amalgamate and synthesize most of this information in one place, utilizing the studies of others, our own surveys, and unpublished data, to arrive at an overall synthesis of this critical region. The focus is on the continental shelf and its deposits. It is designed to serve as (1) a core of information for modern environmental studies, (2) a springboard for future marine geological research, and (3) a solid foundation upon which to build sedimentary facies and sequence stratigraphic models that are applicable to the interpretation of the older rock record.

This research has been funded by grant agencies from two countries, specifically the Natural Sciences and Engineering Research Council of Canada (NPJ), the Australian Research Council (YB), the Commonwealth Scientific and Industrial Research

Organization Division of Oceanography ship funding program, Geoscience Australia, and the University of Adelaide.

This science would not have been possible without the ceaseless efforts of officers and crews of CSIRO and Geoscience Australia vessels often under extremely difficult conditions. We are particularly grateful to Captain Neil Cheshire who skillfully guided us through many trying times and raging seas.

Our colleagues at sea and in the laboratory, Tom Boreen, Lindsay Collins, David Feary, Vic Gostin, Steve Hageman, Lisa Hobbs, Kurt Kyser, Jeff Lukasik, John Marshall, and Chris von der Borch are all silent partners in this endeavour.

The research at sea would have been impossible without the tireless efforts of Tony Belperio, Phil Bock, Kirsty Brown, Frank Brunton, Ric Daniels, Vicky Drapala, Margaret Fuller, Kieth Gaard, Paul Gammon, Karen Gowlett-Holmes, Graham Heinson, Alexandra Isern, Andrew Levings, Bobby Rice, Sam Ryan, Paul Scrutton, Rolf Schmidt, and Tony White. The exacting laboratory analyses were carefully performed by Christina Bruce, Elizabeth Campbell, Morag Coyne, Alexandra Der, Christa Kobernick, Heather Macdonald, and Rowan Martindale. Special thanks go to Isabelle Malcolm whose attention to detail, editing, analysis, and photographic skills helped greatly during the final stages of book production.

We are indebted to Peter Davies, Qianyu Li, Brian McGowran, Paul Taylor, and John Rivers for continuing discussions about our interpretations. Seafloor images from NW Tasmania were acquired with the help of Alan Williams and Bruce Barker.

The original manuscript was kindly read and criticized by Vic Gostin, Brian Jones, Andrew Levings, and John Middleton, to whom we are very grateful for their careful, insightful, and helpful suggestions.

Contents

Chapter 1
Introduction

1.1 Scientific Approach

Global carbonate sedimentation is partitioned into discrete marine realms whose character is determined by seawater temperature (Fig. 1.1). The latitude 25°S, which bisects the Australian continent, is the boundary between tropical, warm-water deposits in the north (e.g. the Great Barrier Reef) and temperate, cool-water sediments in the south (Fig. 1.2). Southern Australia is a classic area of shallow and marginal marine carbonate sedimentation (see review in Gostin et al. 1988). As such, it is one of a suite of modern settings that has been repeatedly utilized, along with Florida, the Bahamas, Caribbean islands, Pacific atolls, the Persian Gulf, and western Australia, as modern analogues for the interpretation of the carbonate rock record throughout geologic time. Thus, southern Australia is a touchstone for those scientists wishing to understand how carbonate sedimentation takes place today but also how it took place throughout geologic history. What has been lacking to date is a synthesis of the modern neritic carbonate sedimentary system that lies offshore from the well-studied marginal marine settings of southern Australia.

The purpose of this book is to document and interpret the origin, distribution, and diagenesis of surficial sediments on this immense cool-water carbonate shelf. It is perhaps useful to recall that the length of this environment is the same as the distance between New York to San Francisco or from the English Channel to the Caspian Sea.

The deposits across this shelf are, to a first degree, the products of modern oceanography, the Pleistocene prehistory of the region, and the organisms that produce the sediment. The first part of the book, comprising four chapters, is devoted to each of these aspects. In order to place the resultant neritic carbonates in a holistic context it is also necessary to describe the numerous coeval marginal marine depositional systems, most of which have been documented by others and are here the subject of a separate following chapter. The core of the book is, however, the chapters on neritic facies and depositional environments, many of which are unique to this continental margin. Yet, despite the universality of these facies and environments, each segment of the shelf has a unique suite of attributes. Thus, later chapters are devoted to analyzing the three major sectors of the southern Australian margin. Although the focus is on deposition, the sediments do not enter the rock record as simple biogenic particles, they undergo profound alteration on and just below the modern sea floor; early diagenesis plays an important part in this system and so a separate chapter is devoted to such matters. Finally, all of these aspects are assessed and discussed, both in terms of the modern sedimentary system and the applicability of our findings to global carbonate sedimentation.

1.2 Scope

The character of the sediments is interpreted in light of our current perception of the modern and late Quaternary biota, climate, oceanography, and geohistory of the area. The vast, latitude-parallel continental margin described herein extends some 4000 km from Cape Leeuwin, Western Australia to South West Cape at the southern tip of Tasmania (Fig. 1.3a). Over this distance there are a myriad of marine environments, each of which has a distinctive array of organisms and carbonate deposits. At the broadest scale, it is made up of a Southwestern

N. P. James, Y. Bone, *Neritic Carbonate Sediments in a Temperate Realm,*
DOI 10.1007/978-90-481-9289-2_1, © Springer Science+Business Media B.V. 2011

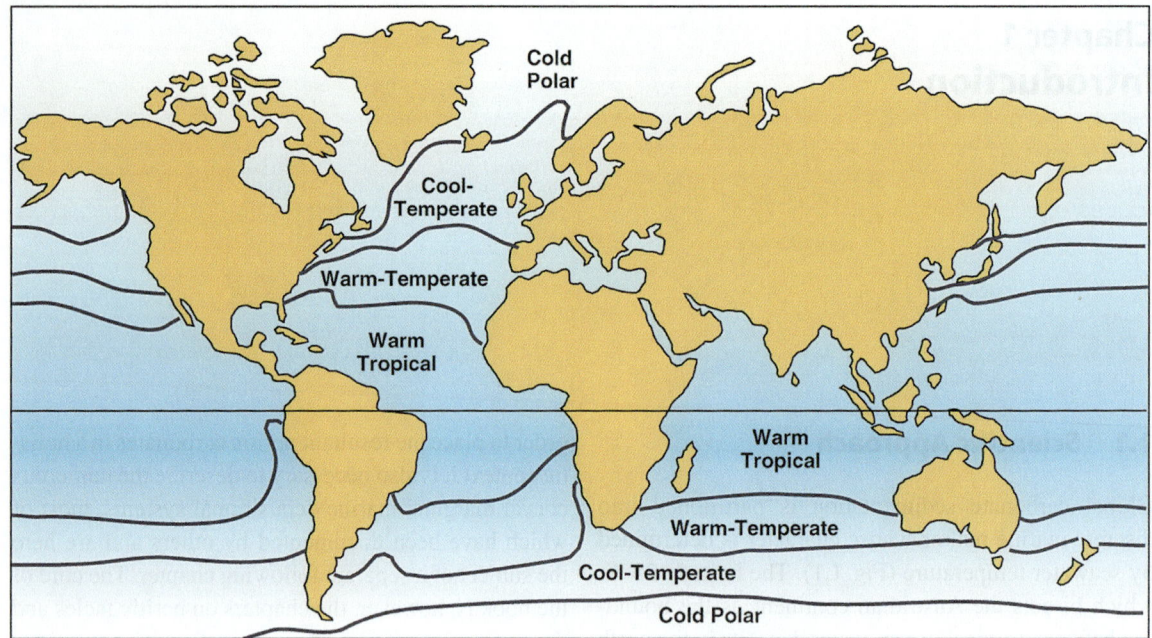

Fig. 1.1 A global map of average surface seawater isotherms delineating different modern carbonate depositional realms. The cool-water realm is differentiated into warm-temperate (15–20°C) and cool-temperate (5–15°C). (After James and Lukasik 2010)

Continental Margin and a Southeastern Continental Margin that pass from one to the other across a complex region of islands and large embayments called the South Australian Sea (Bye 1976) (Fig. 1.3b).

The Southwestern Continental Margin extends from Cape Leeuwin to southern Eyre Peninsula and includes the somewhat narrow Albany Shelf in the very west, but is dominated by the extensive Great Australian Bight. The Bight is, for purposes of documentation, divided into, from west to east, the Baxter, Eyre, and Ceduna sectors. The South Australian Sea is made up of Spencer Gulf, Gulf St. Vincent, Investigator Strait,

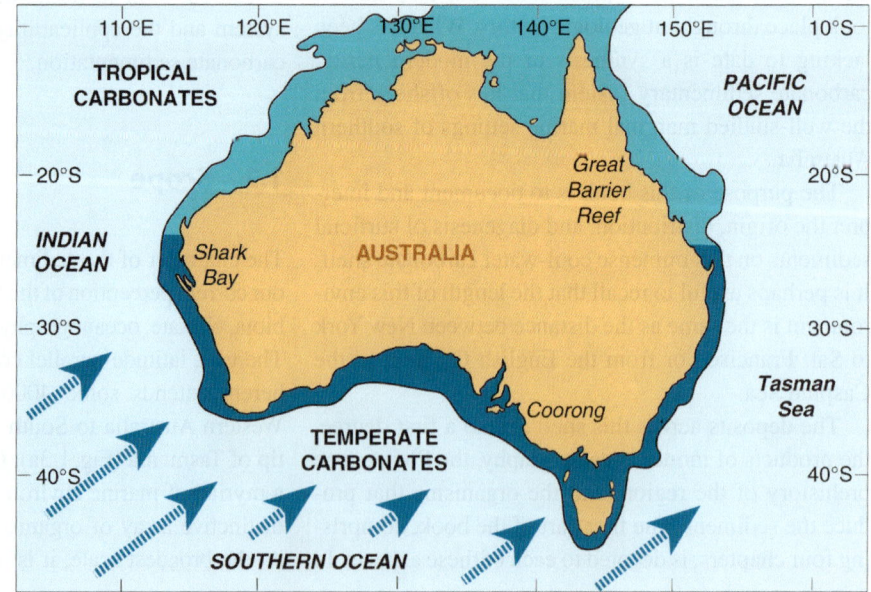

Fig. 1.2 A map of Australia illustrating the surrounding oceans, different neritic carbonate depositional realms (separated by the ~25°S latitude) and the direction of main wave approach (*dashed arrows*). Shark Bay, the Great Barrier Reef, and the Coorong are well known and intensively studied regions of carbonate deposition

Fig. 1.3 (**a**) A map of Australia and surrounding oceans illustrating the states, major cities, and locations along the southern coast that are noted in this book. (**b**) Map of Australia showing the major sectors of the southern Australian continental shelf discussed in this book. Image courtesy of Geoscience Australia

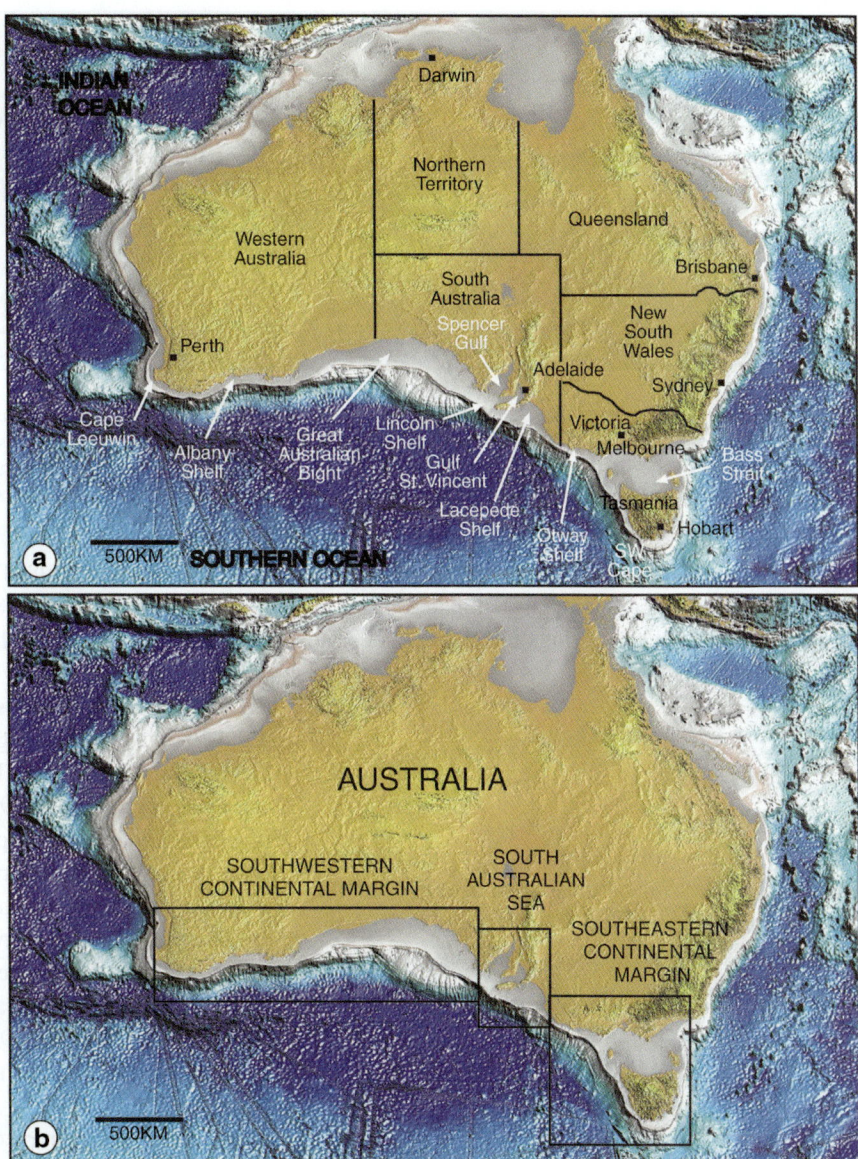

the Lincoln Shelf, and the expansive Lacepede Shelf. The relatively narrow Southeastern Continental Margin fringes the west coast of Victoria as the Otway Shelf and swings southward along Bass Strait to flank Tasmania as the Western Tasmania Shelf. The scope does not specifically include the continental margin off Western Australia that faces the Indian Ocean or the eastern continental margin off eastern Tasmania, eastern Victoria and New South Wales that face the Tasman Sea.

The environments of sediment accumulation although numerous and varied, can be detailed and integrated under a few broad headings. At the core of this system is the subtidal, open marine seafloor and overlying water column. It is in this neritic sediment factory that stretches from the shoreface to the mid-slope, that most of the carbonate sediment is produced and redistributed. In spite of its largely latitude-parallel orientation, the shelf traverses nearly 15° of latitude, from ~31.5° to ~46° S, and ranges from warm temperate in the west to cool temperate in the east. This sediment factory is also active in the large gulfs and embayments, but because of local oceanographic and climatic constraints it is somewhat different from the open shelf factory. The important marginal marine settings range from spectacular high-energy cliffs to

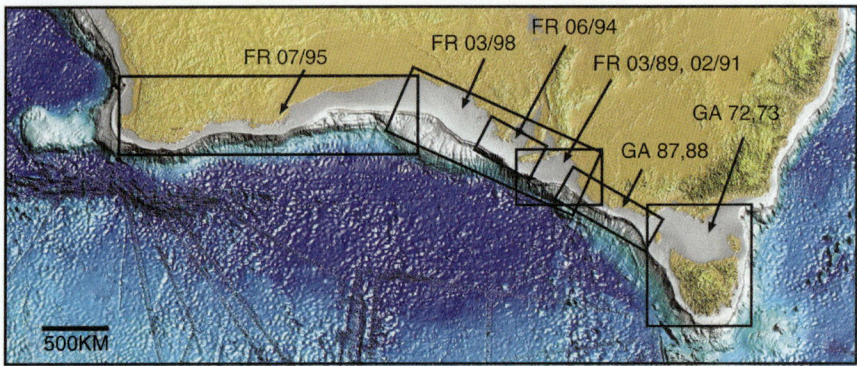

Fig. 1.4 A map of southern Australia. Research cruises (no. = month per year), mainly aboard *RV Franklin*, that enabled complete coverage of the southern continental margin of Australia. Some cruises overlapped previous areas so that seasonal changes in distribution and diversity of the major carbonate producers could be assessed. It also allowed sampling of sites that were not sampled due to weather extremes on previous cruises. Image courtesy of Geoscience Australia

protected muddy tidal flats to strandline dunes and associated saline lakes to fluvial-dominated beaches. Sediment in all of these settings undergoes moderate to significant diagenesis.

1.3 Data Base

The information in this volume comes from documentation of a series of local areas by ourselves, our students, and others over a period of more than 20 years. Research was mainly undertaken during a series of research cruises (Fig. 1.4). The open shelf environment has been documented by James et al. (1992), Boreen et al. (1993), James et al. (1997, 2001, 2008). The slope and associated mounds were described in Passlow (1997), James et al. (2004), von der Borch and Hughes-Clarke (1993), James et al. (2004), Hill et al. (2005), and Exon et al. (2005). The large embayments and their associated marginal facies have been intensively studied (see review for South Australia in Belperio 1995). More terrestrial settings, especially the saline lakes, are also reviewed in Belperio (1995) and Last (1992).

1.4 Data Acquisition and Methodology

The synthesis is principally based on analysis of seafloor sediment samples obtained using CSIRO *RV Franklin* (Fig. 1.5a). These research cruises took place between 1989 and 1998. The first expeditions were to the Lacepede Shelf in 1989 and again in 1991, with the latter focusing on the shelf margin and upper slope. This work was followed by a cruise to the Lincoln Shelf and southeastern Great Australian Bight in 1994. Work in the Great Australian Bight proper began in 1995 by documenting the western half of this huge area. This study was augmented in 1998 by a cruise that collected information from the Lacepede and Lincoln shelves but focused on the eastern part of the Bight. We have not collected samples from the eastern continental shelf ourselves but have generously been given material from the Tasmania continental margin and the Otway Shelf by Geoscience Australia who acquired sediments there in 1971–1972 and 1987–1988 respectively.

Research is based on a total of 1096 sediment samples (Table 1.1). Most of our material was obtained using either a simple pipe dredge (Bleys Dredge) with a volume of ~20 l (Fig. 1.5b), a large epibenthic sled with a 15 cm gape and volume of ~220 l (Fig. 1.5c), a beam trawl (Fig. 1.5d), or occasionally a Smith-McIntyre grab sampler (Table 1.1). Water depths are mostly >30 m, the shallowest operating depth for *RV Franklin*. The pipe dredge and sled were set on the bottom and towed at a speed of 2 knots for 3–5 min, at which time the vessel was stopped and the device retrieved. All sediment samples are, therefore, a mixture of surface and subsurface material to a depth of ~10 cm. A minor amount of the mud fraction may have washed out during retrieval, but enough samples with significant amounts of mud were recovered to indicate that such loss was minimal.

Fig. 1.5 (**a**) *RV Franklin*, the platform used to collect most of the samples and other marine information used in this book, (**b**) the pipe dredge (Bleys Dredge) used to collect most of the bottom sediment samples; hammer is 15 cm long, (**c**) the epibenthic sled being retrieved after sampling; the bag in the center is half full of sediment, (**d**) the material, seafloor biota, and sediment recovered by a beam trawl, (**e**) sediment being analyzed on the stern deck of *RV Franklin*. (Paul Gammon and Kirsty Brown)

Table 1.1 Samples

	Tasmania	Otway Rig Seismic 1987–1988 FR 02-91	Lacepede FR 3-89 FR 2-91	Eucla Rig Seismic 1992	Lincoln FR 06-94	Gab West FR 07-95	Gab East FR 03-98	Total
Grab	272	116	10	11	2	7	1	419
Bleys Dredge	0	0	149	0	52	91	19	311
Epibenthic Sled	0	0	0	0	15	103	91	209
Vibracore	0	24	0	11	0	0	0	35
Garvity Core	0	119	1	0	0	0	0	120
Piston Core	0	0	2	0	0	0	0	2
Total	272	259	162	22	69	201	111	1096

Underwater video images and direct underwater observation via SCUBA confirm these observations. Gravity cores, piston cores and vibracores were taken locally. The materials recovered were, because of operational restrictions, deeper than 30 m water depth (mwd). This information is locally supplemented by shallow-water samples taken using *RV Ngerin* and by materials collected using SCUBA and snorkeling.

Navigation was by Global Positioning System (GPS), transit satellite, radar, and dead reckoning. The average accuracy of navigational fixes varied from meters to several tens of meters. Samples and bottom profiles were widely spaced in order to characterize the entire region. Bathymetry was determined by a precision depth recorder. Surface temperatures and salinities on some cruises were recorded every 10 min, and vertical temperature profiles for selected positions were determined by a conductivity-temperature-depth (CTD) profiler.

Sediment attributes were logged on board (Fig. 1.5e). Images of most samples were taken on deck to ascertain living versus dead biota as well as the size of biota and rocks retrieved. Size fraction separation was done on board. Total samples were taken for further analysis whereas the coarse fraction (>2 mm) was sieved and archived separately. Bulk sample splits are now archived at Geoscience Australia in Canberra and available for future research. Detailed sediment composition was subsequently determined onshore by visual examination, under binocular microscope where appropriate, of sample splits of mud, sand, gravel, and coarser material. Results and facies assignments are tabulated in Appendices A, B, C, and D. Volumetric estimates of various components were made by comparison with standard visual percentage diagrams. Thin sections were prepared as necessary to solve particle identification problems. Scanning electron images were obtained of very fine-grained materials.

Bottom camera stations were located at specific sites after sediment recovery. Images were obtained using either a frame-mounted EG&G camera-flash unit or a 50 mm single lens reflex camera in a watertight housing. Photography during this 10 min drift period enabled coverage over areas that varied from meters to decameters, depending upon the drift rates. Analog videos of sites in the eastern Great Australian Bight were obtained using a camera in a specially designed underwater housing. Images from northern Tasmania were taken from digital videos taken by CSIRO and so are of specific sites.

Chapter 2
Setting

2.1 Geology & Tectonics

2.1.1 Introduction

The southern part of Australia is an ancient terrane that was welded into Gondwana, encompassed within Pangaea, and divorced from the supercontinent during the Mesozoic (Veevers 2000; Johnson 2004). The original tectonic grain, a relict of this long Gondwanan and Pangean prehistory, is predominantly north-south. These structures are, however, truncated by an east-west Mesozoic rift system that has in turn evolved into the modern passive continental margin (Fig. 2.1).

2.1.2 Pre-Mesozoic Craton

Southern Australia is anchored by the Yilgarn Craton and the Gawler Craton (Fig. 2.1). These two Archean-Proterozoic massifs are composed of igneous, metavolcanic, and metasedimentary rocks. The eastern margin of the Gawler Craton is flanked by the Tasman Fold Belt System. This 1,200 km-wide belt (Drexel et al. 1993) comprises, from west to east, the Delamerian Fold Belt, the Lachlan Fold Belt, and the New England Fold Belt (Fig. 2.1). The Delamerian Fold Belt is composed of continental margin Proterozoic and Cambrian sedimentary, metasedimentary, and volcanic rocks that were deformed in the Middle to Late Cambrian and then intruded by Late Cambrian granites (Foden et al. 1990). Today it forms the topographically high Flinders Ranges, Mt. Lofty Ranges, and Olary Arc east and north of Adelaide, and cores much of Kangaroo Island (Foden et al. 2006). The fold belt also underlies much of the Murray Basin.

The eastern part of the continent (Fig. 2.1) is a collage of accreted terranes, linear meridonal sedimentary basins, and volcanic arcs. The western part of the Lachlan Fold Belt is a terrane formed during a protracted period of sedimentation and westward thrusting resulting from repetitive collision of exotic terranes (Birch 2003). Although cratonization of southeastern Australia was largely complete by Middle Devonian, it was quickly followed by intense Late Devonian folding and granite intrusion (Willman et al. 2002). Subsequent rapid unroofing led to erosion and the deposition of middle Paleozoic terrestrial sediments together and post-orogenic acid volcanism. This landscape was scoured by repeated late Paleozoic continental glaciations and locally covered by glacigene and associated cold marine sediments (Fielding et al. 2008).

2.1.3 Australian Southern Rift System

The extensive divergent, passive continental margin of southern Australia, the Australian Southern Rift System (Stagg et al. 1990; Willcox and Stagg 1990; Drexel et al. 1993), is the result of Jurassic to Tertiary rifting and spreading between the Australian and Antarctica plates (Jensen-Schmidt et al. 2002; Duddy 2003; Holdgate and Gallagher 2003; Totterdell and Bradshaw 2004). The initial series of extensional basins formed during Middle-Late Jurassic breakup of eastern Gondwana via extension along the southern margin of Australia as one arm of a triple junction (Fig. 2.2a). This failed rift remained largely quiescent until late Cretaceous when seafloor spreading between Australia and Antarctica began. A major uplift event

N. P. James, Y. Bone, *Neritic Carbonate Sediments in a Temperate Realm*, DOI 10.1007/978-90-481-9289-2_2, © Springer Science+Business Media B.V. 2011

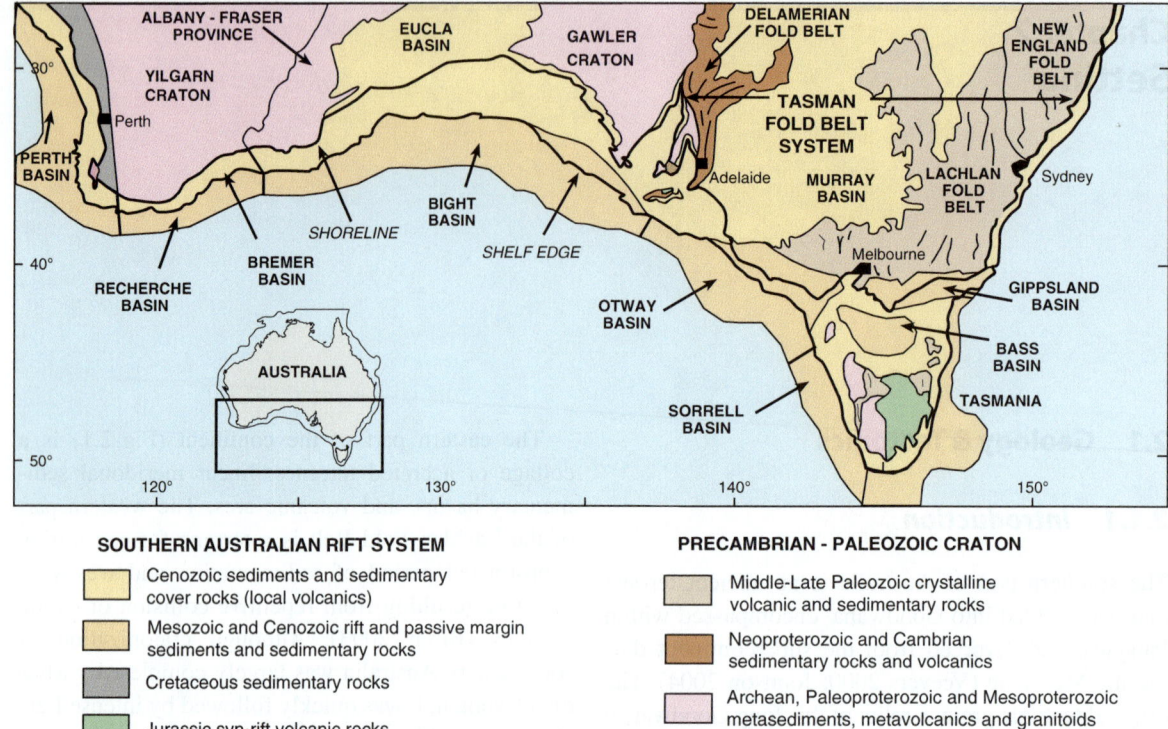

SOUTHERN AUSTRALIAN RIFT SYSTEM

- Cenozoic sediments and sedimentary cover rocks (local volcanics)
- Mesozoic and Cenozoic rift and passive margin sediments and sedimentary rocks
- Cretaceous sedimentary rocks
- Jurassic syn-rift volcanic rocks

PRECAMBRIAN - PALEOZOIC CRATON

- Middle-Late Paleozoic crystalline volcanic and sedimentary rocks
- Neoproterozoic and Cambrian sedimentary rocks and volcanics
- Archean, Paleoproterozoic and Mesoproterozoic metasediments, metavolcanics and granitoids

Fig. 2.1 Geological map of southern Australia. Archean-Paleo-proterozoic massifs of the Yilgarn and Gawler Cratons domi-nate the western half of the continent, and are the sources of quartz sediment on the continental margin in this area. The Tas-man Fold Belt System, comprising the Delamerian Fold Belt, the Lachlan Fold belt, and the New England Fold Belt comprise most of the eastern part of the continent. These Precambrian-Paleozoic terranes with a predominant north-south tectonic grain are cut by the east-west Australian Southern Rift sys-tem. This system comprises local Jurassic volcanics, extensive continental margin rift basins, and Cenozoic cover rocks that accumulated in a series of separate basins. (Recherche Basin to Sorrell Basin)

at ~95 Ma (Cenomanian) probably corresponds to the initial formation of ocean crust south of Australia and initial separation of Australia and Antarctica (Veevers 1986; Veevers and Eittreim 1988) (Fig. 2.2b). Early spreading rates were slow (~9 mm year^{-1}) and largely oriented NW-SE (Fig. 2.2c) but nevertheless resulted in the formation of a narrow, rapidly subsiding seaway between Australia and Antarctica (Fig. 2.2d). Resultant thick Mesozoic sedimentary successions of terrestrial and marine siliciclastic sedimentary rocks are now sequestered in three large roughly east-west, fault-bounded Jurassic-Cretaceous extensional troughs, the Recherche, Bight, and Otway, basins (Figs. 2.1, 2.3) that lie mostly beneath the modern continental shelf and slope in water 200–4,000 m deep.

Spreading rates dramatically increased to ~45 mm year^{-1} during the Middle Eocene coincident with a change of the spreading direction from NW-SE to N-S. Actual continental disconnection was, however, tem-porally protracted, with final detachment of Tasmania from Antarctica not happening until the Oligocene (~33.7 Ma). Although Tasmania also began to separate from continental Australia, it never completely broke away such that Bass Strait is underlain by a thick sedimentary succession on stretched continental crust; the Bass Basin. This west-to-east, scissors-like open-ing resulted in a gradually widening Cenozoic marine embayment, the Australia-Antarctica Gulf (Exon et al. 2004). The modern shelf morphology is, how-ever, largely controlled by a series of major Cenozoic basins (Recherche Basin, Eucla Basin, Bight Basin, Otway Basin, and Sorrell Basin) (Lowry 1970; Fra-ser and Tilbury 1979; Bein and Taylor 1981; Davies et al. 1989; Hocking 1990; Stagg et al. 1990; Hill and Durrand 1993; Totterdell and Bradshaw 2004) whose deposits extend onto the continent proper in a series of epicratonic basins (Bremer, Eucla, Murray, Bass) (Fig. 2.1b).

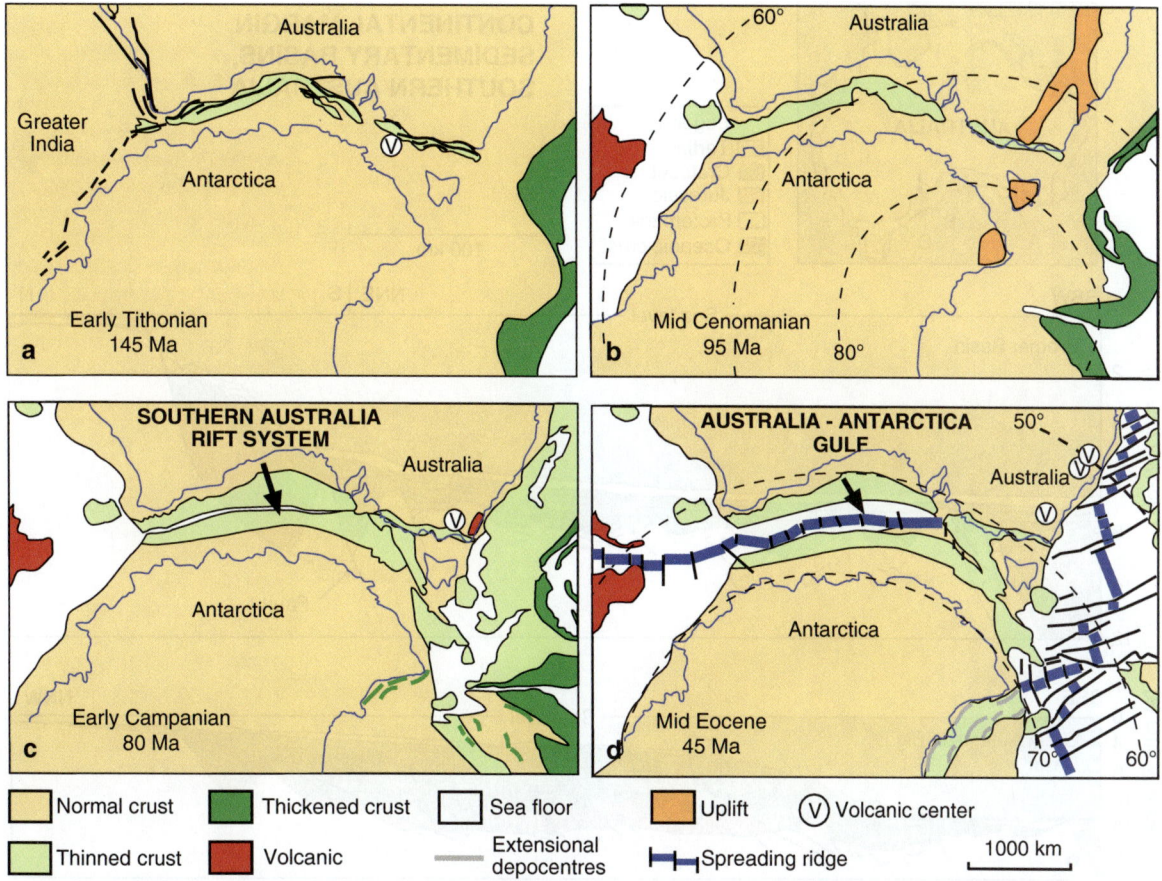

Normal crust | Thickened crust | Sea floor | Uplift | Ⓥ Volcanic center
Thinned crust | Volcanic | Extensional depocentres | Spreading ridge | 1000 km

Fig. 2.2 Images depicting different stages in the separation of Australia from Antarctica (after Norvick and Smith 2001). Separation (**c**) began in the west and resulted in the Australia–Ant- arctica Gulf (**d**) a Paleocene and Eocene embayment that was closed at its eastern end until complete opening and formation of the modern Southern Ocean in the Oligocene

2.1.4 Cenozoic Continental Margin Wedge

The change in drift direction and increase in spreading rate during the Middle Eocene began in the west and was accompanied by left-lateral strike slip as Australia pulled away from Antarctica (Fig. 2.2) until final generation of oceanic crust in the Late Eocene. This change coincided with the first appearance of widespread continental margin and epicratonic carbonate sediments, which continued to be the dominant style of shelf deposition throughout the Cenozoic. The carbonates remained largely cool-water in aspect because, in spite of the fact that Australia had drifted equatorward during the Cenozoic (Fig. 2.4), the surrounding ocean waters remained cool. These epicratonic basin and continental shelf wedge strata have been divided into several discrete successions or sequences (Fig. 2.5), each of which has its distinctive style (Quilty 1977; McGowran et al. 2004). The Middle Eocene to Early Oligocene portion (Succession 2—Fig. 2.5) is characterized by cool-water carbonate and spiculite biosiliceous facies in the west with coeval terrigenous clastic and voluminous coal deposits in eastern Victoria deposited within the elongate Australia–Antarctica Gulf (Fig. 2.2d). The eventual complete separation of Australia, South America, and Antarctica in the early Oligocene led to establishment of the cool Circumantarctic Current and West Wind Drift, isolating Antarctica and profoundly cooling the ocean south of Australia. Following a major eustatic sea level fall due to initial Antarctic glaciation, and profound canyon cutting at the prograding shelf edge (Bernecker et al. 1997), succeeding Late Oligocene to Middle Miocene deposits (Succession 3—Fig. 2.5) are

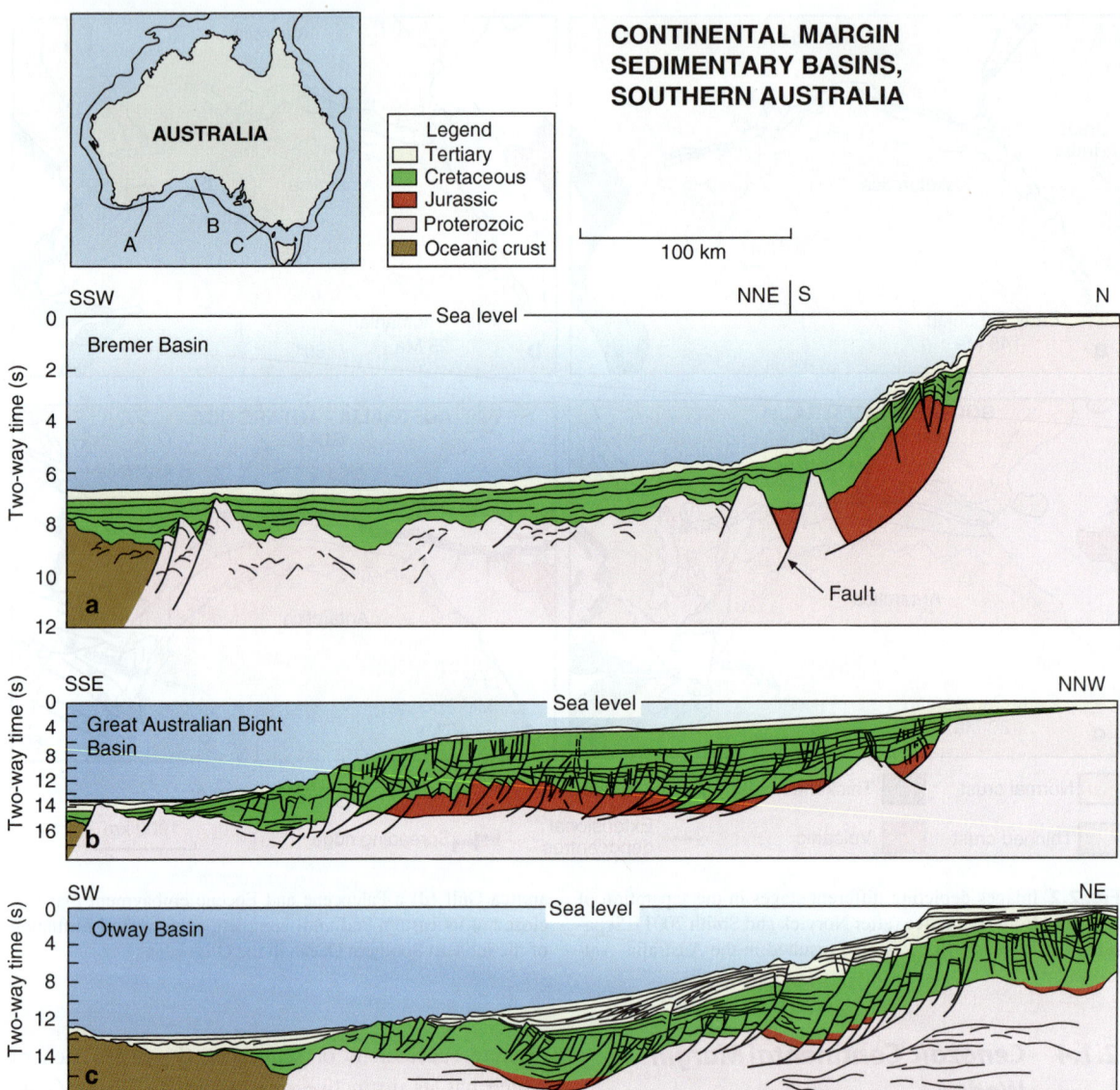

Fig. 2.3 Interpreted seismic sections across the southern Australian continental margin (locations on inset map). Large passive margin basins of the Southern Australian Rift System are filled with Jurassic and Cretaceous siliciclastic sediments, some of which are hydrocarbon rich. The Cenozoic is a rela-tively thin succession of mainly carbonate sediments and sedimentary rocks in sections (**a**) and (**b**), whereas the section in the east (**c**) contains significant siliciclastic deposits at the base. (Modified from Stagg et al. 1990)

typified by temperate neritic carbonates, with local coals in the far east (McGowran et al. 2004).

2.1.5 Tectonic Inversion

The quiescent Mesozoic-Paleogene passive margin history was dramatically interrupted in the late Middle Miocene by complete tectonic inversion, wrench-

ing, compression, and associated tectonics (Hill et al. 1995; Sandiford 2003a, b). Large segments of the inner continental margin and most epicratonic basins were uplifted, locally deformed, and exposed; they have remained so to the present day. This event was due to collisions on the northern and eastern margins of the continent. Such deformation was strongest in the east, in Victoria, and progressively less intense westward. Current thinking is that the Otway Ranges were uplifted (for a second time) by late Miocene tectonism at 8–6 Ma

Fig. 2.4 Movement of Australia during the Cenozoic (from Feary et al. 1992). The Great Australian Bight was located at ~60°S during the early Cenozoic and is now positioned at ~32°S. Initial slow northward drift was followed by a change in direction and accelerated movement to the NNW in the middle Oligocene. The continent is still moving in this direction at a rate of ~7 cm year^{-1}

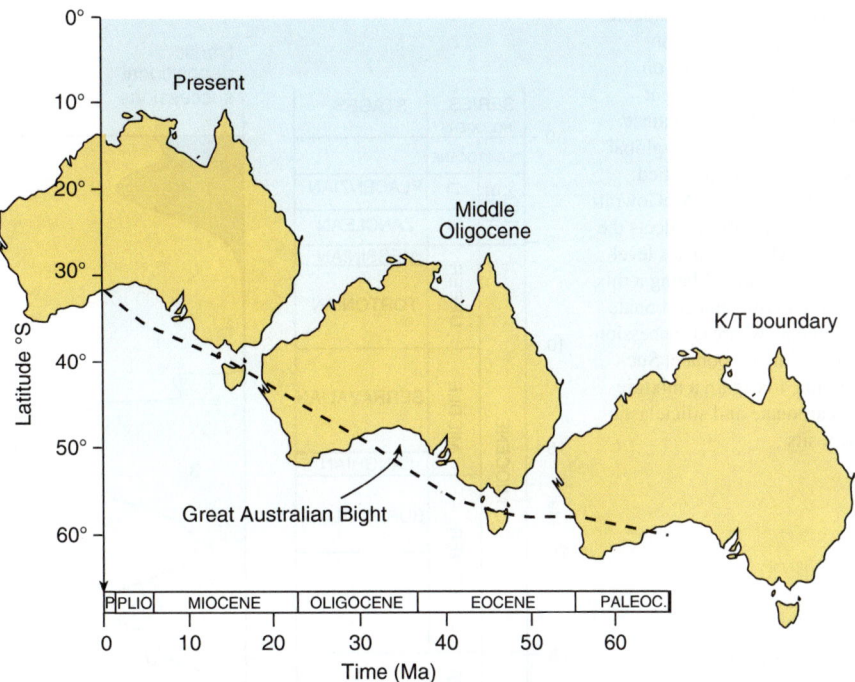

(Dickinson et al. 2002; Sandiford 2003a), either the result of compression between the entire Australian plate and plates of SE Asia or a change in relative plate motion between the Australian and Pacific Plates. All of Victoria is currently under strong E-W and SE-NW compression with evidence of tectonic activity through Pliocene to the Holocene wherein Pliocene strandlines are displaced upward 200–250 m on the west side of the Otway Ranges. Basin inversion was accompanied by Plio-Pleistocene volcanism (the Older Volcanics) (Price et al. 2003), again largely restricted to the east. Such tectonic activity and attendant seismicity continues in South Australia (Greenhalgh et al. 1994) and Victoria with volcanism (the Newer Volcanics) in South Australia documented by aboriginal oral tradition as recently as 1500 BP (Sheard 1986).

The major episode of uplift that created the Otway Ranges in Victoria, exposed Eocene-Miocene sediments in the epicratonic basins, and resulted in sporadic, ongoing volcanism in the east, was succeeded by Plio-Pleistocene deposition in epicratonic basins and on the outer shelf, whereas little sediment accumulated on the inner shelf. The Pliocene (Succession 4—Fig. 2.5) is typified by mixed siliciclastic–carbonate deposits (Belperio et al. 1988) from oyster-rich estuarine to inboard marine shoreface and grassbed deposits (Brown and Stephenson 1991; Pufahl et al. 2004; James et al. 2006; James and Bone 2007), to local

spectacular prograding outer shelf and slope sediments (Feary et al. 2004) and extensive periods of laterization inland. The Pleistocene is, by contrast distinguished by spectacular eolianites that are plastered along many coastlines, especially those that are SW-facing into the prevailing winds and seas (Wilson 1991; Belperio 1995). These sets of aeolianite are locally prograding with evaporites and lacustrine dolomites in interdune corridors (Alderman and Skinner 1957; von der Borch et al. 1975; von der Borch 1976; von der Borch and Lock 1979). Coeval sediments in gulfs and basins are marine (Shepherd and Sprigg 1976; Blom and Alsop 1988; Gostin et al. 1988; Fuller et al. 1994). Outboard the shelf edge and upper slope is (1) gently dipping as a series of impressive prograding clinoforms, (2) steep with numerous submarine canyons or (3) erosional and subject to mass wasting (von der Borch and Hughes-Clarke 1993; Passlow 1997; Feary and James 1998; James et al. 2004; Exon et al. 2005; Hill et al. 2005).

2.2   Meteorology & Climate

2.2.1   Introduction

Meteorology and climate, summarized by Gentilli (1971) have a profound influence on the nature of

Fig. 2.5 A plot of Cenozoic depositional successions in southern Australia, on the shelf and in adjacent epicratonic basins, against geologic time and the global sea level curve (modified from Quilty 1977; McGowran 1997). Deposition reflects the eustatic changes in sea level with succession 2 being a mix of siliciclastic and carbonate sediments whereas succession 3 is mainly carbonate. Succession 4 is again a mixture of carbonate and siliciclastic deposits

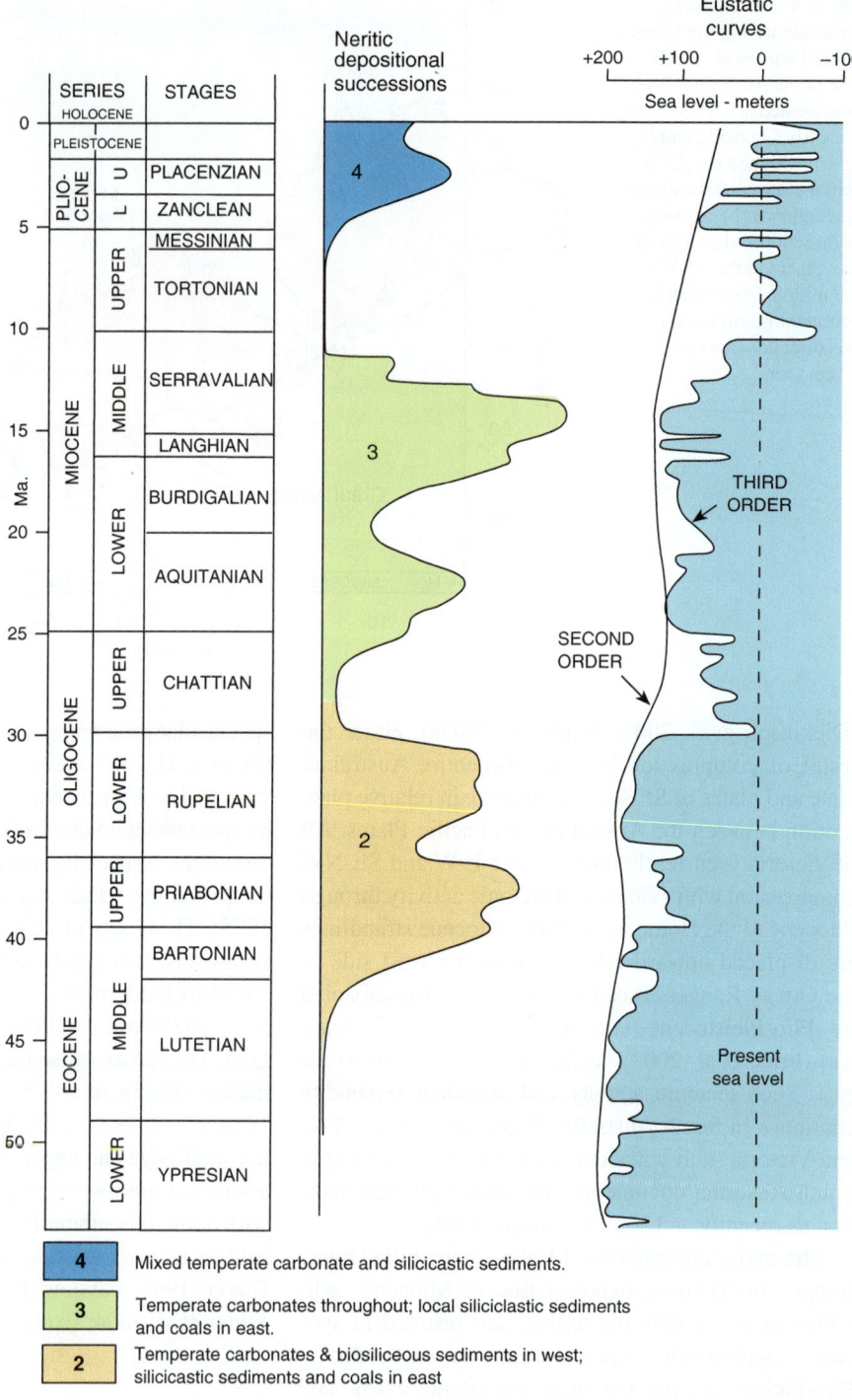

4 — Mixed temperate carbonate and silicicastic sediments.

3 — Temperate carbonates throughout; local siliciclastic sediments and coals in east.

2 — Temperate carbonates & biosiliceous sediments in west; silicicastic sediments and coals in east

sedimentation across the southern Australian marine environment. The continent lies between ~11°S and ~43°S and at this location weather is dominated by the ridge of high pressure produced by descending air at the boundary between the Hadley Cell and the Ferrel Cell.

During summer, this boundary moves southward some 5–8° of latitude, at times allowing the tropical belt of low pressure (intertropical convergence) to penetrate the northern coast of Australia (Fig. 2.6b). In winter the high-pressure belt moves northward (Fig. 2.6a) and

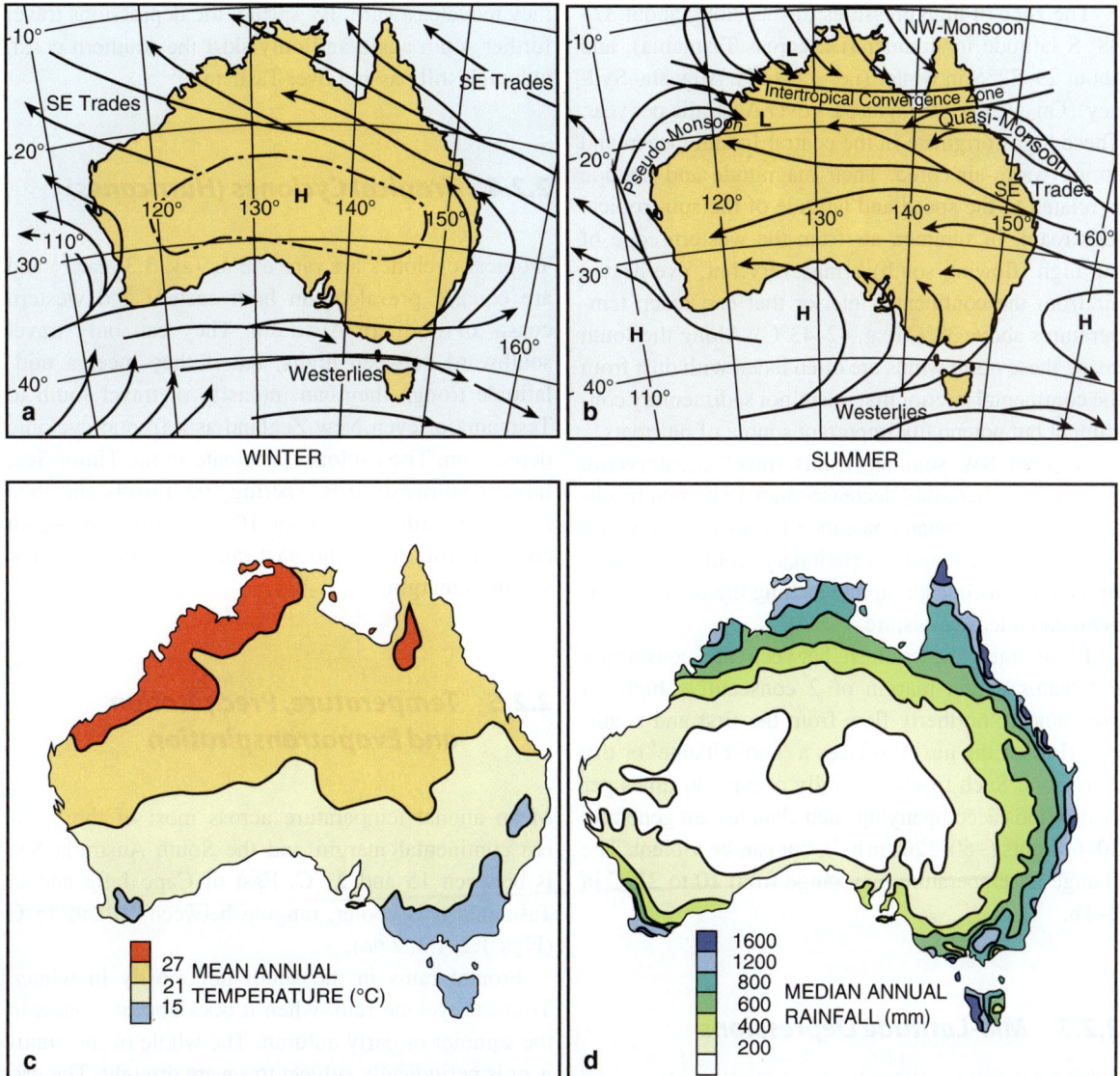

Fig. 2.6 Maps of Australian meteorology (modified from Gentilli 1971). (**a**) A plot of the major intracontinental high and associated wind patterns during winter. (**b**) A plot of the southward movement of the intertropical convergence zone into northern Australia, southward shift of the high pressure system, and associated wind patterns during summer. (**c**) A plot of mean annual temperature. (**d**) A plot of mean annual rainfall illustrating the humid character of the southwestern and southeastern coasts and the semi-arid character of the Great Australian Bight and South Australian Sea

permits the mid-latitude jet stream to meander over the land. Thus, summer circulation is meridonal (N-S) and winter circulation is zonal (E-W).

2.2.2 Anticyclonic Highs

The high-pressure ridge is really a series of highs that move eastward, especially during summer (Figs. 1.2,

1.3). Circulation around each cell is counterclockwise with winds on the south side irregular and often violent. Summer highs bring tropical continental air to the SW and maritime tropical air to the SE. Between highs in summer, there is a quasi-stationary front and light clouds, whereas in winter there is lower pressure, multiple cold fronts, and stormy weather. This winter pressure trough between highs is due to the formation of upper air depressions that often bring copious rain.

The zone of high pressures travels along about 37–38° S latitude in summer (i.e. across Tasmania), and about 29–32° S in winter (i.e. across Pt. Augusta–Sydney). On average, 40 highs pass over Australia per year. These highs originate in the central Indian Ocean and so are warm-air cored. Their magnitude and latitude is related to the speed and latitude of the sub-tropical jet stream. In summer, air from the western edge of the high (flowing south) brings very hot, overheated air from the continental interior that can reach temperatures above 38°C (e.g. 42–43°C). Along the south coast, these north winds are often laden with dust from the continental interior that is a minor sedimentary constituent but potentially important source of nutrients.

As cool SW summer winds travel equatorward, their relative humidity decreases such that upon reaching the hot Australian coast they fail to yield any rain (although there may be a stationary front and cloud). In winter maritime air, upon reaching the colder continent, can release moisture.

In summer, the contrast between the converging airstreams at the margin of 2 consecutive highs is pronounced; northerly flow from the first and southerly flow of the next produces a 'cool change' or dry cold front. Such fronts generally occur ~30 times per year. Winds accompanying such changes are generally 30–60 knots (~60–120 km h^{-1}), but can be violent. The change in temperature may range from 10 to 22°C in 2–4 h.

2.2.3   Mid-Latitude Depressions

These intense lows occur well south of Australia, but they have a strong effect on Australian climate. In summer, they are too far south and weak to affect Australia. During autumn, the frequency of heat-cored highs decreases considerably and the mid-latitude lows travel slightly further north than in summer. By late autumn, zonal circulation is more intense, the jet stream flows faster and enters Australia at a lower latitude, and depressions go past much closer to the southern coast. Well-developed fronts appear across southern Australia, bringing rain. During winter, the closeness of mid-latitude depressions increases rapidly to its mid-July maximum, with an average of three cyclonic centers per month skirting the southwestern coast and passing over Tasmania. Their intensity increases as they move eastward. By spring, the depressions travel further south again and only skirt the southern coast, although still passing over Tasmania.

2.2.4   Tropical Cyclones (Hurricanes)

Tropical cyclones are rare events (av. 3.3 year^{-1}) and are equally prevalent on both eastern and western coasts of northern Australia. They can only travel southward between highs, but if they meet a mid-latitude trough they can intensify or travel south to Tasmania or even New Zealand as a frontal cyclonic depression. The cyclones originate in the Timor Sea, travel southwestwards, veering southwards and then southeastwards with about 10% reaching the south coast. Those along the east coast do not affect the southern margin.

2.2.5   Temperature, Precipitation and Evapotranspiration

Mean annual temperature across most of the western continental margin and the South Australia Sea is between 15 and 21°C. East of Cape Jaffa and in Tasmania it is cooler, ranging between 10 and 15°C (Figs. 1.2, 1.3, 2.6c).

Frontal rains in the south fall mostly in winter. Tropical cyclone rain, when it does appear, comes in the summer or early autumn. The whole of the continent is periodically subject to severe drought. The driest area is in the center of the continent.

In summer, all of the area from Albany to Portland is arid, west of Albany and east from Portland to Melbourne it is semi-arid whereas most of western Tasmania is perhumid (wet) (Fig. 2.6d). In winter, the balance ranges from perhumid west of Esperance and across all of Tasmania, to humid from Esperance east to Cape Pasley and from Streaky Bay east to Melbourne. It is semiarid from Cape Pasley across the Great Australian Bight to Streaky Bay.

In January (summer), mean evaporation across the region ranges from ~200 mm per month in Eucla and Adelaide to ~100 mm per month in Tasmania. In July (winter), it ranges from ~75 mm per month in Eucla to ~25 mm per month in Tasmania.

2.3 Oceanography

2.3.1 Introduction

The continental margin along southern Australia is one of the world's longest, latitude-parallel, zonal, shelves; there are few other such shelves in the modern world. Most are meridonal and so transect major oceanographic boundaries. The shelf faces the continent of Antarctica, with its cold and isolated water masses that drive much of global ocean circulation. The overall region is, from an oceanographic (but not a geological) perspective, called the South Australian Basin.

The region between Australia and Antarctica is commonly referred to as the Southern Ocean, both in the scientific and non-scientific literature. The 2,500–3,500 km–wide body of water is, however, composed of two very different water masses, polar waters surrounding Antarctica and sub-tropical waters adjacent to southern Australia. The Subtropical Convergence Zone, a complex region of fronts, separates these water masses (Fig. 2.7).

The Southern Ocean is not recognized as such by some oceanographers but is instead treated as individual southern segments of the other three oceans (Tomczak and Godfrey 1994). In this view, only the ocean south of the Subtropical Convergence around Antarctica is called

the Southern Ocean (Schodlok et al. 1997). It is here that the permanent thermocline reaches the surface. The region is characterized by unimpeded zonal circulation around the polar continent leading to perpetual strong westerly winds and associated strong currents (West Wind Drift–Antarctic Circumpolar Current) (Fig. 2.7), and some of the highest sea states on the globe. Since water circulation is largely from the west, the waters are mostly influenced by the Indian Ocean; Pacific Ocean waters are not present to any significant extent in the region. As a result, waters north of the Subtropical Convergence, and which cover the southern Australian shelf, are treated as part of the SE Indian Ocean. For ease of discussion, the whole region herein is called the Southern Ocean, recognizing other more stringent definitions. The following discussion is a combination of observations, summary articles, and numerical studies.

2.3.2 Sea State

The southern Australian continental shelf overall is swell- and storm-dominated with high (>2.5 m) modal deep-water wave heights (Davies 1980; Short and Hesp 1982; Short and Wright 1984; Hemer and Bye 1999). The dominant wind and swell wave approach is from

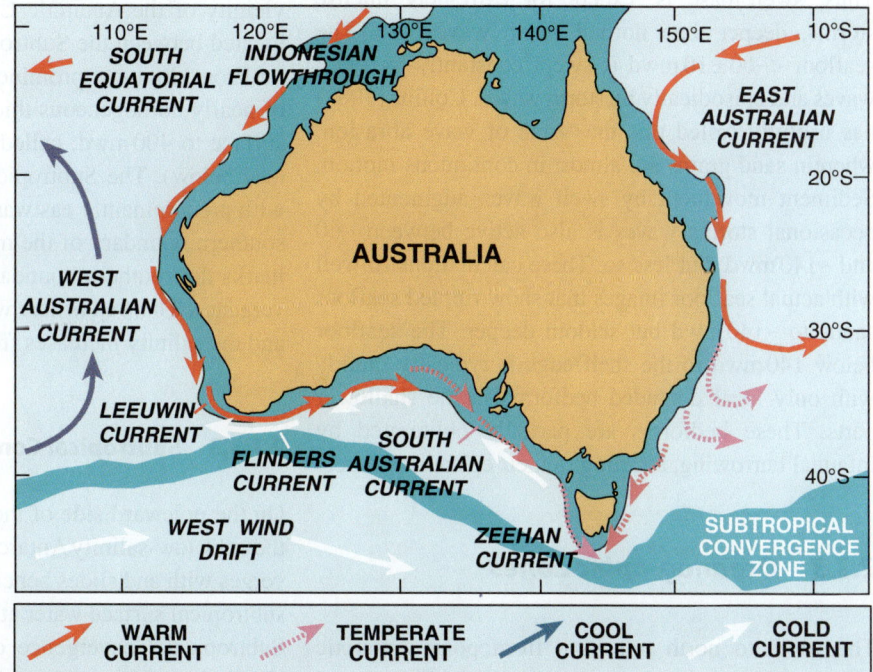

Fig. 2.7 A map of Australia and surrounding oceans highlighting the major current systems

the southwest, year-round. Wave drift is onshore with a mean speed of $0.15\,cm\,s^{-1}$ (over a period of a month, organisms would be advected onshore ~400 km). The importance of this wave drift conveyor belt is not yet known. Wave height will exceed 3 m for 30–60 days of the year; will exceed 6 m for 0–10 days; will exceed 8 m 1.3% of the time.

For typical swells there is refraction around Kangaroo Island. A westerly swell penetrates directly into Investigator Strait and is refracted northward into Spencer Gulf along the western shore of Yorke Peninsula; models predict 6 m swells; swells in the surf zone of Adelaide beaches would be 0.4–0.7 m with velocities of 1.0–$0.6\,m\,s^{-1}$ (Bye 1976).

The two critical sedimentological interfaces in this hydrodynamically energetic region are storm wave base and swell wave base. Swells move onshore, generally from the southwest year round whereas storm waves are most intense during winter and spring (June to November). Swell period off southwest Australia ranges between 10 and 20 s with peak energy at 14 s and so waves would interact with the seafloor to depths of ~160 mwd Collins (1988). Storm wave periods are somewhat shorter, ranging between 8 and 10 s with peak energy at 8.5 s and so they come in contact with the seafloor at depths shallower than ~60 mwd. This relationship is more or less the same ±10 mwd across most of the southern Australian continental margin. Thus, swell base is, except for extremely intense storms, deeper than normal storm wave base. The seafloor <~60±10 mwd is swept constantly by swell waves and episodically by storm waves. Collins (1988) has usefully called this the 'zone of wave abrasion' wherein sand grains are almost in continuous motion. Sediment movement by swell waves augmented by occasional storms waves is also active between ~60 and ~140 mwd, but less so. These calculations fit well with actual seafloor images that show rippled seafloor sands to ~140 mwd but seldom deeper. The seafloor below 140 mwd to the shelf edge is typically muddy with only local degraded bedforms in the shallower parts. These bedforms are partially obliterated by infaunal burrowing, attesting to their episodic nature.

2.3.3   Oceanographic Zones

The south to north transition from polar Antarctic to subtropical Australian waters is complicated and is marked by several fronts or narrow areas of rapid spatial temperature and salinity change. These are the Polar Front, Subantarctic Front and Subtropical Front (Fig. 2.8a). It is the water masses generated and subducted at these fronts that are potential upwelling waters along the south Australian margin.

2.3.3.1   Antarctic Region or Zone

The Antarctic Zone lies south of the Polar Front. The Polar Front Zone, where water temperatures change northward from 4 to 10°C is complex and composed of the Subantarctic Front and the Polar Front. The Polar Front is where surface temperatures change northward from ~4 to 8°C and salinities change from ~34.2 to <33.9‰. The Subantarctic Front is where temperatures change from 8 to 10°C and salinities increase from 34.3‰ to 34.7‰.

2.3.3.2   Subantarctic Region or Zone

This zone is bounded by two fronts, the Subtropical Convergence (~40°S) and the Polar Front (~50°S) and contains another, the Subantarctic Front (~45°S). It has also been called the Subantarctic Water Ring (Longhurst 1998). The northernmost zone is in the vicinity of the Antarctic Circumpolar Current and is located between the Subtropical Front and Subantarctic Front. Its most prominent feature is a water layer of nearly homogeneous thickness, extending from the surface to 400 mwd, called Subantarctic Mode Water (see below). The Subtropical Front is a current band with predominantly eastward flow associated with the southern boundary of the major subtropical gyres and marks the southern boundary of the Subtropical Convergence. The waters here warm rapidly from 8 to10°C and the salinity increases from 34.2 to 34.8‰.

2.3.3.3   Subtropical Convergence Zone

On the poleward side of the Subtropical Convergence the cold, low-salinity Antarctic Intermediate Water converges with and slides beneath the warm, high salinity, subtropical surface water at the Subtropical Front. The Subtropical Convergence divides anticyclonic water circulation of the southern Atlantic, Pacific, and Indian

Fig. 2.8 Isometric blocks illustrating the major water masses described in the text **(a)** off Antarctica (after Sverdrup et al. 1942) and **(b)** off Southern Australia (synthesized from McCartney 1982; Condie and Dunn 2006; Middleton and Bye 2007). The Leeuwin Current System flows from west to east whereas the Flinders Current system flows from east to west

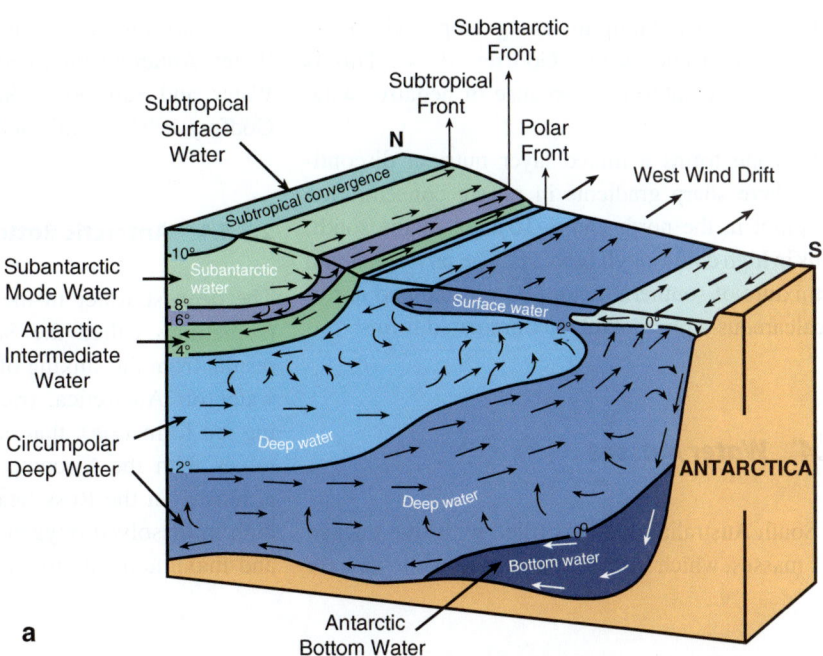

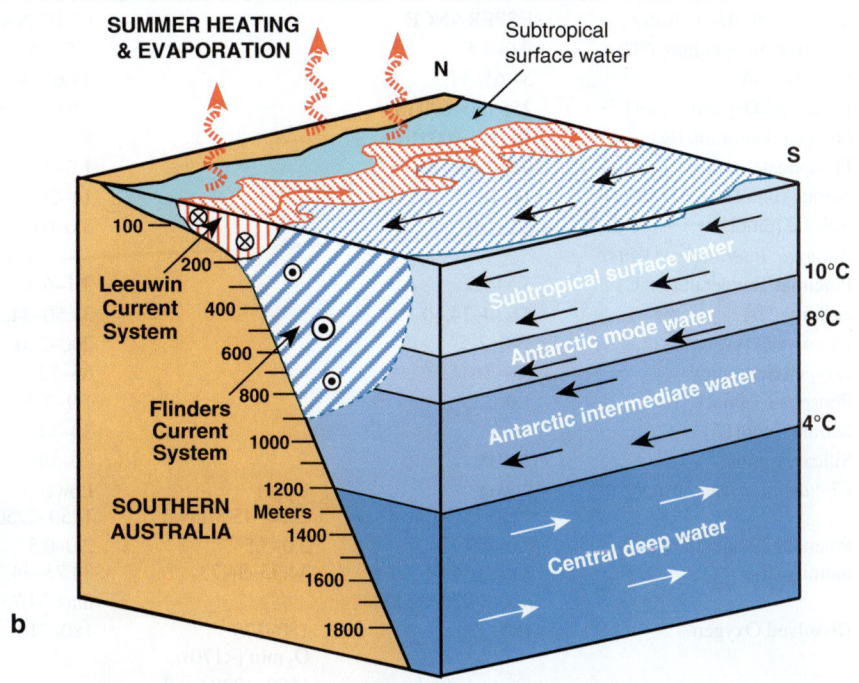

oceans from the cyclonic circulation of the Antarctic Circumpolar Current and can be traced almost continuously around the globe.

The Subtropical Convergence in the Australian-Antarctic sector lies north of the Subtropical Front and between ~35° and 45°S and is where surface waters increase in temperature from ~10°C to locally 18°C over a distance of <5 km. Increasing sea surface temperature and salinity occurs with decreasing latitude. The water spreads northward along isopicnals (a process called subduction). Subduction proceeds by the injection of surface water into intermediate depths by Ekman Pumping along isopicnals of its own density (Tomczak and Godfrey 1994). This is common in the subtropics because of negative wind curl.

The zone forms a mixed layer nutrient discontinuity where sharp gradients in nitrate concentration (<0.5 μmol in the north and 8–10 μmol in the south) results in high chlorophyll biomass. The phytoplankton is a mixture of southern diatom-dominated and northern calcareous—phytoplankton dominated forms.

2.3.4 Water Masses

The South Australian Basin is filled with five stacked water masses, which from the deepest to shallowest are

(Fig. 2.8a) Antarctic Bottom Water, Circumpolar Deep Water, Antarctic Intermediate Water, Antarctic Mode Water, and Subtropical Surface Water (Tomczak and Godfrey 1994; Condie and Dunn 2006) (Table 2.1).

2.3.4.1 Antarctic Bottom Water

The deepest water in the Australian-Antarctic Basin (sometimes called the South Indian Ocean Basin), results from the sinking of cold and dense nearsurface water off Antarctica, (rich in solutes excluded during ice formation), that descends to >4,000 mwd and mixes with the Circumpolar Deep Water. Its origin is mostly in the Ross Sea and off Adelie Land. It is high in dissolved oxygen and low in nutrient content and maximum salinity and spreads to the east along

Table 2.1 Intermediate and deep water masses of the Southern Ocean

Subantarctic Mode Water	ESPERANCE		CEDUNA	
Potential Temperature (°C)	9.0–9.5		8.5–9.5	
Salinity (‰)	34.65–34.7		34.60–34.65	
Dissolved Oxygen (μmol l^{-1})	255 (N)–290 (S)		250(N)–285 (S)	
Oxygen Saturation (%)	89 (N)–100 (S)		88 (N)–100 (S)	
Phosphoate (μmol l^{-1})	0.9–1.2		1.0–1.2	
Nitrate (μmol l^{-1})	14–18		14–20	
Silicate (μmol l^{-1})	2.0–6.0		3.0–6.0	
Antarctic Intermediate Water				
Potential Temperature (°C)	3.5–5.5		3.1–6.3	
Salinity (‰)	34.30–34.40		34.50–34.32	
Dissolved Oxygen	190–2-5		200–220	
Oxygen Saturation	65–70		66–72	
Phosphate (μmol l^{-1})	1.9–2.2		1.9–2.3	
Nitrate (μmol l^{-1})	28–34		30–34	
Silicate (μmol l^{-1})	22–35		20–39	
Circumpolar Deep Water	Lower 1250–2250 mwd	Upper 2250–4500 mwd	Lower 1250–2250 mwd	Upper 2250–4500 mwd
Potential Temperature (°C)	2.0–0.5	2.0–3.5	2.0–0.5	2.0–3.0
Salinity (‰)	34.73–34.71 max 34.75 @3 km	34.55–34.73	34.73–34.71 max 34.75@2.7 km	
Dissolved Oxygen	180–220	180–170 O$_2$ min (<170) 1500–1700 mwd	180–210	180–200 O$_2$ min (<175) 1200–1500 mwd
Oxygen Saturation	55–62	53–55 O$_2$ min <51	53–62	53–60 O$_2$ min <51
Phosphate (μmol l^{-1})	2.2–2.4	2.2–2.5 O$_2$ min >2.5	2.4–2.5	2.4–2.5 O$_2$ min >2.5
Nitrate (μmol l^{-1})	36–38	36–38	34–36	34–36
Silicate (μmol l^{-1})	90–125	50–90	100–120	60–100

Compiled from Schodlok et al. (1997)

the sea floor. The water temperature is <0.5°C and the salinity <34.7‰. Dissolved oxygen is >220 µmol l^{-1}, >62% saturation. Nutrients are about the same as the overlying Circumpolar Deep Water whereas silicate is ~125 µmol l^{-1}. This water mass has little effect on the southern Australian continental slope or shelf.

2.3.4.2 Circumpolar Deep Water

Low in dissolved oxygen, this is the most prominent water mass in the Southern Ocean. In this region, it originates as the North Indian Ocean Deep water and spreads southward from the western Indian Ocean at depths of 1,000–2,000 m to mix with North Atlantic Deep Water east of Africa and form Central Deep Water that is characterized by low dissolved oxygen and high nutrients. In this area it can be divided into two parts, upper (1,250–2,250 mwd) and lower (2,250–4,500 mwd) (Table 2.1). The upper part has nutrient levels higher than anywhere else in the water column.

2.3.4.3 Antarctic Intermediate Water

For many years it has been known that this water mass is formed as high salinity North Atlantic Deep water and Circumpolar Deep water together move upward from depths greater than 2,000 m, come to within 200 m of the surface in the Antarctic Zone where they mix with nearsurface waters. Recent studies have also found much evidence that Subantarctic Mode Water is also another precursor of Antarctic Intermediate Water (Carter et al. 2009). The mixture of these waters is warmed and diluted by rain and snow, resulting in low salinity, and is then subducted northward near the polar front between the Polar Front and the Subantarctic Front.

This relatively nutrient-rich water forms an intermediate depth temperature and salinity minimum layer with high oxygen content that flows northward at depths of ~800–1000 m (Lynch-Stieglitz et al. 1994). It is slightly denser than Subantarctic Mode Water (see below). The winter deep mixed layers are isolated from the surface during the summer by a seasonal thermocline. The sinking occurs in a stepwise path northward. This water intersects the south Australian slope at ~600–1200 mwd (Middleton and Cirano 2002; Wood

and Terray 2005) and with Sub Antarctic Mode Water feeds into upwelling circulations over the shelf during the Austral summer. It has a temperature of 4–8°C and salinities of 34.30–34.50‰, fresher than the Subantarctic Mode Water positioned above it, although in upwelled circulations is sometimes reheated to 14°C (Levings and Gill, in press).

2.3.4.4 Subantarctic Mode Water

This large, nearly homogeneous layer formed between the Subantarctic Front and the Subtropical Front is called a thermostad or pycnostad because temperature and salinity variation with depth is very small. The water body results from deep winter convection due to erosion of the seasonal thermocline and exposure of this layer to the cold atmosphere. Convective overturning and thus ventilation results in high dissolved oxygen (~95%–McCartney 1977, 1982). It is tracked throughout southern hemisphere oceans as water with an oxygen maximum.

Adjacent to the Australian continental slope this water has descended to ~450–700 mwd but is much thicker and extends to the surface south of the subtropical front. It has a temperature of ~8–10°C and salinity of 34.60–34.70‰ (Table 2.1). In the Subantarctic Ring Longhurst (1998) referred to this as an oligotrophic regimen even though it is a relatively high-nitrate but low chlorophyll signature, probably because of low Fe supply.

2.3.4.5 Surface Waters

Surface waters, those shallower than ~400 mwd are complex, especially on and adjacent to the continental shelf proper where they are a combination of Subtropical Surface Water, waters introduced from offshore Western Australia, evaporated waters, and upwelled waters. Subtropical Surface Water is also called Southern Subtropical Surface water, Tropical Surface Waters, Subtropical Surface Water and Indian Central water (Condie and Dunn 2006; Woo and Pattiaratchi 2008). Herein called Subtropical Surface Water, it is between 10 and 22°C, has intermediate relatively high salinity (35.1–35.9‰) and intermediate oxygen content (220–245 µmol l^{-1}) and a weak minima of dissolved N, Si, and P. Details of the water in different areas are discussed in subsequent chapters.

2.3.5 Current Systems

2.3.5.1 General Aspects

Waters that overlie the shelf and upper slope, because they are relatively shallow, are strongly affected by seasonally changing climate, both in terms of their composition and movement. In summer, the high-pressure ridge that is maintained over the area induces a consistent pattern of southeasterly winds; in winter when the anticyclone moves north over central Australia there is a predominantly westerly wind regimen. The autumn change from an easterly to westerly wind pattern causes, (1) a change in the character of Ekman transport, and (2) a switch in the direction of coastal currents from east to west. Southeasterly summer winds along the northern edge of the high pressure cells drive waters offshore via Ekman transport, lower coastal sea level by ~25 cm, induce a westward coastal current, and locally upwell waters towards the coast (Herzfeld and Tomczak 1997; Middleton and Platov 2003; Levings and Gill in press).

This situation reverses in winter as the westerly winds force downwelling to depths of 600 mwd over the shelf break, raising coastal sea level ~25 cm, and generating an easterly nearshore current. This is a rare example of downwelling currents driven by seasonally reversing winds (Middleton and Cirano 1999; Cirano and Middleton 2004)

2.3.5.2 Open Ocean

Water movement south of the Subtropical Convergence Zone is perpetually to the east in the form of the West Wind Drift or Circumpolar Current (Figs. 2.7, 2.8). The Circumpolar Current is the largest mass transport of all ocean currents. Driven by a circumpolar belt of westerly winds and associated frequent storms, this current comprises all water masses south of the Subtropical Front and travels eastward at a velocity of $0.05–0.15 \, m \, s^{-1}$ (~0.2 knots). North of the Subtropical Front water movement in the upper 1,000 m is largely westward. Movement of Subantarctic Mode Water

adjacent to the Australian continental margin is westward (McCartney 1977; Schodlok et al. 1997). To the south it may be anticyclonic; east in the south, north adjacent to Tasmania, and west along the shelf edge (McCartney 1977, 1982) or west north of the Subtropical Front, east in the centre, and west adjacent to the shelf (Schodlok et al. 1997).

2.3.5.3 Continental Margin

There is a considerable amount of information available concerning water masses and circulation across the continental margin. The two dominant features are a warm, mixed surface layer underlain by cooler Antarctic Intermediate and Subantarctic Mode Water (Newell 1961, 1974; Wyrtki et al. 1971; Bye 1972, 1983; Callahan 1972; Rochford 1977; Lewis 1981; Godfrey et al. 1986; Hahn 1986; Harris et al. 1987; Schahinger 1987; Cresswell and Peterson 1993; Hufford et al. 1997). The mixed surface layer flows in a generally east-southeast direction and is known as the Leeuwin Current off west Australia, the South Australian Current off South Australia and Victoria, and the Zeehan Current off west Tasmania. The underlying counter current of Antarctic Intermediate Water and Antarctic Mode Water flows in a generally north westward direction at a depth of about 400–600 m (Wood and Terray 2005) and is known as the Leeuwin Undercurrent off west Australia and the Flinders current off western Tasmania, western Victoria and South Australia (Middleton and Cirano 2002). This current feeds into shallower shelf circulations during summer when alongshore winds from the southeast in combination with coriolis force advects the mixed surface layer offshore and triggers a compensatory upwelling of Antarctic Intermediate Water and Antarctic Mode Water from greater depths (Hahn 1986; Schahinger 1987; Levings and Gill in press).

2.3.5.4 Flinders Current System

Monthly mean wind stress curl south of Australia is positive during summer and winter and leads to Ekman pumping and downwelling throughout the region. Calculations indicate that such downwelling ought to result in northward transport and that this should be deflected westward into an upwelling favourable boundary current that flows westward along the southern shelf; the Flinders Current (Bye 1972, 1983; Middleton and Cirano 2002; Middleton and Bye 2007), a northern boundary current. This may be the only such current on the globe. There are but a handful of real observations about the Flinders Current (Cresswell and Peterson 1993; Wood and Terray 2005) and most of what is understood comes from an integration of these measurements and mathematical modeling.

The Flinders Current is sourced from the Subantarctic Zone where Subantarctic Mode Water (cool, high oxygen, moderate nutrient levels) and Antarctic Intermediate Water (very cold, relatively fresh, high nutrient levels) flow north across the Subtropical Front (Fig. 2.8a) and then west along the Australian continental slope (McCartney and Donohue 2007; Currie et al. in press). The Flinders Current is also thought to be fed by the Tasman outflow, that flows westward around the southern tip of Tasmania (Cirano and Middleton 2004). Upwelled waters would therefore come mostly from Antarctic Intermediate Water (and Subantarctic Mode Water).

The Flinders Current flows throughout the year, even though coastal winds and currents reverse. The current runs beneath the Leeuwin Current during winter (Cirano and Middleton 2004), but during summer locally upwells onto the shelf and is present as a shallow current. Flow is strongest in summer due to increased Sverdrup transport. The Current intensifies from east to west, can extend from the surface to 800 mwd, and flows year-round with a velocity of ~16 cm s^{-1} (0.3 knots). Maximum westward speed is 20 cm s^{-1} (~0.4 knots) off Cape Pasley between 400 and 600 mwd; decreasing to zero at 1000 mwd. It should be thickest in summer with the bottom boundary layer extending some 50 km out from the shelf.

2.3.5.5 Leeuwin Current System

The western and southern margins of the continent are characterized by a unique series of strong warm shallow-water (<200 mwd) currents that flow south and east along the shelf edge in a narrow band for almost 5500 km (Ridgway and Condie 2004) from Northwest Cape to the southern tip of Tasmania (Fig. 2.8).

Leeuwin Current The wind-dominated circulation pattern generates a shelf-edge jet that changes with the season. The Leeuwin Current (Godfrey and Ridgway 1985; Cresswell 1991) is a narrow (<100 km), shallow (<200 m) stream of comparatively warm, (17–19°C), low-salinity (35.7–35.8‰) nutrient-depleted oceanic water of tropical origin that flows southward at relatively high velocity (0.1–1.4 m s^{-1}; 0.2–3.0 knots, ~20 km per day) along the western Australian continental slope and eastward into the Great Australian Bight (Pearce 1991). Temperature decreases from 26.3°C at the northern origin to 21.5°C at Cape Leeuwin to 18°C in the Great Australian Bight.

The Leeuwin Current originates off northwest Australia where the Indonesian Throughflow creates a warm-water pool. This in turn drives a meridonal pressure gradient that through geostrophy drives an onshore transport towards the western Australian coast. This situation results in a strong alongshore pressure gradient that provides the driving mechanism for a poleward flowing boundary current. The flow is driven by a large-scale meridonal pressure gradient that generates onshore geostrophic transport that is sufficient to exceed Ekman transport induced by equatorward wind stress. This is a unique current because it flows against the prevailing winds and transports warm, low-salinity, oligotrophic tropical water southward along the western Australian coast and then eastward across the Great Australian Bight. The strong alongshore pressure gradient is usually sufficient to overwhelm the effects of coastal, wind-forced upwelling. It has a strong seasonal cycle with greatest strength in the autumn and early winter, associated with weakening of alongshore winds from the southwest. It flows into the Great Australian Bight to about 120° E (off Esperance), although in many years it may penetrate further east.

There is also a substantial alongshore steric height gradient eastward from Cape Leeuwin capable of driving the Leeuwin Current eastward (it is here that the Leeuwin Current flows fastest). The influence of the Leeuwin Current diminishes eastwards such that off the eastern Great Australian Bight (135° E) it only drives ~15% of the flow. Ridgway and Condie (2004) concluded that fortuitously, the west coast pressure gradient delivers the Leeuwin Current to the southern margin just in time for the winds to change to winter westerlies and drive it eastwards.

As the Leeuwin Current propagates southward along the western coast of Western Australia it passes over the high-salinity subsurface core of the South Indian Central Water and the associated mixing makes it progressively more saline and cooler (Webster et al. 1979). Thus, there is a seasonal increase in the salinity of the Leeuwin Current as it approaches Cape Leeuwin (due to this drawing up of the South Indian Central Water), such that after rounding Cape Leeuwin the Leeuwin Current is more saline than surrounding waters. The saline water is, however, slowly lost through energetic mixing with fresher offshore waters. The Leeuwin Current appears to reach a longitude of ~138° E (south of Kangaroo Island) before its temperature-salinity signature is swamped by the Great Australian Bight outflow onto the slope.

South Australian Current The Leeuwin Current merges with the South Australian Current in autumn (Black 1853) as a plume of warm (2–3°C above the surrounding waters), saline water from the Great Australian Bight. The South Australian Current water forms in summer and moves east and off the shelf just as the winter winds start to blow from the west, driving it eastward. The current reaches its maximum flow during May, June, and July. Wind stress curl that leads to the anticyclonic gyre in the Great Australian Bight is intensified off Eyre Peninsula. In the western Great Australian Bight, seaward surface Ekman transport and Flinders Current shoreward transport converge to form a ridge over the shelf edge resulting in an eastward current over the shelf break and downwelling to 100 mwd. Temperature patterns of the Leeuwin Current indicate that the saline pulse from the Great Australian Bight travels 3000 km in ~2.5 months (speed of 0.4 cm s^{-1}, ~1 knot).

Zeehan Current Further east, there is a strong southeasterly flow off western Tasmania, called the Zeehan Current (Baines et al. 1983). The current is seasonally reversing, and steric height measurements show that the strongest flow is in winter and spring (June to November), although there is a southward flow all year round. Once the full winter state has been established with a positive sea level anomaly, the jet of warm water moves progressively eastward until it reaches the coast of Tasmania in July. Surface water decreases in temperature as it moves east from 18.3°C in the Great Australian Bight to 14°C along the Tasmania coast. The higher salinity water that is entrained into the flow comes from the Great Australian Bight and not from Spencer Gulf, Gulf St. Vincent, or the Indian Ocean. The winter outflow of cool and very saline waters from Spencer Gulf intrudes

onto the shelf and slope and finds its own density level at 250–300 mwd; too deep to be incorporated into the shallower flow.

2.3.6 Tidal Currents and Internal Waves

Tidal currents on the southern Australian shelf are small (<10 cm s^{-1}, <0.2 knots) but are locally amplified somewhat in the gulfs. The nature and role of internal tides and solutions is unknown but inertial wave currents off the Bonney Coast are relatively large (20 cm s^{-1}; 0.4 knots) (Middleton and Bye 2007).

2.3.7 Seasonal Variability and Trophic Resources

Prevailing westerlies reverse during summer months and this together with coastal heating leads to the formation of warm waters in the Great Australian Bight and South Australian Sea. These east and southeasterly winds result in weak coastal currents that flow westward. Wind stress curl can lead to an anticyclonic gyre in the Great Australian Bight as well as the formation of the South Australian Current. For over 25 years it has been known that summer coastal upwelling occurs along the narrow continental shelf adjacent to the Bonney Coast between Cape Jaffa and Portland (Rochford 1977; Lewis 1981; Schahinger 1987; Griffin et al. 1997). Simultaneous upwelling also occurs along wider segments off western Victoria (Levings and Gill in press), south of Kangaroo Island, at the mouth of Spencer Gulf (Hahn 1986), and along the western coast of Eyre Peninsula (Kämpf et al. 2004; Middleton and Bye 2007) but does not usually shoal to the surface and is often masked in sea surface temperature images by the warm surface layer.

Strong winter westerlies and cooling lead to intense downwelling throughout to 200 mwd or more as well as the formation of dense coastal waters in the Great Australian Bight and the South Australian Sea. There is no upwelling during winter. Dense waters that have been evaporated during the summer and subsequently cooled exit the Great Australian Bight and Spencer Gulf as gravity currents to equilibrium depths of ~200 m. Winter cooling also leads to the formation of cold and dense waters (14–15°C, 36‰) in shallow waters of the Lacepede Shelf.

2.4 Synopsis

The southern part of Australia is a passive continental margin presently covered by cool-water carbonate sediments. The strongly seasonal climate varies from arid to humid. The ocean that covers the shelf lies north of the subtropical convergence zone and is a high-energy, swell dominated hydrodynamic regimen.

1. Southern Australia is composed of two major Precambrian cratons bounded on the east by the Tasman Fold Belt System, a series of Neoproterozoic to late Paleozoic continental margin and accreted terranes. The tectonic grain is predominantly north-south. This fragment of Gondwana is truncated by the Southern Australian Rift System, an east-west, Mesozoic–Cenozoic structure involving rift and drift coincident with separation of Australia from Antarctica. Marginal rift basins and subsequent shelf deposits in this system comprise Jurassic to Cretaceous siliciclastic sedimentary rocks overlain by Cenozoic limestones. The entire region underwent tectonic inversion and uplift in the middle Miocene resulting in reduced shelf accommodation and exposure of the Paleogene and early Neogene limestones in a series of shallow epicontinental basins.
2. A strong, mid-latitude, high-pressure cell dominates weather in Australia today. The high lies over the southern Australian coastline during summer resulting in an overall semi-arid climate. It shifts northward during winter, allowing a succession of low-pressure systems with accompanying rains and strong westerlies to dominate weather across the region. This pattern results in an overall climate that is arid to semi-arid across the Great Australian Bight and South Australian Sea but becomes progressively more humid southward in Victoria and around Tasmania.
3. Waters on the continental shelf are largely sub-tropical and separated from cold Antarctic waters by the Subtropical Convergence Zone. There is unimpeded zonal water circulation from the west (West Wind Drift). The sea state is swell–dominated with high wave heights and long period swells. Storm wave base is estimated to lie on average at ~60 m water depth (mwd) whereas swell wave base extends to ~140 mwd.
4. The ocean south of Australia comprises five stacked water masses:

   - *Antarctic Bottom Water:* very cold, high salinity waters with high dissolved oxygen and high nutrient levels—it has little affect on the Australian shelf or slope.
   - *Circumpolar Deep Water:* 4500–1250 mwd, cold, moderate salinity waters with low dissolved oxygen but very high nutrient levels—it has little effect on the Australian shelf or slope.
   - *Antarctic Intermediate water:* 1250–800 mwd, intermediate temperature, nutrient-rich water with high oxygen contents but low salinity levels that are recognizable as a salinity minimum throughout the Southern Ocean.
   - *Subantarctic Mode Water:* 800–450 mwd, a homogeneous, intermediate temperature (8–10°C), low oxygen, intermediate salinity water mass with moderate nutrient levels adjacent to the southern Australian continental margin.
   - *Subtropical Surface Water:* <450 mwd, a relatively warm (10–22°C) water mass with intermediate oxygen content and low nutrient levels that is complex and locally saline because of intense seasonal evaporation.

5. Current systems are strongly affected by climatic seasonality. The autumn change from an easterly to westerly wind pattern results in a change of Eckman transport and a switch in coastal currents. The regimen is over all downwelling with local cool summer upwelling. Nearsurface environments are affected by two currents systems:

   - The *Flinders Current System* is an upwelling-favourable northern boundary current that extends from the surface to ~800 mwd and flows from south of Tasmania westward and northward, generally outboard of the shelf edge to Cape Leeuwin. It is composed of Antarctic Intermediate Water and Subantarctic Mode Water that is cool, well oxygenated, and with moderate nutrient levels.
   - The *Leeuwin Current System* is a seasonal phenomenon that flows eastward along the

entire length of the continental shelf edge. It is composed of relatively warm, nutrient-depleted subtropical surface waters, and generally prevents upwelling. The system comprises Leeuwin Current waters in the west, South Australian Current waters in the center, and Zeehan Current waters in the east.

6. Upwelling of waters from the Flinders Current occur in summer but are localized to the Bonney Shelf, off Kangaroo Island, near the mouth of Spencer Gulf, and along the eastern coast of Eyre Peninsula. There is no upwelling during winter because of the strong westerlies.

7. Spencer Gulf and Gulf St. Vincent are inverse estuaries in which strong summer heating and evaporation form saline waters at the head of each gulf. These waters cool in winter and flow oceanward as dense saline bottom currents.

Chapter 3
The Pleistocene Record

3.1 Introduction

The focus of this book is on sediments that currently cover the continental margin seafloor across southern Australia. It quickly became obvious during the early stages of our research, however, that these deposits are not simply a product of modern processes but that they are inexorably linked to the Pleistocene prehistory. The influence of the Pleistocene has been two-fold; (1) antecedent topography created by Pleistocene deposition has formed a template upon which modern sediments are accumulating–this control is especially important in shallow marine environments, (2) today's seafloor sediments contain older particles that were produced in the Pleistocene.

Glacial-driven eustatic cyclicity on a 20,000–100,000 year scale (Fig. 3.1) resulted in interglacial highstands at or near modern sea level and glacial lowstands that exposed much of the continental shelf (Figs. 3.2, 3.3, 3.4). Australia, like other continents in the mid latitudes under Hadley Cell circulation (the intertropical zone), was as much as 10°C colder, significantly drier, and also much windier than today (Bowler 1976; Zheng et al. 1998; Williams 2000; Ruddiman 2001) during the last glacial maximum (LGM). There was nearly 25% greater land area (Figs. 3.2, 3.3, 3.4) and land bridge links to Tasmania and New Guinea (Fig. 3.5) at that time. The axis of winter storms moved south and considerable amounts of aeolian dust blew off the arid center towards the south and southeast (Williams et al. 1998; Williams 2001). Sea surface temperatures along the southern continental margin are estimated to have been 3–5°C colder than today (Williams et al. 1998).

As ice caps melted during the latest Pleistocene, sea-level rose, evaporation from the intertropical oceans increased, the land bridge connections were severed, and the climate became warmer and wetter (Williams et al. 1998; Williams 2000). The dune systems in southern Australia became rapidly vegetated as exemplified on the Roe Plains (Lowry 1970; James et al. 2006).

The record of Pleistocene marine deposition is, however, not straightforward. This is largely because there was little accommodation on the shelf and so most sediment generated there was swept outboard onto the slope or inboard into beach-dune complexes (Fig. 3.6). As a result, the record lies in (1) continental slope sediments, (2) sediments flooring the large gulfs, and (3) aeolianite complexes, but not on the open shelf. Slope deposits contain a record that extends through the entire Cenozoic but is particularly well developed in the Pleistocene (see below). The sedimentary record in the gulfs and the aeolianites is restricted to sea-level highstands.

Sediment on the shelf does, nevertheless, contain evidence of late Pleistocene deposition. Such evidence is not stratigraphic but is compositional. Current seafloor sediments are an assortment of late Pleistocene and recent grains (Fig. 3.7). The oldest Pleistocene particles are *relict grains*, skeletons and intraclasts that were generated during Marine Isotope Stages 3 and 4 (MIS 3 and 4) when the shelf was partially flooded (Figs. 3.1, 3.8, 3.9, 3.10). The younger Pleistocene grains were formed during early stages of the post-last glacial maximum (LGM) sea level rise (MIS 2). These *stranded grains* accumulated in shallow-water environments and are now marooned on a seafloor that is much deeper than when they were produced. The importance of these two late Pleistocene components cannot be overemphasized because they are the most numerous constituents in

N. P. James, Y. Bone, *Neritic Carbonate Sediments in a Temperate Realm*, DOI 10.1007/978-90-481-9289-2_3, © Springer Science+Business Media B.V. 2011

Fig. 3.1 A sea level curve for the southern Australian continental margin, (adapted from Chappell and Shackleton 1986; Hails et al. 1984a, b; Cann et al. 1988). The upper curve illustrates the position of sea level and pertinent marine isotope stages for the last 150,000 years. The lower curve details the last 75,000 years, highlighting the periods when relict, stranded, and Holocene skeletal grains formed as well as the position of sea level during the last glacial maximum (LGM)

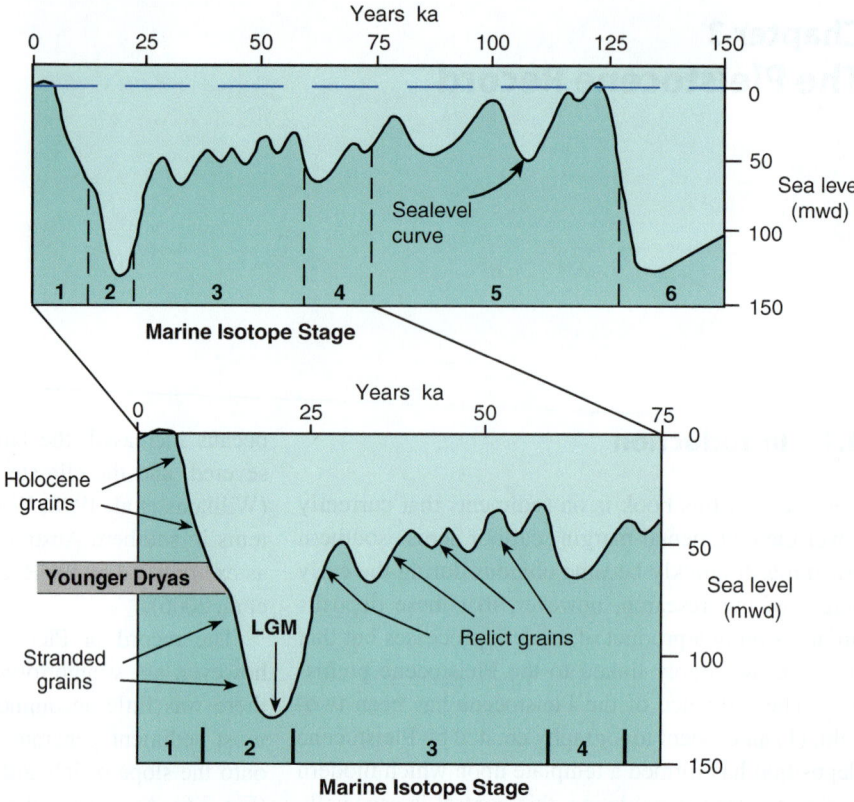

many surface sediments on the modern shelf. The relict and stranded particles are now being mixed with Holocene carbonate gains to form a palimpsest sediment apron across the neritic part of the continental margin.

3.2   The Continental Slope Record

The continental slope ranges from a relatively gentle incline locally incised by numerous submarine canyons (von der Borch 1968; Exon et al. 2005; Hill et al. 2005) to a broad progradational sediment wedge (Feary et al. 2000b) to a relatively steep slope marked by fault detachment (von der Borch and Hughes-Clarke 1993). The clearest record is in the prograding wedge in the central Great Australian Bight. There the slope laps down onto the Eyre Terrace as an impressive ~550 m-thick sediment wedge (Fig. 3.11) composed of fine-grained carbonate derived largely from the shelf, together with bryozoan reef mounds (James et al. 2000, 2004; Feary et al. 2004).

3.2.1   Slope Sedimentation

Most of the shelf edge and upper slope in the Great Australian Bight has a pronounced progradational seismic geometry (Fig. 3.11), with calculated accumulation rates for the Pliocene–Pleistocene succession of ~26 cm ky^{-1}. The succession in the Eucla sector is, however, an unusually thick, sigmoid-shaped sedimentary package with spectacular clinoform geometry. At its thickest point, this prograding unit is more than twice as thick as the entire underlying succession of upper Miocene to middle upper Eocene cool-water carbonates. The highest calculated accumulation rate is 62.5 cm ky^{-1}. This rate compares favorably and even exceeds accumulation rates estimated for warm shallow-water carbonate sedimentary environments (30–100 cm ky^{-1}) (Schlager 1981; James and Bone 1991). It is clear that the vigorous off-shelf transport of carbonate sediments from the shallow-water carbonate factory is of critical importance for the development of this prograding wedge (James et al. 1994).

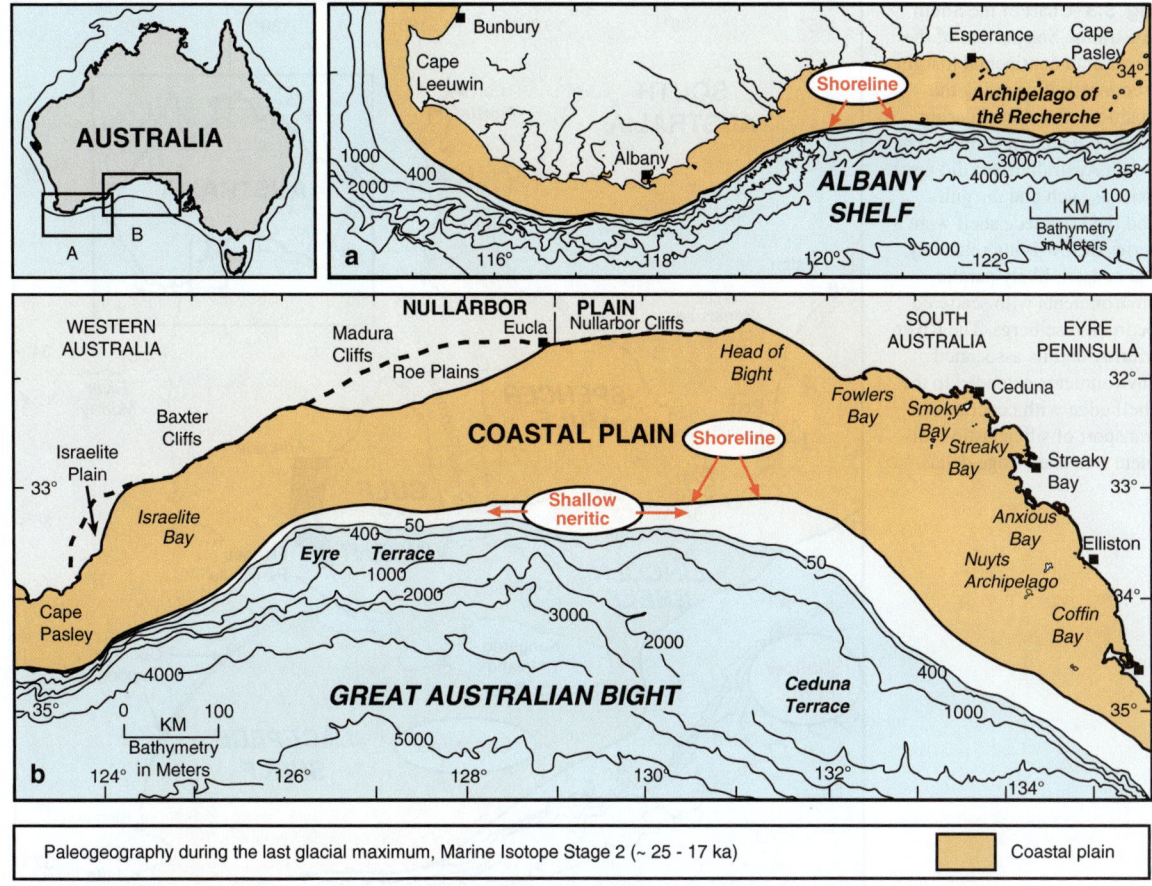

Paleogeography during the last glacial maximum, Marine Isotope Stage 2 (~ 25 - 17 ka)   Coastal plain

Fig. 3.2 Charts of two sections of the southern Australian continental margin (see inset), illustrating the position of sea level during the last glacial maximum. In most cases the modern neritic zone was a wide coastal plain. The Albany Shelf (**a**) was mostly exposed with a narrow shallow neritic zone whereas the Great Australian Bight (**b**) was a vast low, coastal plain more than 200 km wide in places adjacent to ~ a 20 km wide shallow neritic environment

The Plio-Pleistocene clinoform succession shows pronounced cyclicity (Feary et al. 2000b). All the global Marine isotope stages (Shackleton et al. 1990) can be recognized when the data are compared using the age models developed for each site and the SPECMAP stack (Imbrie et al. 1984; Holbourn et al. 2002; Andres et al. 2003; Andres and McKenzie 2004). The Great Australian Bight oxygen isotope signals have unusually low amplitudes when compared to other oxygen isotope curve based on planktonic foraminifers. This expression could be due to the contribution from shelf-derived sediment or to only subtle water temperature changes at the seafloor during sea level fluctuations (Holbourn et al. 2002).

Mid-Pleistocene cycles (~10 m thick) can be correlated to sea level changes arising from obliquity-scale (41 ky) orbital forcing (Saxena and Betzler 2003). Fine-grained sediment (wackestone) at the bottom of each cycle grades upward into coarse-grained sediment (packstone) at the top. Tunicate spicules, brown bioclasts, bryozoan fragments, and coralline algal debris were shed to the slope during sea level rises whereas sponge spicules and carbonate mud were shed during sea level falls. Such particle constituent partitioning produced distinct mineralogic cyclicity. High magnesium calcite (HMC) and aragonite are most abundant during sea level rises and highstands (Saxena and Betzler 2003; Andres and McKenzie 2004; Swart et al. 2004). Siliciclastic input, in the form of clay minerals, was highest during sea level highstands, reflecting higher rainfall and runoff.

Late Pleistocene cycles comprise alternating olive gray (coarse grained and aragonite and HMC rich) and light gray (fine grained and LMC-rich) sediment (Brooks et al. 2004; Simo and Slatter 2004). Coarse

Fig. 3.3 Chart of the South Australian Sea portion of the Australian continental margin (see inset), illustrating the position of sea level during the last glacial maximum. The South Australian Sea did not exist as such and the gulfs and the Lacepede shelf were a series of low-lying relatively flat coastal, likely paralic environments with scattered bedrock inselbergs. The River Murray and its associated environments extended to the shelf edge with concomitant transport of siliciclastic sediment into shelf-edge deltas.

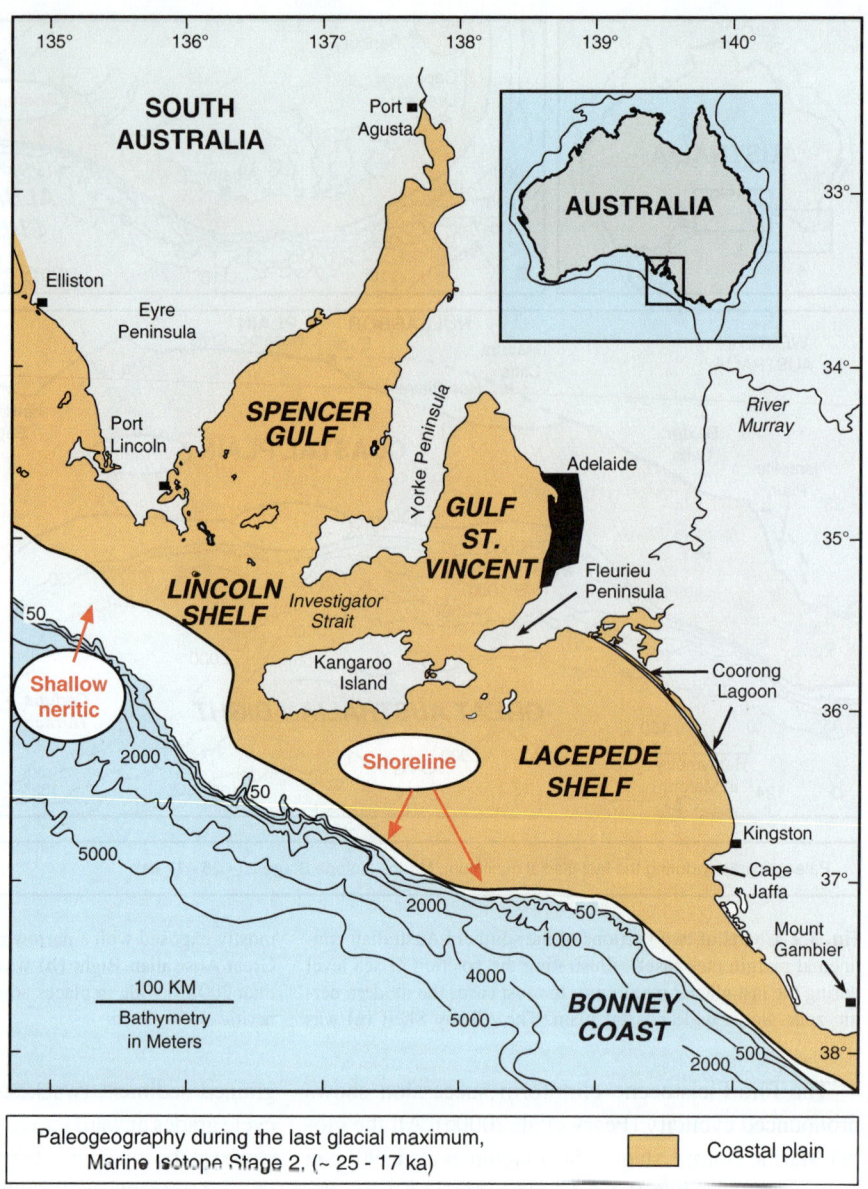

material is interpreted to represent the lower sea level part of the cycle with the finer material representing the higher sea level component.

Hine et al. (2004) determined that the most rapid sediment accumulation rates occurred during the fastest transgressive component of the sea level cycle. Calculated sediment accumulation rate between 19 and 13.95 ka, for example, was 656 cm ky^{-1}. This indicates that when the vast adjacent shelf is initially flooded, there is a pulse of sediment production and off-shelf export that decreases toward peak flooding.

3.2.2 Biogenic Mounds

Bryozoan-rich, biogenic mounds (Figs. 3.11, 3.12) grew periodically on the prograding carbonate slope of the central Great Australian Bight throughout Pliocene–Pleistocene time (James et al. 2000, 2004). Cores from three ODP Leg 182 drill sites provide a record of mound growth during the LGM. These mounds grew between paleodepths of 100 and 240 m with the upper limit of growth limited by swell wave base, and the lower boundary fixed by an oligotrophic

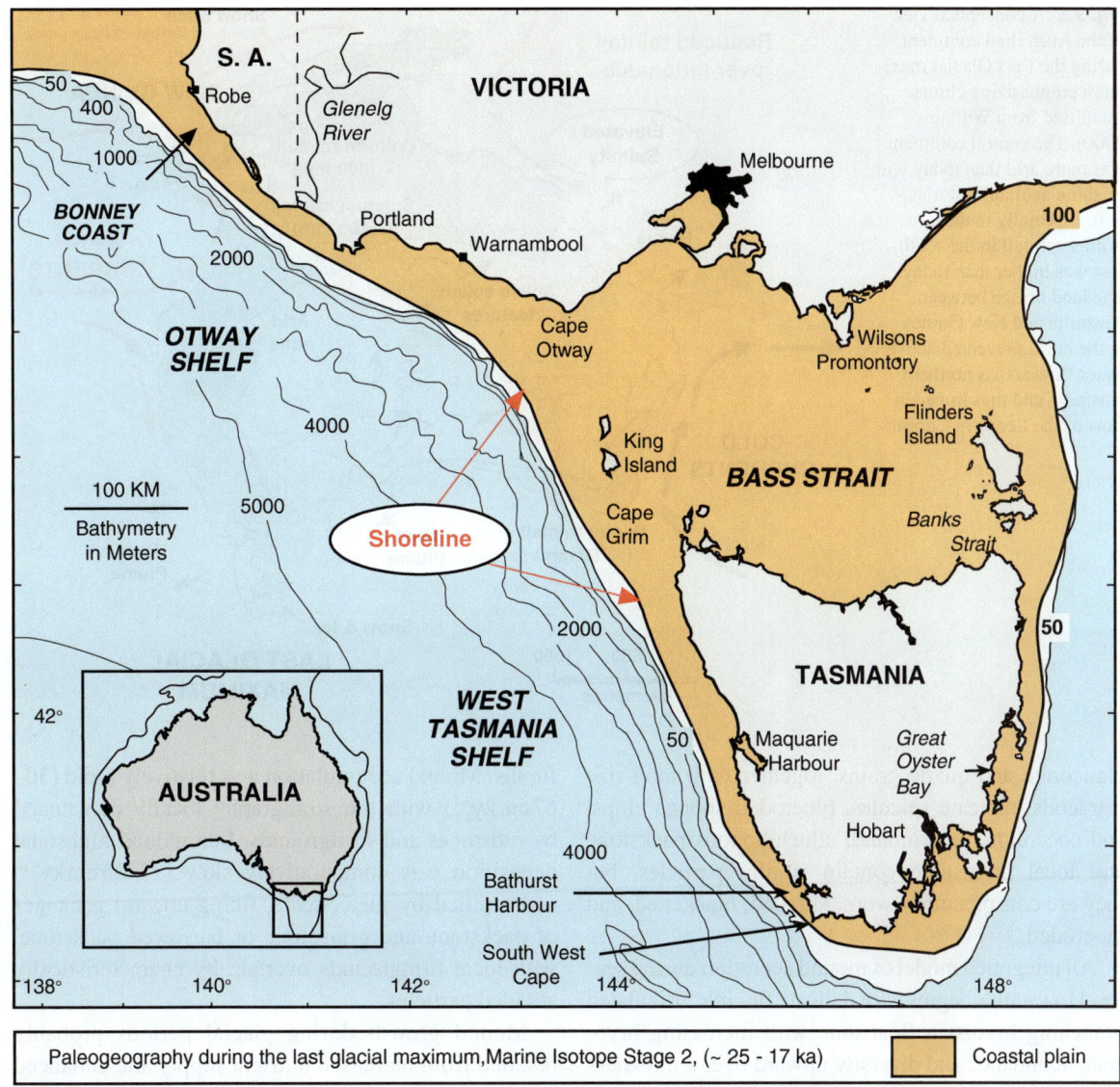

Paleogeography during the last glacial maximum, Marine Isotope Stage 2, (~ 25 - 17 ka)   Coastal plain

Fig. 3.4 Chart of the Southeastern portion of the Australian continental margin (see inset), illustrating the position of sea level during the last glacial maximum. The southeast coastal plain was relatively narrow but Bass Strait was a wide shallow depression, with hills and lakes along both sides

water mass. Detailed chronostratigraphy, based on radiometric and U-series dating, benthic foraminifer stable-isotope stratigraphy, and planktic foraminifer abundance ratios, confirm that buildups flourished during glacial lowstands (even-numbered MIS) but were largely moribund during interglacial highstands and are not forming today. The most recent mounds, whose tops are 7–10 m below the modern seafloor, flourished during the last glacial lowstand but perished during transgressive sealevel rise.

Mound floatstones (Fig. 3.12) are compositionally a mixture of *in situ* bryozoans comprising 96 genera (Bone and James 2002) and characterized by fenestrate, flat rigid robust branching, encrusting multilaminar, domal, globose and arborescent, robust rigid fenestrate and branching, delicate flexible articulated branching growth forms. The packstone matrix comprises autochthonous and allochthonous sand-size bryozoans, benthic and planktonic foraminifers, serpulids, coralline algae, sponge spicules, peloids, and variable

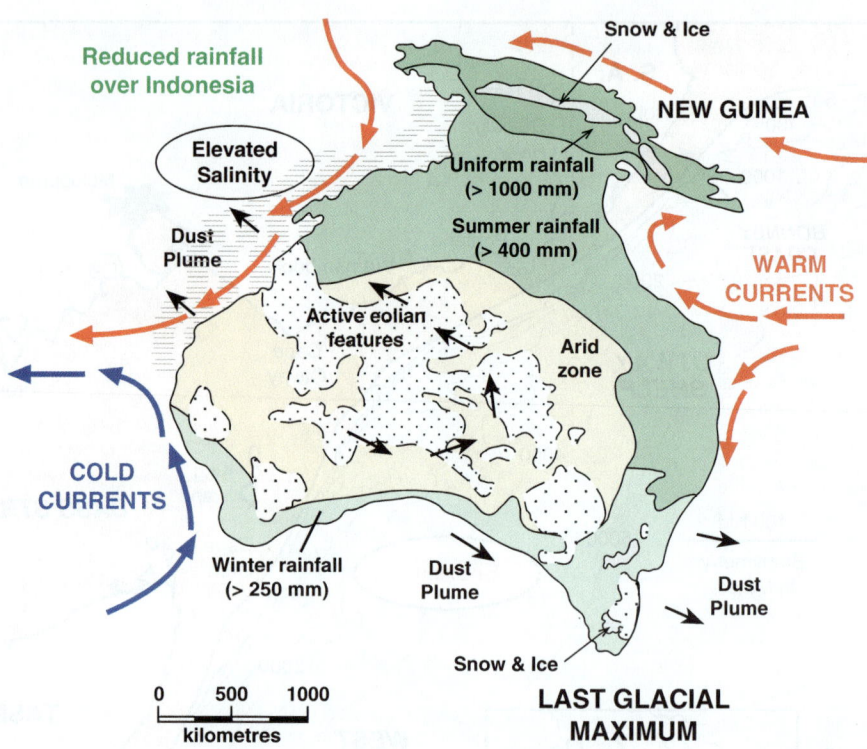

Fig. 3.5 A conceptual view of the Australian continent during the Last Glacial maximum emphasizing climate (modified from Williams 2000). The central continent was more arid than today with offshore aeolian dust transport, especially to the east. Winter rainfall in the southeast was higher than today. The land bridge between Australia and New Guinea in the north prevented active water flow across northern Australia and thus impeded flow of the Leeuwin Current.

glauconite and quartz grains, together with mud-size ostracods, tunicate spicules, bioeroded sponge chips, and coccoliths. Intermound, allochthonous packstone and local grainstone contain similar particles, but they are conspicuously worn, abraded, blackened, and bioeroded.

An integrated model of mound accretion during sea-level lowstands begins with delicate flexible articulated branching bryozoan floatstone with increasing bryozoan abundance and diversity upward over a thickness of 5–10 m, culminating in thin intervals of grainstone characterized by reduced diversity and locally abraded

fossils. Mound accumulation was relatively rapid (30–67 cm ky^{-1}) with the stratigraphy locally punctuated by rudstones and firmgrounds. Intermound highstand deposition was comparatively slow (17–25 cm ky^{-1}) and typified by meter-scale, fining-upward packages of packstone and grainstone or burrowed packstone, with local firmgrounds overlain by characteristically abraded particles.

Mound growth during glacial periods probably resulted from increased nutrient supply and enhanced primary productivity. Large specimens of benthic foraminifers restricted to the mounds confirm overall

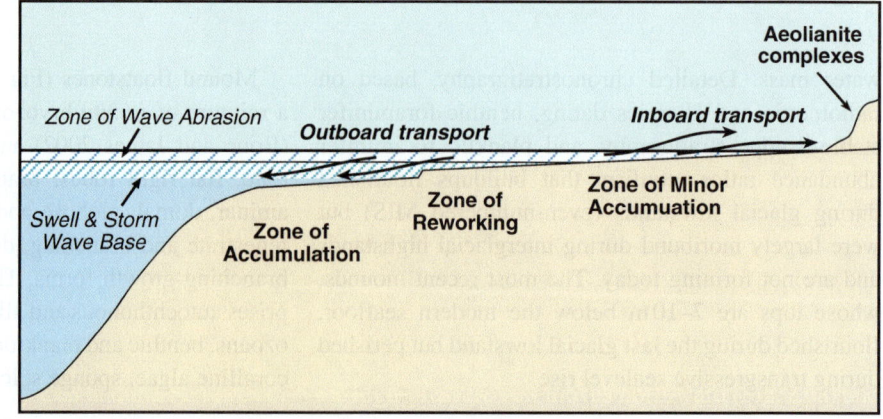

Fig. 3.6 A simplified sketch of the southern Australian continental shelf illustrating the main zones of sediment movement as related to modern ocean hydrodynamics

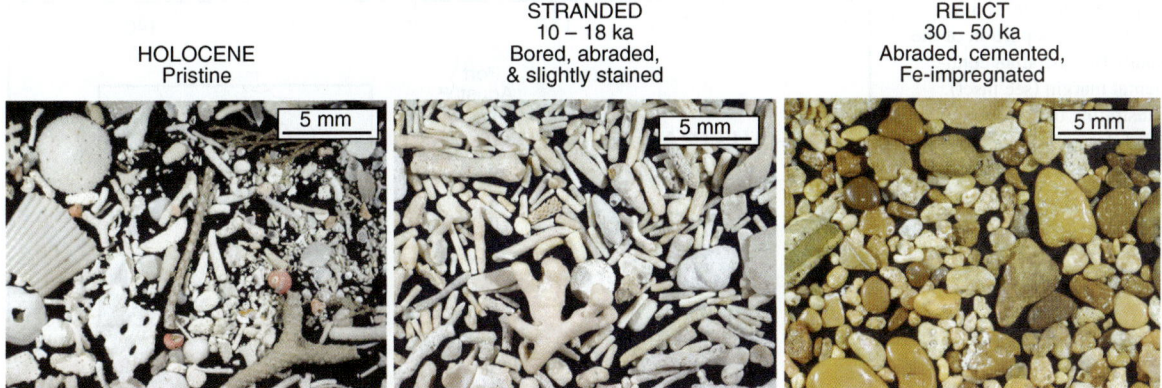

Fig. 3.7 Images of the grain types produced in the neritic zone at different times during the late Quaternary. (Modified from James et al. 2005)

mesotrophic growth conditions. Such elevated trophic resources were both regional and local, and thought to be focused in this area by cessation of Leeuwin Current flow, together with northward movement of the subtropical convergence and related dynamic mixing.

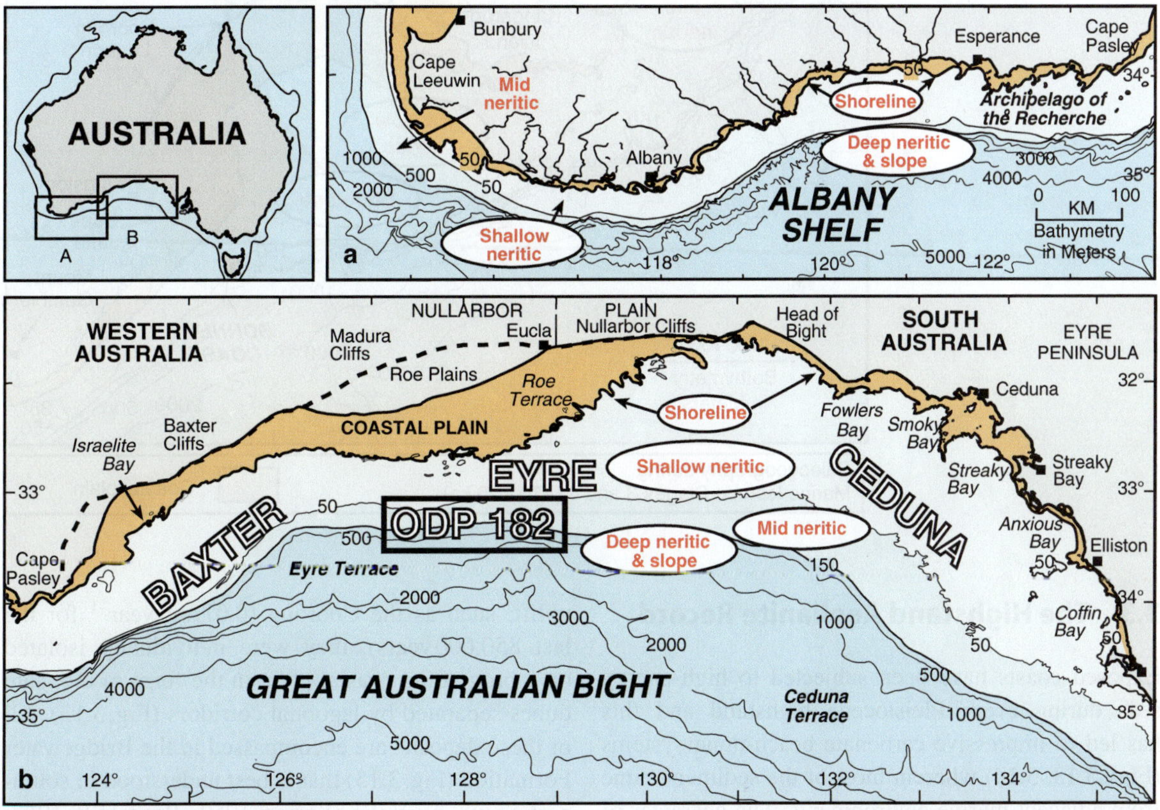

Fig. 3.8 Charts of the two western sections of the southern Australian continental margin (see inset), illustrating the position of sea level during Marine Isotope Stages 3 and 4 (see Fig. 3.1) when sea level was roughly 50 m below what it is today. The shallow neritic zone extended across most of the Albany Shelf (**a**) and the Great Australian Bight (**b**) except for the Roe Terrace, which was exposed

Fig. 3.9 Chart of South Australian Sea portion of the southern Australian continental margin (see inset), illustrating the position of sea level during Marine Isotope Stages 3 and 4 (see Fig. 3.1) when sea level was roughly 50 m below what it is today. The gulfs were likely lacustrine during this period

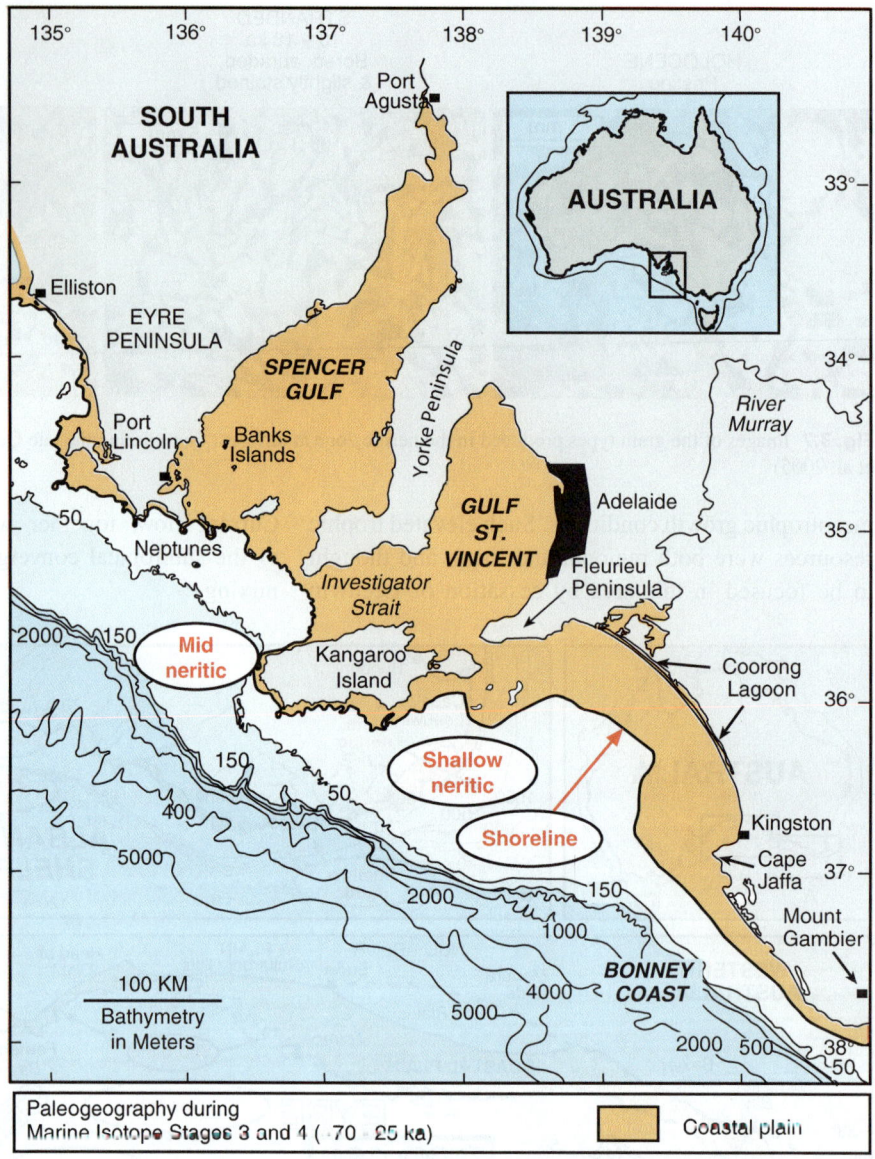

3.3 The Highstand Aeolianite Record

Exposed coasts have been subjected to high-energy seas during every Pleistocene highstand and this has led to impressive carbonate beach-dune systems (Figs. 3.13, 3.14) wherein most of the sediment came from adjacent marine environments. The character of the aeolianites, in turn, depended on the local tectonics. Aeolianites on tectonically stable coasts adjacent to Precambrian massifs were stranded in the form of vertically stacked complexes with eroded seaward margins (Fig. 3.14). By contrast, in areas of gradual

uplift, such as the Coorong (0.07 mm year^{-1} for the last 850,000 years), they were individually isolated on a prograding coastal plain in the form of elongate dunes separated by lagoonal corridors (Fig. 3.13). All of these deposits are encompassed in the Bridgewater Formation (Fig. 3.15) that is best understood in southeast South Australia (Sprigg 1952; Boutakoff 1963; Cook et al. 1977; Schwebel 1983; Gostin et al. 1988; Wilson 1991; Belperio 1995). Whereas the stacked aeolianites have not yet been isotopically dated, the prograding beach/dune-lagoon systems in the Coorong have, via a variety of techniques, including

Fig. 3.10 Chart of the eastern part of the southern Australian continental margin (see inset), illustrating the position of sea level during Marine Isotope Stages 3 and 4 (see Fig. 3.1) when sea level was roughly 50 m below what it is today. Bass Strait was likely lacustrine whereas the shelves were shallow neritic throughout

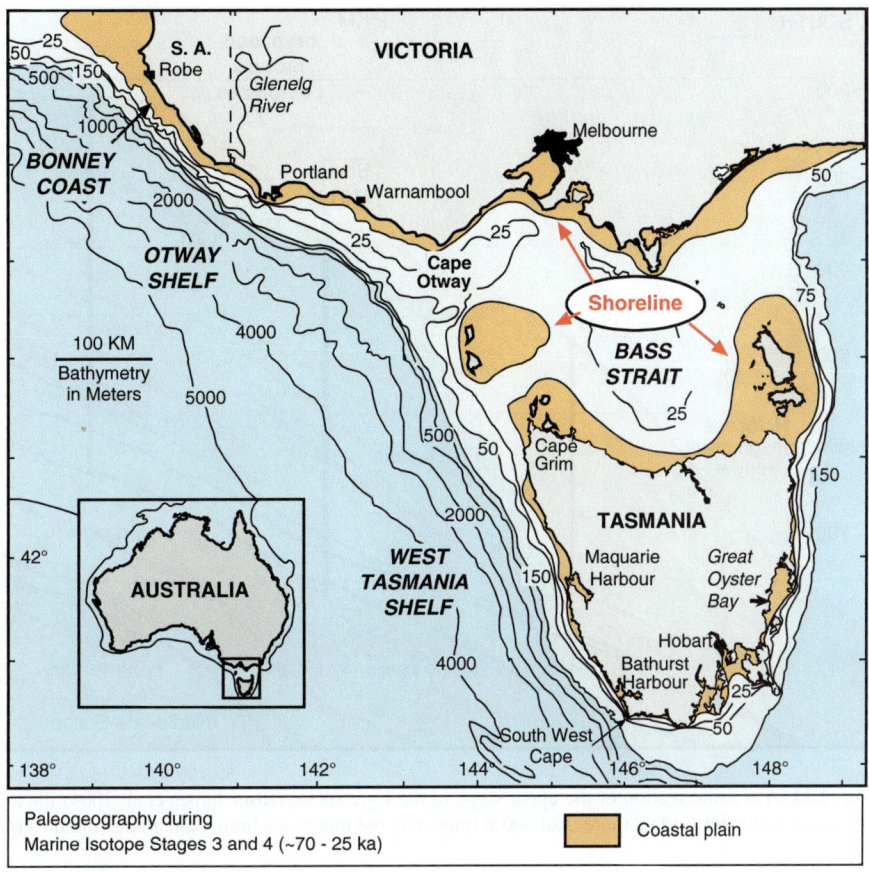

thermoluminescence and amino acid racimization been correlated with the global Quaternary sea level curve (Fig. 3.13) such that each dune ridge corresponds to a sealevel highstand and insolation maximum (Sprigg 1952, 1979; Cook et al. 1977; Schwebel 1983; Murray-Wallace and Belperio 1991; Murray-Wallace et al. 1991, 2001; Huntley et al. 1993; Belperio 1995; Murray-Wallace 2002).

3.4   The Spencer Gulf and Gulf St. Vincent Record

The Late Pleistocene history of the central part of the region has been largely determined via extensive coring in the large gulfs (Belperio et al. 1984; Hails et al. 1984b; Cann and Gostin 1985; Cann et al. 1988, 1993). Poorly sorted, mollusc-rich peritidal Glanville Formation sediments (MIS 5e –~125,000 ka) form an extensive blanket beneath younger sediments in Spencer Gulf and Gulf St. Vincent and in many places outcrop along the shore (Figs. 3.14, 3.15). These deposits are rich in molluscs that live in the gulfs today whereas MIS 5e age deposits also contain several warmer water neritic organisms that no longer populate southern Australian waters, in particular the coral *Goniopora somalienesis*, the blood cockle *Anadara trapezia*, the pearl oyster *Pinctada carchariarum*, the gastropod *Euplica bidentata*, and the large benthic foraminifer *Marginopora vertebralis* (Lindsay and Harris 1975; Ludbrook 1976; Bone 1978). Overall shallow seawater temperatures are calculated to have been 7–9°C warmer than today in similar modern environments. It also is estimated that at that time, sea level stood ~2 m higher than it does today (Bone 1978; Hails et al. 1984a; Cann et al. 1988).

Younger MIS 5 interglacial deposits, which were deposited during highstands that did not reach modern levels, are also present in the gulfs. The 10 m-thick False Bay Formation (Fig. 3.15) accumulated as a series of restricted gulf, beach, and estuary-lagoon,

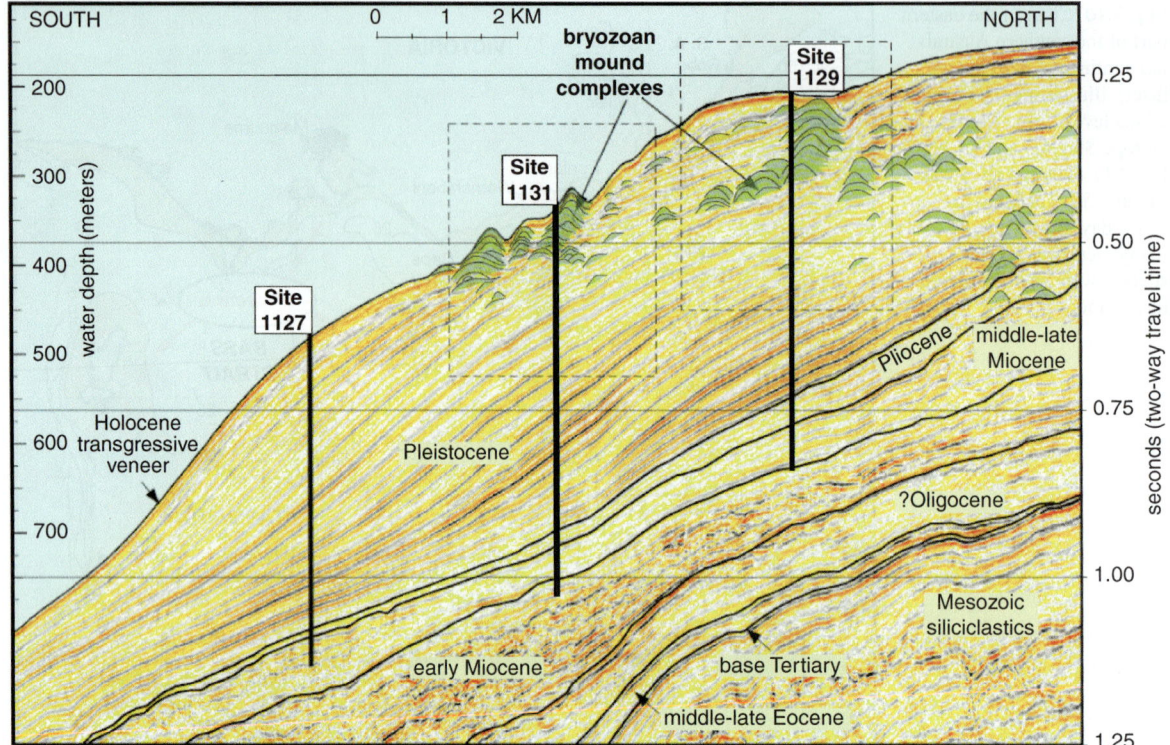

Fig. 3.11 A seismic section of the upper slope in the Eyre sector (from James et al. 2004) illustrating the Pleistocene prograding carbonate sediment wedge, more than 500 m thick, and the numerous bryozoan-sponge mounds (green)

sabkha-like peritidal environments during MIS 5c (~110,000 ka) when sea level rose to within 8 m of modern levels. The overlying Lowly Point Formation (Fig. 3.15), which is mostly supratidal, laminated clays with gypsum and aeolian sands, formed in protected lagoons some 14 m below modern sea level during MIS 5a (~80,000 ka) (Hails et al. 1984a).

3.5 The Continental Shelf Record

3.5.1 Overview

Much of the shelf has virtually no Pleistocene record. The impression of sea level rise and fall is in the form

Fig. 3.12 Slabbed core from a bryozoan reef mound illustrating large *Celleporaria* sp. bryozoan colonies (arrows) in floatstone; ^{14}C age = 48,000 years BP (James et al. 2004); Ocean Drilling Program Core 1132B-03H-3 -20.33 to 20.83 cm (~22 mm below sea floor; unit 2B); cm scale

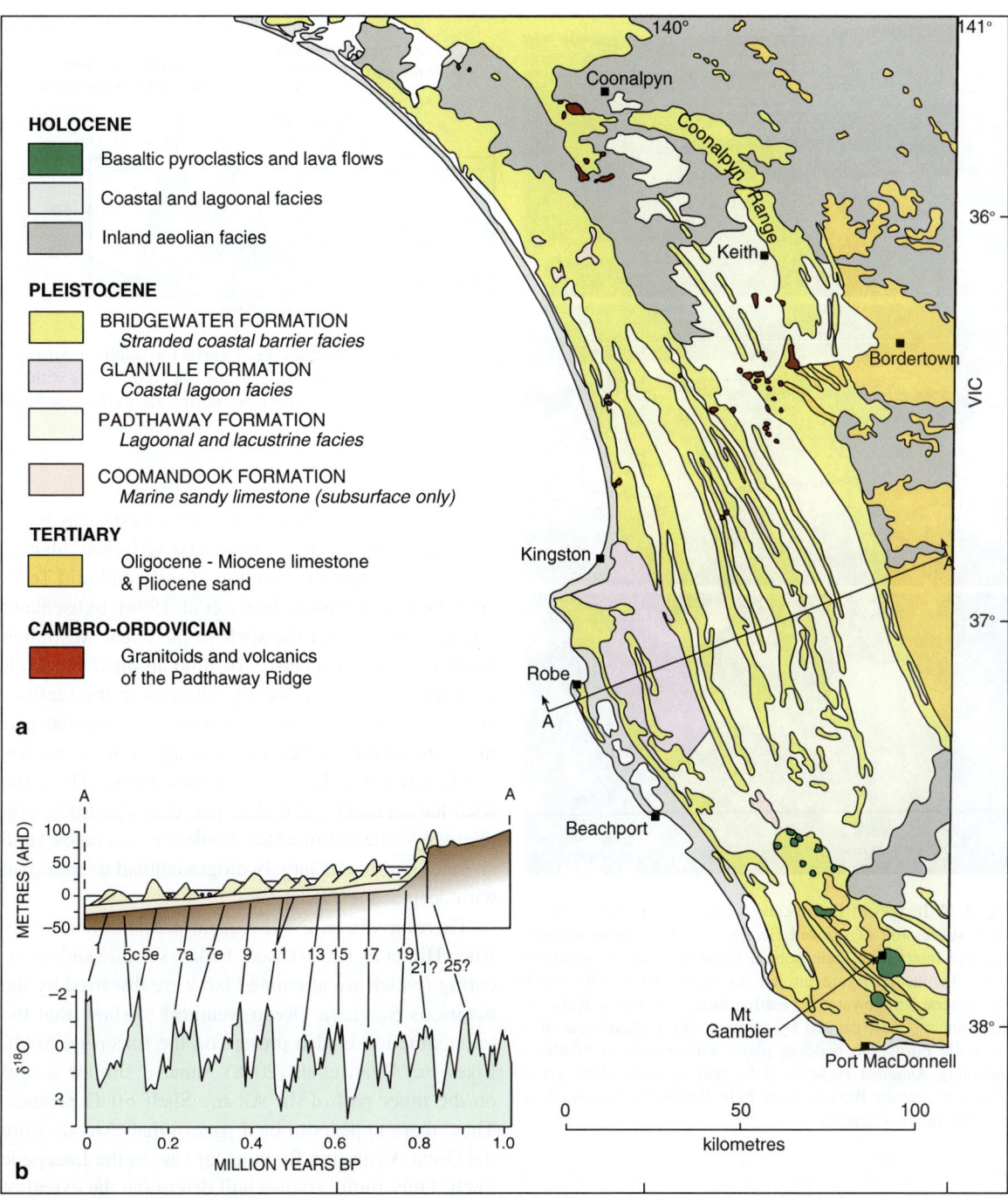

Fig. 3.13 A geological map of the southern part of the Lacepede Shelf, the Bonney Coast and the prograding coastal plain inland from the shore-hugging Younghusband Peninsula. The series of Pleistocene aeolianite complexes (called Ranges) are highlighted. Cross-section A-A is depicted at lower left with each dune complex correlated to a known sea level highstand. (After Belperio 1995)

of recurring shelf terraces and a patchy sediment cover. Bathymetric profiles universally record a series of terraces or ridges on most shelves, many of which recur at similar, but not exact, water depths. The most prominent is the terrace and small submarine cliff at 110–130 mwd that probably represent the Late Glacial Maximum sea level lowstand (~20 ka). The other prominent terrace or ridge at ~50 mwd, (the 25 fathom

Fig. 3.14 Images of Pleistocene carbonates. **(a)** The ~100 m-thick succession of stacked Bridgewater Formation aeolianites and intervening paleosols at Cape Spencer on Southern Yorke Peninsula, figure circled for scale, **(b)** an 18 m-thick Pleistocene Bridgewater Formation dune complex at Robe on the Bonney Coast capped by calcrete, **(c)** a close view of a Glanville Formation bedding plane with numerous bivalves, including *Anadara trapezia* (left) and zooxanthellate coral (right) at Streaky Bay, western Eyre Peninsula. Camera lens cap = 50 mm in diameter

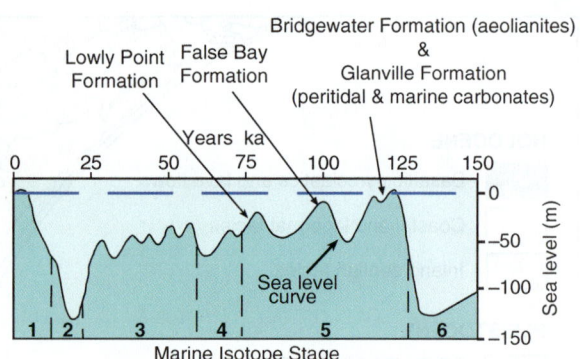

Fig. 3.15 The eustatic sea level curve for southern Australia (adapted from Belperio 1995) for the last 75,000 years with the different formations that developed during Marine Isotope Stage 5 highlighted

Vibracoring on the central Great Australian Bight shelf revealed a patchy Holocene sediment blanket only a few meters thick on top of early and mid Tertiary limestone bedrock (James et al. 1994). Subsequent seismic showed that the shelf in the Great Australian Bight is partitioned (Fig. 3.16) into an inner pre-Middle Miocene part that locally outcrops on the seafloor and a younger, outer progradational or erosional part that is mainly Plio-Pleistocene in age (James and von der Borch 1991; Feary and James 1998). Thus, the shelf has an inner and middle part that is bedrock with relatively little sediment accumulation (see below) and an outer part that is largely progradational or erosional with deep accommodation.

There have been no subsequent studies until recently when Hill et al. (2009) used shallow seismic and vibracoring (which we attempted but were thwarted by the notorious Southern Ocean weather) to show that the same situation is also present on the Lacepede Shelf. Likewise, Ryan et al. (2008) found a similar record on the inner part of the Albany Shelf off Esperance. Thus, there appears to be a pattern that extends from the Great Australian Bight as far east as the Lacepede Shelf. Only future studies will determine the extent of this theme.

Regardless, the nature of sea-level fluctuations across the shelf, at least in the late Pleistocene, can be deduced because of coring in the gulfs (see above). The shelf was flooded and exposed several times between 80 ka and 20 ka, before sea level fell to ~120 mwd during the intervening MIS 2 (20–17 ka) lowstand (Figs. 3.2, 3.3, 3.4, 3.5). Most highstands during MIS

terrace of Sprigg (1979)), is thought to be one of several paleoshorelines associated with Marine Isotope Stages 3 and 4 (Fig. 3.1). The seafloor between these two depths is usually relatively smooth or exhibits several small terraces (e.g. Figs. 8.10–line 2, 8.11–line 5, 9.9–line 6). The deepest terrace at ~150 mwd, is most prominent on the shelf of the South Australian Sea and the Otway Shelf.

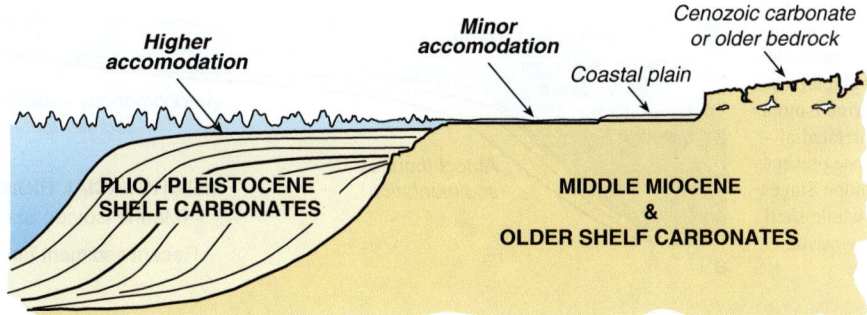

Fig. 3.16 A simplified sketch of the southern continental margin, particularly as represented in the Great Australian Bight. The shelf comprises two sectors, an inner part that has resulted mainly from late Cenozoic uplift wherein the rocks have been beveled by northward progressing shoreline erosion creating a shelf with a bedrock surface and little accommodation. This part of the neritic environment is veneered with minor recent sediment. The outer part, which appears to be mostly post mid-Miocene, is a prograding wedge of late Cenozoic carbonates with higher accommodation

3 and 4 were between 50 and 80 m below modern sea level. Although the shelf was largely underwater during this time (Figs. 3.8, 3.9, 3.10, 3.17), shelf carbonate sediment production took place in much shallower water environments than are there today. It was during these periods (Figs. 3.15, 3.17) that most of the relict sediment was produced and altered (James et al. 1997). An implication is that only those parts of the shelf now deeper than 50 m were sites of deposition during this time. Modern shallow neritic environments less than 50 mwd have not been sites of sedimentation since MIS 5. By contrast, modern middle neritic environments have been inundated, drained, and exposed several times.

The only record of shelf sedimentation between 80 ka and the Holocene is from the relict and stranded grains (Fig. 3.7). These particles are either biofragments or intraclasts (sensu Folk 1959). *Biofragments* are coarse-sand to granule size skeletal particles. Intra skeletal pores are typically empty. Most are bryozoans, bivalves, coralline algae, azooxanthellate corals, and benthic foraminifers. *Intraclasts* (Fig. 3.18) include; (1) biogenic intraclasts, wherein the pores are filled with fine-grained carbonate, and (2) lithic intraclasts, which are a mixture of biogenic grains and intergranular micrite cement, locally containing conspicuous dolomite crystals.

3.5.2   Relict Particles

Shallow-water relict grains formed during MIS 3 and 4 when sea level was 50–80 m lower than it is today

(Figs. 3.1, 3.8, 3.9, 3.10, 3.17) and the shelf was a vast shallow-water environment. Repetitive rise and fall of sea level (Figs. 3.1, 3.15, 3.17) resulted in repeated surf zone reworking, abrasion and chemical alteration over a period of ~30,000 years. Such grains are (1) highly abraded, reduced in size, rounded, and to some degree polished, (2) stained brown to more rarely black due to the presence of Fe-oxides or other diagenetic phases and, (3) predominantly filled (when intraskeletal pores are present) with fine- sand to silt- size biogenic carbonate particles as well as micrite and smectite clays with a mudstone to wackestone texture (Fig. 3.18). It is commonly difficult, without thin section petrography, to correctly identify the original grain composition. Sediments are rich in lithic intraclasts, benthic foraminifers and coralline algal rods; molluscs and bryozoans are not numerous.

Bulk analysis of relict sediments show that they are ~70% 10–13 mol% Mg-calcite, 5% aragonite and 20% LMC; i.e. they contain conspicuously less aragonite than stranded grains and overwhelming less aragonite than modern sediments. Gastropod particles in such sediments are typically steinkerns with the outer wall dissolved and the remaining columella encased in indurated brown-stained micrite that inherited the shape of the original gastropod chamber (Rivers et al. 2008). The relict benthic foraminifer *Marginopora vertebralis* is large and brown-strained in the Great Australian Bight with the pores filled with brown-stained micrite. [14]C ages are mostly between 18 and 35 ka (Rivers et al. 2009).

The relict particles, as demonstrated by relict benthic foraminifers are not, however, pre-Pleistocene (Li et al. 1999). Their polished and abraded character indicates

Fig. 3.17 A sketch illustrating neritic sedimentation during **(a)** interglacial highstands such as today, **(b)** glacial lowstands such as Marine Isotope Stage 2, and **(c)** interstadial periods such as those present during Marine Isotope Stages 3 and 4 when the whole shelf was largely shallow neritic

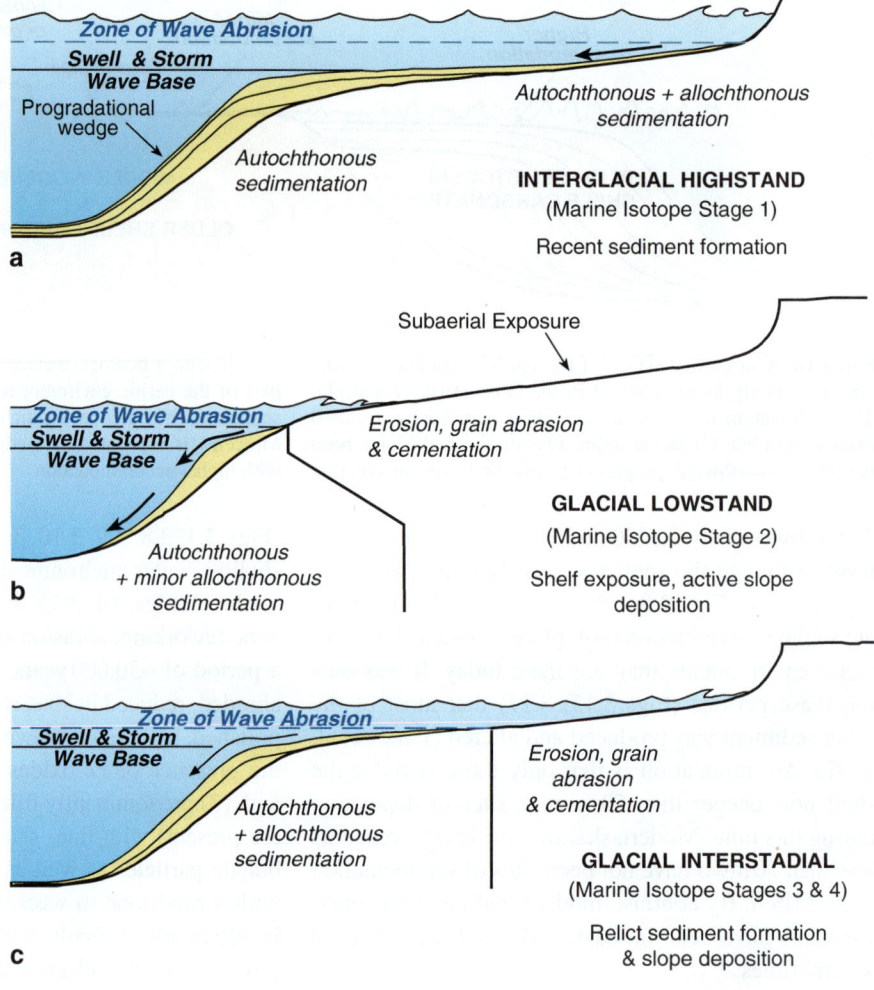

that they passed through the surf zone, whereas the ~10 mol% Mg-calcite micrite fill implies that they resided in the marine environment for an extensive period of time and have not been substantially affected by meteoric diagenesis. They are not common shallower than 50 mwd and are absent deeper than 120 mwd. The situation is complicated somewhat in very shallow water environments because of the transport of nodular ferricrete grains into the environment and local iron-rich fresh or saline groundwaters and associated iron-fixing bacteria in strandline environments.

3.5.3   Stranded Particles

Sea level rise since the LGM lowstand has been relatively rapid (Fairbanks 1989), concomitant with chang-

ing climate (Figs. 3.5, 3.19). Since this is not a rimmed shelf, sea-level rise during the last 17 ky (Fig. 3.1) has gradually flooded the seafloor. What is now a relatively deep mid neritic environment was, during early stages of this inundation, a site of shallow water carbonate sediment production. The grains produced then are now out of equilibrium with modern marine conditions; they were marooned in relatively deep water as sea level rose over the last ~15,000 years. This is especially evident with the numerous articulated coralline algal rods in depths well below where they are forming today (Rivers et al. 2007). Stranded grains (Fig. 3.7) are somewhat abraded, light grey to buff-coloured and can be distinguished from younger Holocene particles by their (1) lack of luster, (2) rounding of skeletal edges, (3) lack of well defined surface structure, (4) discolouration, or (5) missing skeletal elements such as spines and opercula on bryozoans. In some cases

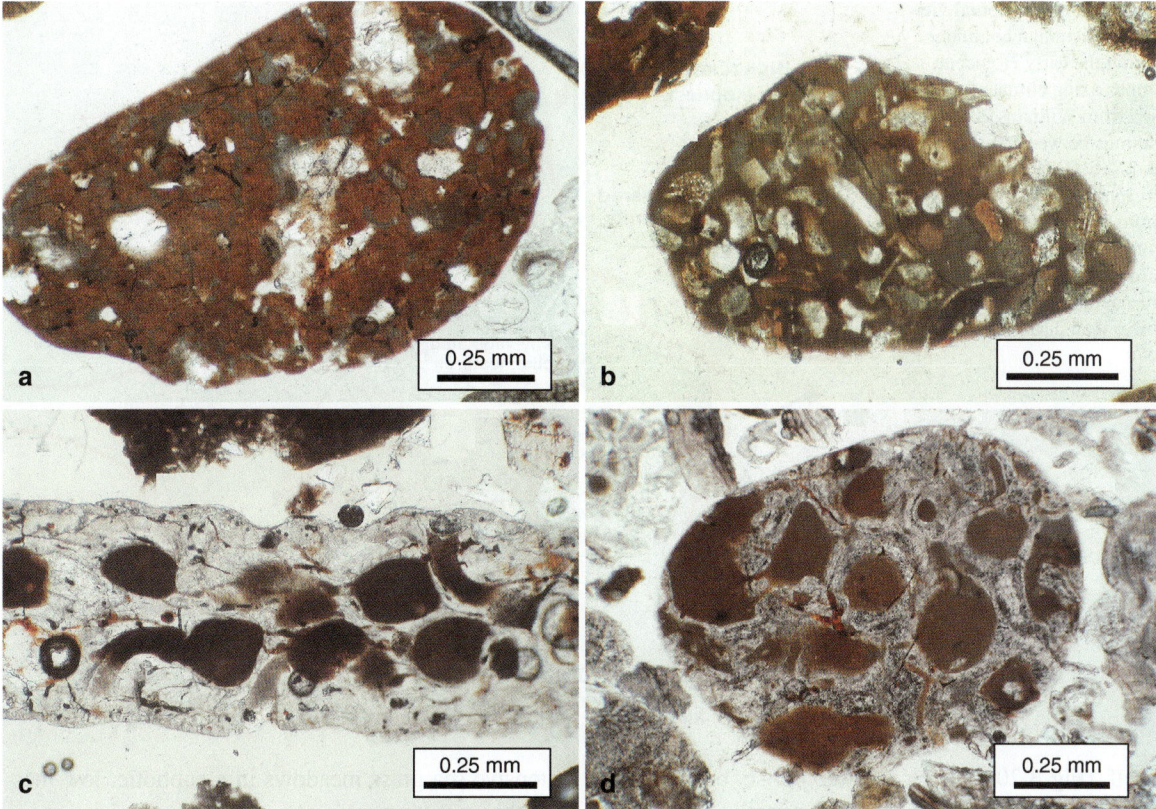

Fig. 3.18 Relict particles, micrographs, plane polarized light, Otway Shelf. (**a**) A rounded relict wackestone lithoclast containing ~20% quartz grains; 67 mwd. (**b**) An angular packstone lithoclast with a bioclastic–quartzose composition; 79 mwd. (**c**) A bryozoan skeletal bioclast that is not significantly abraded and whose zooecia are filled with Fe-oxides, 126 mwd. (**d**) A rounded bryozoan bioclast with microborings and zooecia filled with Fe-oxides, 67 mwd

the separation of stranded and relict grains is difficult because the same process of Fe-impregnation is taking place today, with the intensity of staining in relict grains being due to their longer residence time in seawater.

The bulk mineralogical composition of stranded sediments from the Great Australian Bight is ~65% 10–13 mol% Mg-calcite, 20% aragonite, and 15% low magnesium calcite (<4 mol%). This stands in contrast to modern sediments that have a much higher aragonite content, implying that there has been some dissolution of aragonite on or beneath the seafloor (James et al. 2005; Rivers et al. 2008). Stranded gastropods have a white to light grey colour, a dull luster, lack intricate skeletal features, display dissolution features, are variably filled with grey or buff-coloured micrite, and are conspicuously microbored. In many cases the outer wall is dissolved exposing the columella. Modern snails, by contrast, have a pearly luster, delicate and

subtle skeletal detail, no dissolution features, and no intraskeletal sediment fill. The large benthic foraminifer *Marginopora vertebralis*, although composed of 10–13 mol% Mg-calcite (like living forms), has similar dissolution features. ^{14}C age determinations mostly range between 7.6 and 21.4 ka (Rivers et al. 2007, 2009). The stranded particles are not found in waters deeper than ~120 m.

3.5.4 Late Pleistocene Shelf Paleoenvironments

The composition of relict and stranded particles yields important information about the nature of the shelf between 80 ka and the Holocene (Fig. 3.1). Relict particles are most numerous in the eastern part of the Great Australian Bight and on the outer part of the Lacepede

Fig. 3.19 A conceptual view of the Australian continent during the early Holocene emphasizing climate (modified from Williams 2000). The isthmus between Australia and New Guinea was partially breached and so there was a proto-Leeuwin Current, the continental interior was less arid and winter rainfall along the southern margin was lower than during the LGM

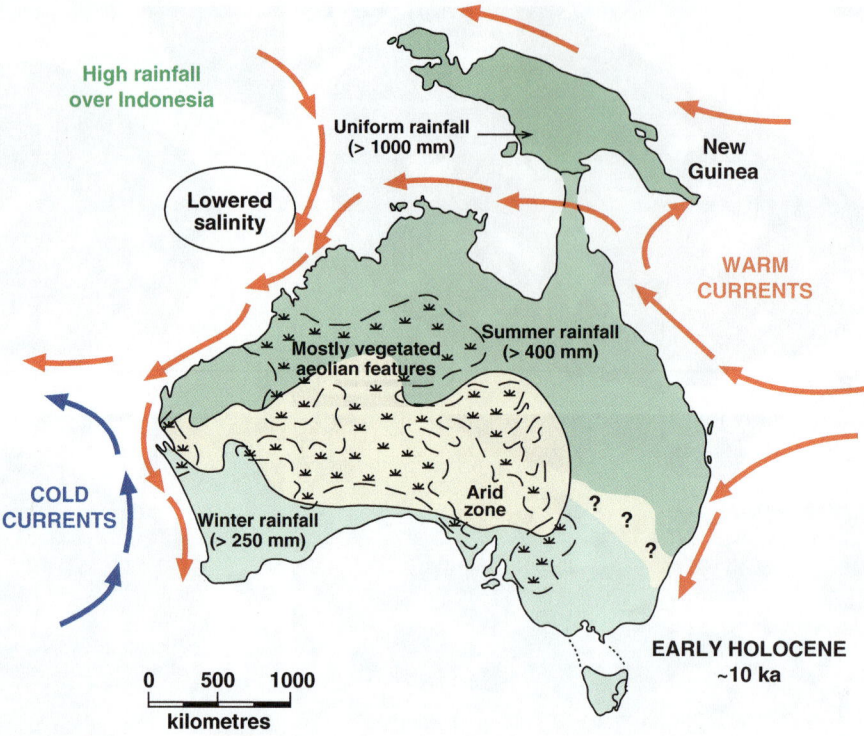

Shelf (Fig. 3.20) where they comprise between 40 and 60% of the sediment. Most particles are coralline algae and benthic foraminifers, particularly large benthic foraminifers (Fig. 3.21). Relict grain facies in the Great Australian Bight have numerous intraclasts inboard but most of the material is dominated by corallines and benthic foraminifers (Fig. 3.21). Relict sediments on the Lacepede Shelf, although having many intraclast or quartz grains inboard and numerous corallines outboard have no large benthic foraminifers and bryozoans are conspicuously more numerous than in the Great Australian Bight.

The relict benthic foraminifer assemblages across the southwestern continental margin are dominated by *Quinqueloculina* spp., *Triloculina* spp., *Discorbis dimidiatus*, and *Elphidium* spp. (>70%). This assemblage, together with the *Marginopora-Sorites* group is generally recognized as a warm-temperate 'marine lagoon' and not open shelf assemblage (cf. Murray 1991) and elements of it are present in the gulfs today. Thus, the relict fauna is interpreted as one that may have formed in vast somewhat saline lagoons but with some open shelf attributes across the shelf (Li et al. 1999). These foraminifers together with articulated coralline algae, and the sea grass associated benthic foraminifer *Nubecularia* sp. implies the presence of

extensive sea grass meadows in a euphotic, low mesotrophic environment.

None of these relict warm-temperate indicators are present on the southeastern continental margin, suggesting more cold-temperate ocean waters east of Kangaroo Island. Furthermore, the inner part of the Lacepede Shelf is characterized by numerous relict *Ammonia beccarii*, a form not found on the southwestern continental margin. These foraminifers would seem to represent a somewhat brackish water environment, (Li et al. 1996b) probably signaling a time when the River Murray was delivering more fresh water to the shelf than it is today (Rahimpour-Bonab et al. 1997).

Stranded sediments are rich in heterozoan elements throughout (Rivers et al. 2007). The grains are particularly abundant in central parts of the Great Australian Bight where they locally form up to 80% of the sediment (Fig. 3.22). There are no obvious trends in composition (Fig. 3.22). Bryozoans are most numerous outboard throughout. Those sediments between 50 and 90 mwd also contain many *Marginopora vertebralis* (which are not there today) together with molluscs and coralline algae, and a few *Heteristegina* sp., a classic warm-temperate seagrass assemblage. Neither the large benthic foraminifers nor the articulated corallines are found east of Kangaroo Island (Fig. 3.22). A typi-

Fig. 3.20 A series of charts showing the percentage of relict (**a**) & (**c**) and stranded (**b**) & (**d**) particles in shelf surface sediments–colours the same for a & c and b & d. (Modified from Rivers et al. 2007)

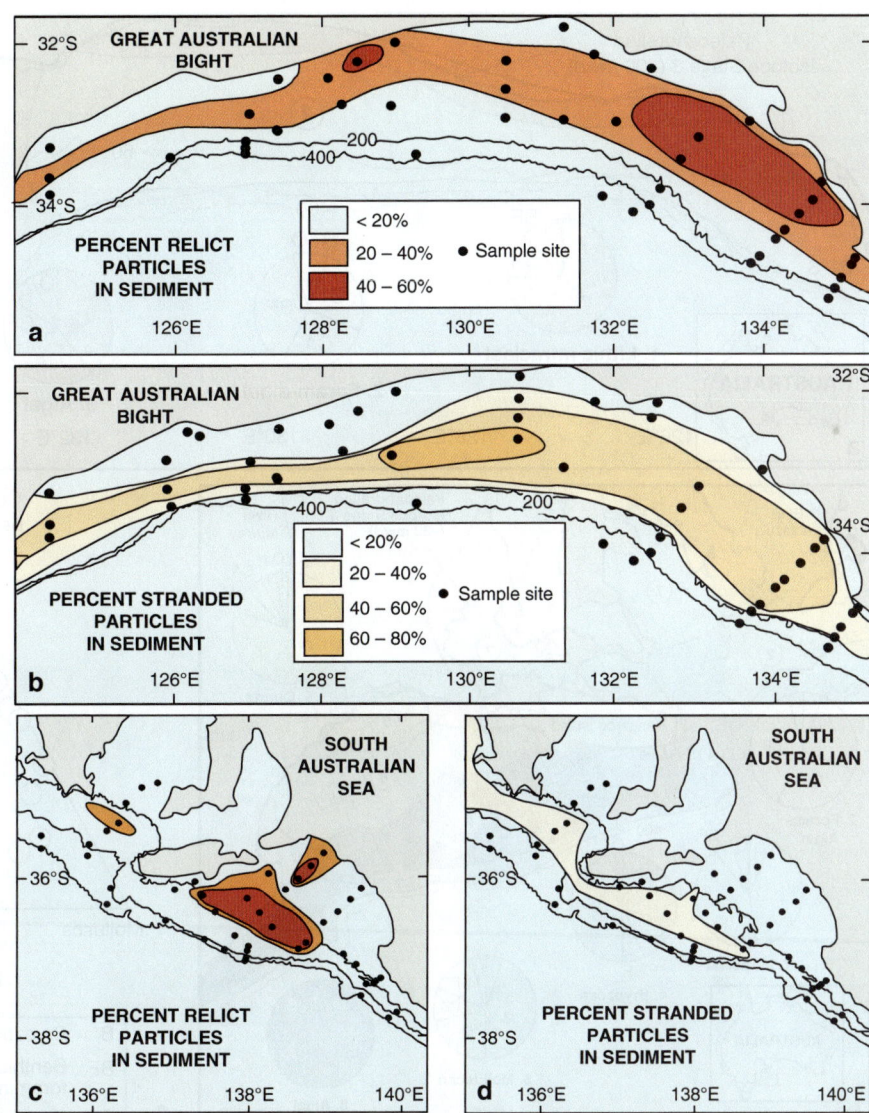

cal cool-temperate, bryozoan-dominated assemblage dominates sediments to the east. These observations support the proposition of Li et al. (1999) that during these times, and even today, the southern margin was divided into a shallow warm-temperate setting with sea grasses in the west and a cool-temperate kelp-dominated setting in the east (cf. James and Lukasik 2010).

Finally, stable isotopic analysis of stranded and relict particles (Rivers et al. 2009) confirm the notion that deposition during these times of lower sea level and shallow waters across the shelf took place in environments of elevated sea water salinity. This interpretation is proposed because *Marginopora vertebralis* tests in particular are enriched by 1–3‰ in both ^{18}O and ^{13}C relative to modern specimens from the same region.

The overall environment is thought to be similar to outer Shark Bay today (Logan and Cebulski 1970), an area of somewhat elevated salinity and enhanced photosynthetic activity whereby dissolved ^{13}C is enriched in calcareous invertebrates.

3.6 Synopsis

Late Pleistocene geomorphology created the framework within which the recent sedimentary system operates. Not only do the Pleistocene deposits form a template upon which recent sediments are accumulating, but Pleistocene sediments are also typically mixed

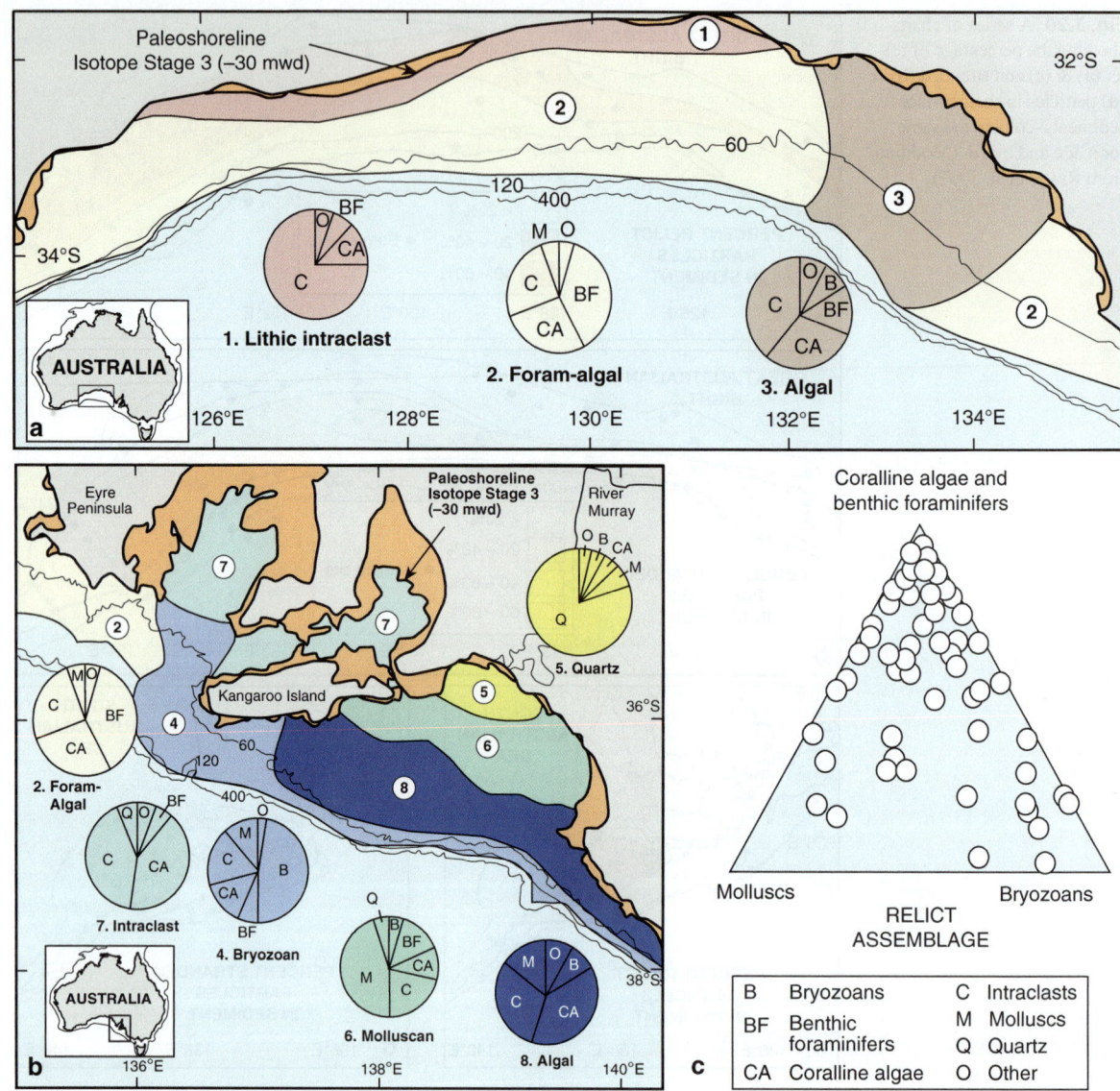

Fig. 3.21 Charts illustrating the different lithofacies of relict particles in (**a**) the Great Australian Bight and (**b**) the South Australian Sea, together with (**c**) a triangular plot of the main relict particle types demonstrating the predominance of coralline algae and benthic foraminifers. (Modified from Rivers et al. 2007)

with modern skeletal deposits to form palimpsest accumulations. The Pleistocene record, because of low shelf accommodation resulting from tectonic uplift, is negligible and in the form of a thin, patchy sediment veneer on top of Cenozoic limestone; there is virtually no late Quaternary stratigraphic record on the shelf. Pleistocene depositional history must, therefore, be deduced from (1) slope sediments, (2) deposition in the large gulfs, and (3) late Pleistocene grains within the modern sediments.

1. The record of Pleistocene sedimentation on the slope is excellent, because of ODP drilling, and can be correlated with the global marine record. Accumulation rates on the slope sediment wedge are locally very high, exceeding carbonate deposition rates in similar tropical marine settings. This is due to active off-shelf sediment transport, especially during initial stages of sea level rise and shelf flooding.

2. Bryozoan–sponge reef mounds are embedded in the slope sediment wedge. The structures appear

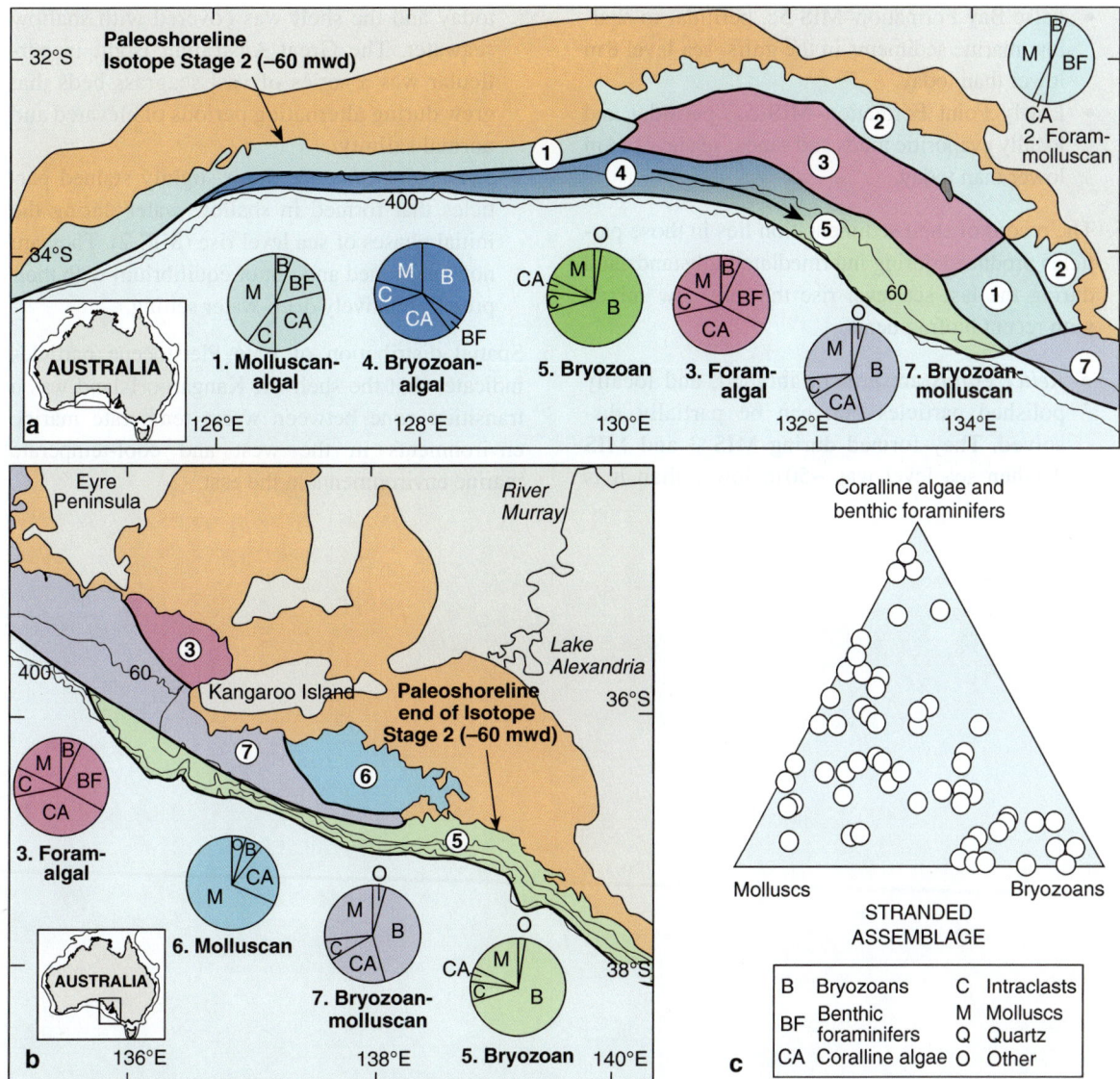

Fig. 3.22 Charts illustrating the different lithofacies of stranded particles in (**a**) the Great Australian Bight and (**b**) the South Australian Sea, together with (**c**) a triangular plot of the main relict particle types demonstrating the predominance of coralline algae and benthic foraminifers. (Modified from Rivers et al. 2007)

to have grown during each glacial sea level lowstand when nutrient-rich Antarctic Intermediate Waters were much closer to the slope than they are today.

3. Marginal marine aeolianites formed during each highstand of sea level when the source of carbonate sediment (particularly seagrass banks) was in adjacent shallow water. They are either aggrading or prograding and when the latter shore-parallel dune ridges have elongate interdune corridors. The

youngest of these corridors are today locally the sites of modern lagoonal sedimentation.

4. Extensive coring in Spencer Gulf and Gulf St. Vincent together with low shoreline outcrops reveals a clear late Pleistocene depositional record, principally;

- Glanville Formation–MIS 5e, marine calcarenites, warmer ocean waters compared to today (7–9°C higher), sea level 2 m above that today.

- False Bay Formation–MIS 5c, peritidal to shallow marine sediments in the gulfs, sea level 8 m lower than today.
- Lowly Point Formation–MIS 5a, peritidal and locally evaporitic muds and sands, sea level 14 m lower than today.

5. The record of shelf sedimentation lies in those particles produced during intermediate highstands and during the last sea-level rise that are now mixed with recent biofragments.

- *Relict grains* are brown, abraded, and locally polished particles that can be partially dissolved. They formed during MIS 3 and MIS 4 when sea level was ~50 m lower than it is today and the shelf was covered with shallow seawater. The Great Australian Bight in particular was a series of vast seagrass beds that grew during alternating periods of elevated and normal salinity.
- *Stranded grains* are those slightly stained particles that formed in shallow water during the initial phases of sea level rise (MIS 2). They are now marooned and out of equilibrium with their present relatively deep-water setting.

6. Spatial distribution of Late Pleistocene particles indicates that the shelf off Kangaroo Island was a transition zone between warm temperate marine environments in the west and cool-temperate marine environments in the east.

Chapter 4
The Neritic Carbonate Factory

4.1 Introduction

Carbonate sediments in the vast cool-water, mid-latitude marine environment of southern Australia are fundamentally different from warm-water neritic, tropical carbonates. They are heterozoan and palimpsest, not photozoan and modern. Heterozoan particles (cf. James 1997) are biogenic (Fig. 4.1), the only phototrophs are red calcareous algae, the only mixotrophs are large benthic foraminifers, and most of the benthic invertebrates from whence the particles came are filter feeders. The sediments are palimpsest because they form relatively slowly and are mixed together with grains generated during earlier periods of deposition in this overall high-energy setting. Only recently has it been possible to recognize, separate, and interpret these particle types (Rivers et al. 2007). The different and progressively older particles formed at separate times are recognized herein as (1) relict (~70–25 ka), (2) stranded (18–10.4 ka), and (3) Holocene (<10.4 ka) grains. Today, the seafloor sediments are a mixture of particles of several ages, each of which reflect a different set of environmental controls (Fig. 4.2). The term recent is used herein to encompass all particles formed since the LGM and during MIS 1 and 2 and therefore includes both stranded and Holocene particles. This chapter is a description of the Holocene sediments, and by implication, stranded materials produced on the modern shelf. Relict particles have been documented in Chap. 3.

4.2 Biogenic Sediment Production

The most important skeletal elements (Fig. 4.1) are calcareous algae, bryozoans, molluscs (scaphopods, bivalves, and gastropods), and foraminifers (free and encrusting, large and small, benthic and pelagic). They form particles across the grain-size spectrum (cf. Bone and James 1993; James et al. 1997, 2000) (Fig. 4.3). Other locally important skeletons are those of calcareous worms, epifaunal and infaunal echinoids, and barnacle plates together with spicules from sponges, ascidians, gorgonians, and other soft corals, ostracods, and fragments of coccoliths. Locally conspicuous but not regionally important particles include corals and brachiopods. Remains of fish, crustaceans and mammals are rare but present.

Mineralogy of these particles is characteristic of the modern marine environment wherein most are $CaCO_3$ polymorphs, namely aragonite or calcite with varying concentrations of magnesium in the calcite lattice. The calcites, however, can be divided into several types depending on their magnesium content. Low-magnesium calcite (LMC) contains <4 mol% $MgCO_3$, intermediate magnesium calcite (IMC) contains between 4 and 12 mol% $MgCO_3$, and high-magnesium calcite contains >12 mol% $MgCO_3$. The majority of biofragments are either aragonite, LMC or IMC.

The carbonate factory is not, however, simply the sum of all calcareous algae and invertebrate animals. Seagrasses and non-calcareous red, green, and brown algae are also an integral part of the sediment-producing system (Figs. 4.5, 4.6, 4.7). They act as substrates for encrusting calcareous animals and plants, profoundly modify the environment by reducing wave and current energy, bind sediments with their roots systems, modify light levels, and are a seafloor habitat for organisms of all types.

As with other modern temperate carbonate areas, there is a general trend of inner shelf sediments rich in bivalve-bryozoan sand and outer shelf deposits dominated by bryozoan sediments (Ginsburg and James 1974;

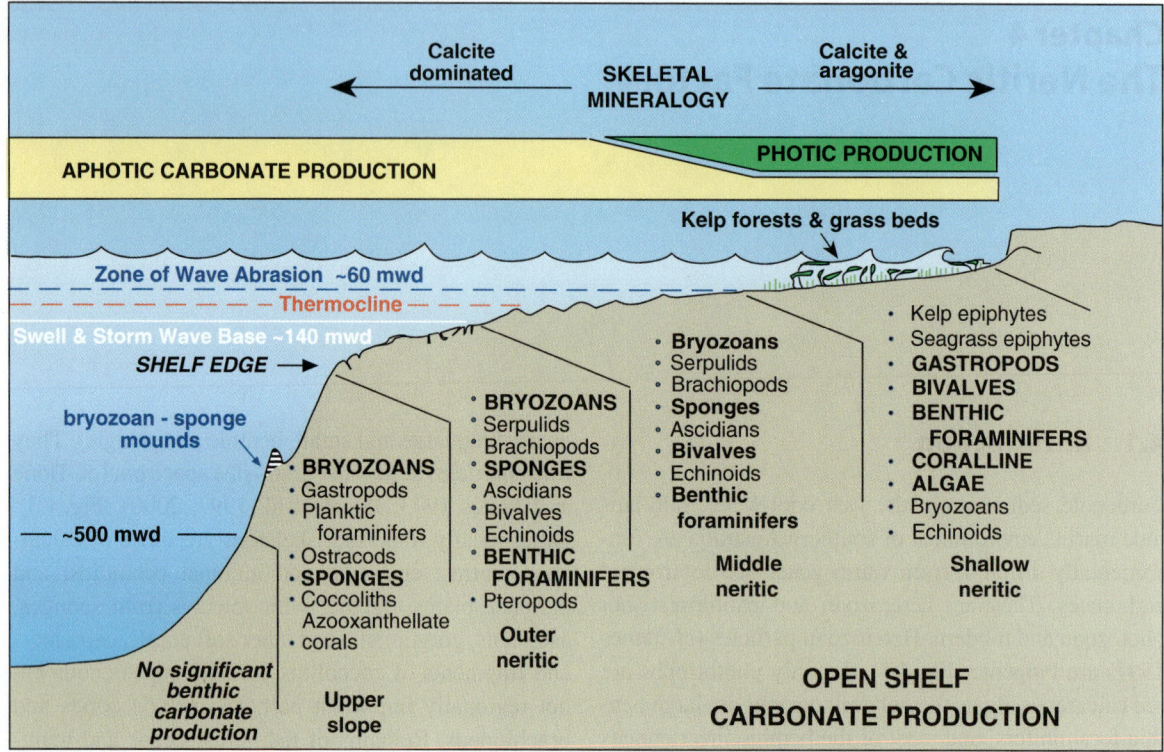

Fig. 4.1 Carbonate sediments produced and deposited on the southern margin of Australia are predominantly a heterozoan assemblage (James 1997). Major producers are highlighted in bold capitals. Biota distribution is controlled by water depth, energy levels, substrate availability, temperature, light and nutrient levels. Most sediments are from aphotic invertebrates

Nelson 1988b). Shelf deposits are all grainy and rippled, with little or no mud anywhere, although there are very fine sands. Texture is controlled by plant and invertebrate skeletal architecture (Fig. 4.3), and the relative mix of Holocene and older components.

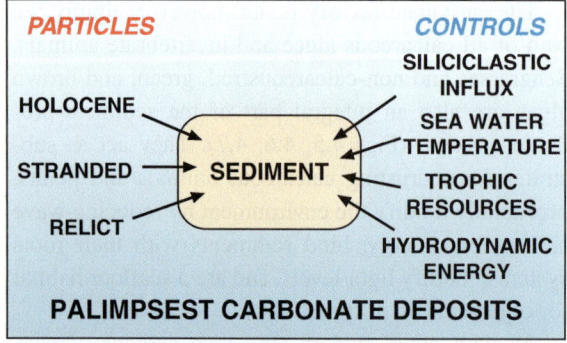

Fig. 4.2 A simplified sketch illustrating the main particle types in the palimpsest sediments and the various controls that determine their distribution

4.2.1   Seagrass and Macroalgae

As in modern tropical carbonate factory, marine seagrass is an important part of the euphotic zone in this temperate system (Shepherd and Robertson 1989; James et al. 2009). Major differences in this temperate setting are the high abundance of seagrasses, and the presence of copious non-calcareous macroalgae. Macroalgae are here recognized as the spectrum of soft algae that range from millimeters to many meters in size. Seagrasses are most diverse in warm temperate environments, red, green, and brown macroalgae are found throughout, whereas large brown macroalgae grow best in cold temperate settings. Thus, southern Australian euphotic environments west of Cape Jaffa (Fig. 4.4) are sites of seagrasses and macroalgae, between Cape Jaffa and Bass Strait seaweeds, macroalgae (especially kelp) and reduced seagrasses, and along the western coast of Tasmania macroalgae, prolific kelp, and sparse seagrass.

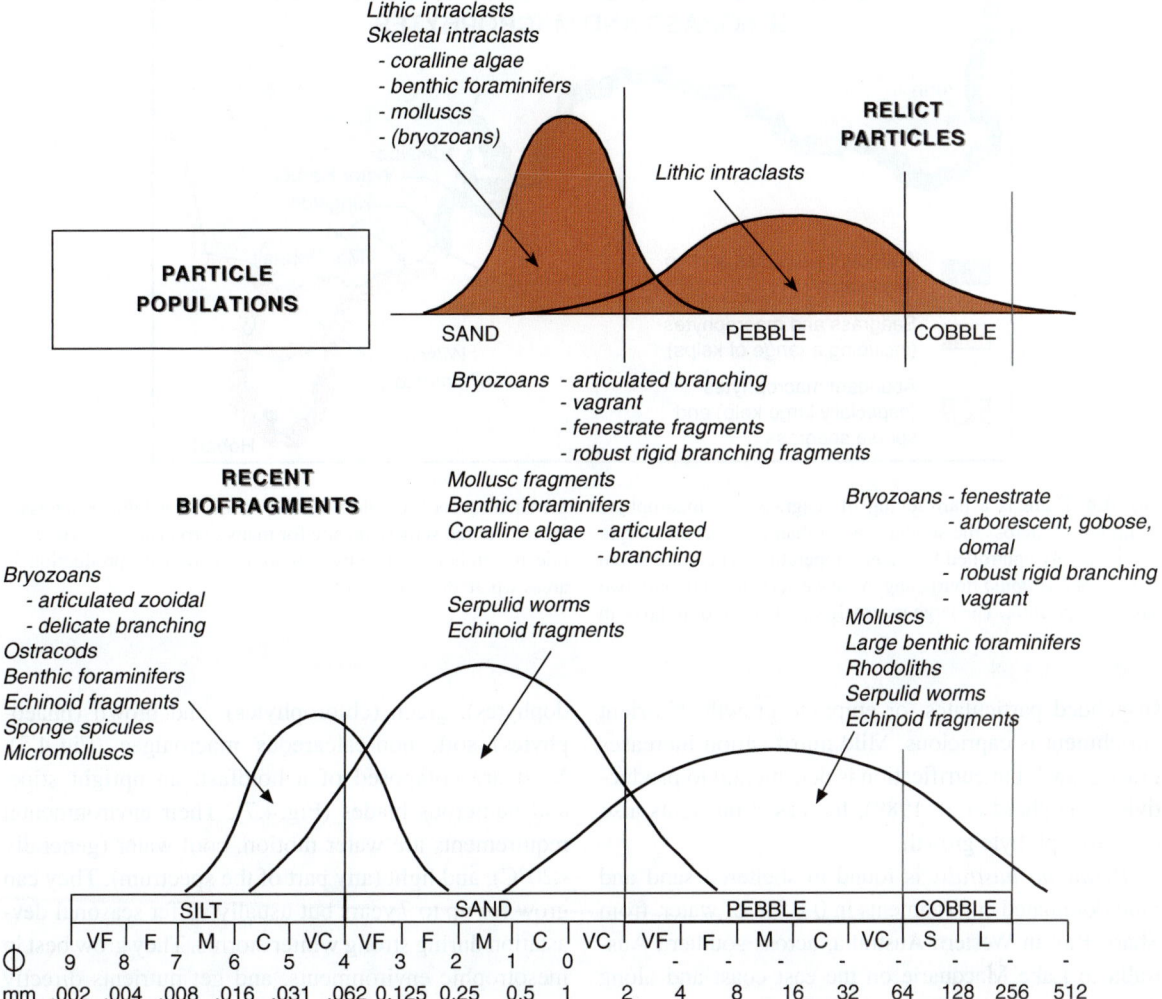

Fig. 4.3 The size of grains or fragments produced by different biota can vary by several orders of magnitude. Recent biota produce silt to cobble-sized pieces, but sponge spicules will always be in the range of medium silt to fine sand whereas rhodoliths will always be coarse sand to cobble sized. Bryozoans range from medium silt to cobble sized, with texture controlled by the architectural style of the genus. Relict particles are sands and cobbles, with a somewhat bimodal range

4.2.1.1 Seagrasses

Most seagrass meadows consist of *Posidonia, Amphibolis, Heterozostera, Zostera,* and *Halophila* with only the first two (Fig. 4.5) significant as hosts for calcareous epiphytes (Figs. 4.5, 4.6) mainly coralline algae, bryozoans, spirorbid worms, and benthic foraminifers (James et al. 2009). The grasses are all rooted in soft sediment with the exception of *Thallasodendrum pachyrhizum* and *Amphibolis antarctica* which can also locally grow on hard substrates (Edgar 2001) (Fig. 4.4).

These marine angiosperms consist of a well-developed rhizome that runs in sediment beneath the seabed and has regularly spaced nodes, each bearing roots below and an erect stem or shoot with strap-like leaves above. The strong root systems form a dense mat in the sediment and both grasses shed their blades over a period of about 2 months. The upper limit of the seagrass growth window is determined by wave energy and exposure to desiccation. The lower limit is imposed by the depth of light penetration where photosynthesis equals respiration (compensation depth). Photosynthesis is reduced by living and non-living

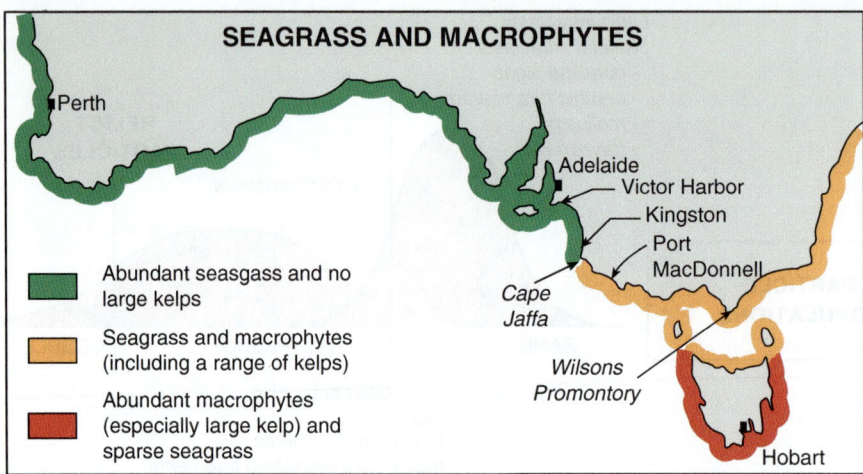

Fig. 4.4 There is a partitioning of seagrass and macrophyte communities across the southern Australian continental margin that is largely controlled by water temperature. There are seven species of *Posidonia* flourishing in the western half, but only two east of Cape Jaffa. The large macroalgae, however, only grow in the shallow cool to cold waters west of Cape Jaffa. Seagrasses are a favoured settlement site for many carbonate epiphytic benthic invertebrates, thereby promoting carbonate production in areas where they are prolific

suspended particulates, or epiphyte growth. Nutrient enrichment is capricious. Mild nutrification increases grass growth but eutrophication is detrimental to productivity (Shepherd et al. 1989). Increased nutrients also increase epiphyte growth.

Posidonia australis is found in sheltered sand and mud dominated environments in 0–15 m of water, from Shark Bay in Western Australia, across southern Australia to Lake Macquarie on the east coast and along the northern coast of Tasmania. *P. angustifolia* grows in moderately exposed sand between 2 and 35 mwd from the Houtman Abrolhos in Western Australia to Port MacDonnell, South Australia (Fig. 4.4). *P. sinuosa* lives in moderately exposed sand as well as sheltered sand from 0 to 15 mwd. It is found from Shark Bay, Western Australia to Kingston, South Australia (Fig. 4.4). *Amphibolis antarctica* grows in moderately exposed sand in 0–23 mwd and extends from Carnarvon, Western Australia to Wilson's Promontory, Victoria and halfway down the coast of Tasmania. *A. griffithii* is found on moderately exposed sand and rock in 0–40 mwd, from Western Australia across to Victor Harbor (Fig. 4.4).

4.2.1.2 Macroalgae

Hard substrates from the intertidal zone to depths of ~40 m are typically colonized by a variety of red (rho-

dophytes), green (chlorophytes), and brown (phaeophytes) soft, non-calcareous macroalgae (Fig. 4.7). Most are composed of a holdfast, an upright stipe, and numerous blades (Fig. 4.7). Their environmental requirements are water motion, cool water (generally <20°C), and light (any part of the spectrum). They can grow for up to 7 years but usually suffer seasonal devastation during strong winter storms. They grow best in mesotrophic environments, and get nutrients directly from seawater. Phaeophytes require lower nutrient levels than chlorophytes and all macroalgae require fewer nutrients than phytoplankton (Fig. 4.8).

Kelp are by definition restricted to large brown algae within the order Laminariales, which in southern Australia includes the genera *Lessonia*, *Macrocystis* (giant kelp) and *Ecklonia* (Womersley 1987). The large phaeophyte *Durvillea*, although not strictly a kelp, is commonly referred to as bull kelp. The general definition is used herein and all are referred to as kelps.

The most conspicuous kelp is *Ecklonia radiata*. All of the macroalgae, with the exception of *Caulerpa* sp., require a hard substrate, but this can be as small as a pebble or a bivalve. Large macroalgae such as the giant kelp (*Macrocystis*) and the bull kelp (*Durvillea*) grow in the colder waters east of Cape Jaffa and are common along the coast of southern Victoria and Tasmania.

Calcareous epiphytes do not usually grow on macroalgae in any numbers because many of these algae,

Fig. 4.5 Sketches of the important seagrasses on the southern Australian continental margin. (**a**) The two major sea-grasses that act as substrates for calcareous epiphytes. (**b**) *Amphibolis* spp. are individual plants with wiry stems, topped by a tuft of leaves that are continuously shed while stems live for 2 years or more. *Posidonia* spp. are runner-like grasses, with sub-seafloor sheaths from which multiple, tightly-packed leaves emerge. These are shed after 3–4 months. (**c**) Large and small foraminifers epiphytic on *Posidonia* leaves. (**d**) Articulated coralline algae epiphytic on *Amphibolis* stems. (Modified from James and Bone, 2007 and James et al. 2009)

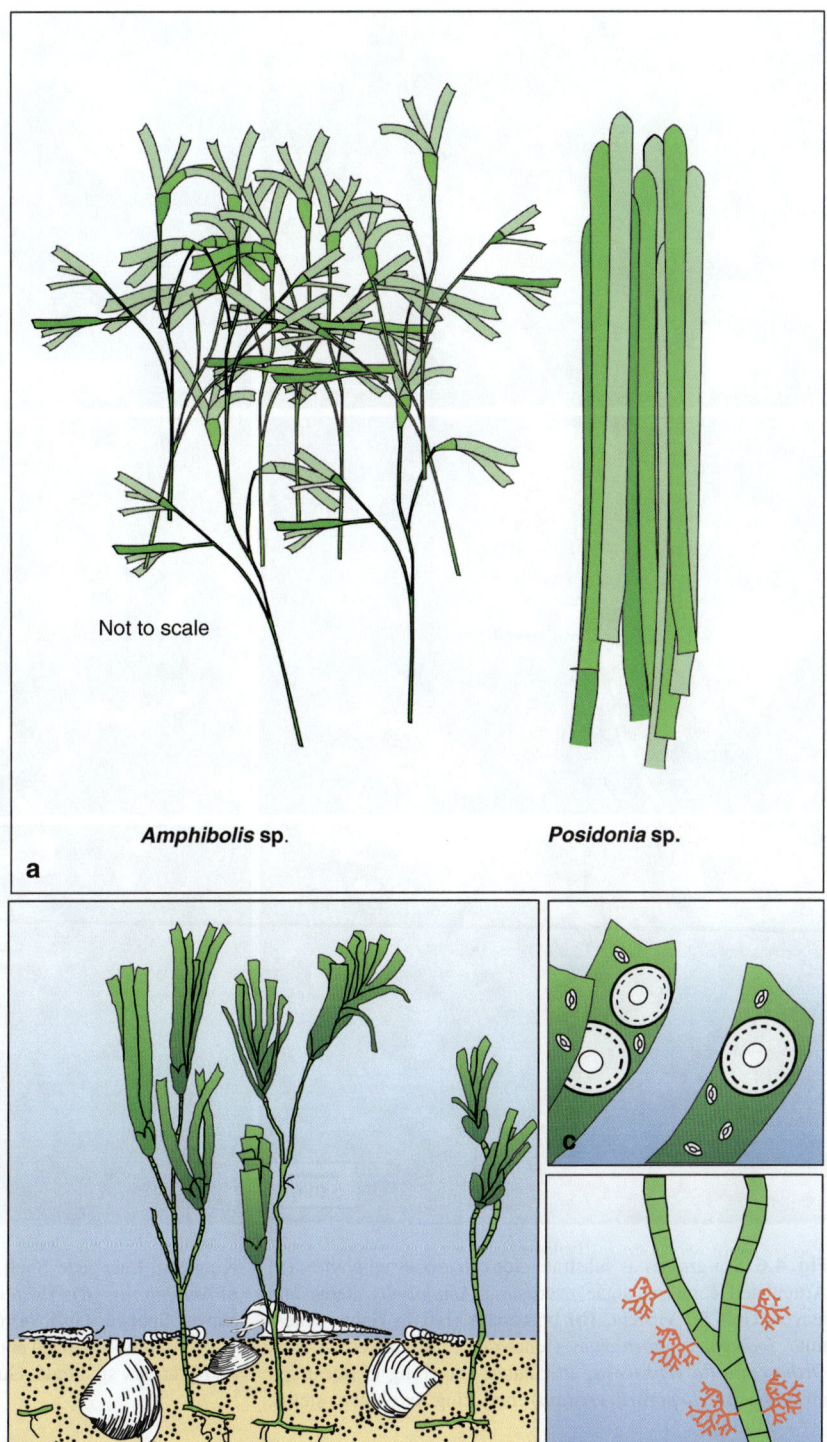

Not to scale

Amphibolis sp.   **Posidonia sp.**

a

b   c

d

morphology, phloroglucinol-type tannin abundance, and inside the thallus, in particular occurring alga-(terpenoids are *Z*-sackene (Fig. 4.4) and several spe-bacteria and spheroids, and brush, smaller modified bivalves of *Zostera* soft, which are encountered? ... the brown. Ostracods live on the algae and feed on the encrust-open *Amphibolis* are attenuation, and or spheroid.   ing organisms, bryozoans grow the algal leaf.

Fig. 4.6 Sea-grasses as substrates for calcareous epiphytes. (**a**) Articulated coralline algae encrusting *Amphibolis* stems and leaves, Gulf St. Vincent, (**b**) bryozoans (left to right—fenestrate *Iodictyum phoenicium* (fenestrate), articulated zooidal *Orthoscuticella ventricosa*, articulated branching *Cellaria* sp., arborescent *Celleporaria cristata*) encrusting *Amphibolis* stems, Kingston, Lacepede Shelf, (**c**) articulated corallines and stems of *Amphibolis*, (**d**) *Posidonia* meadow with encrusting algae on leaves, Spencer Gulf, (**e**) a *Posidonia* seagrass blade encrusted with coralline algae, (**f**) encrusting *Disporella* sp. *Thairopora* sp. on *Posidonia* sp. blade, Gulf St. Vincent

and brown algae in particular, secrete secondary metabolites (phlorotannins) that inhibit attachment. Exceptions are *E. radiata* (Fig. 4.8) and several species of *Sargassum*, which are encrusted by the bryozoan *Membranipora membranacea* and or spirorbid worm tubes. Epiphytes are, however, numerous around and inside the holdfast, in particular coralline algae, bryozoans, spirorbids, and bysally attached bivalves. Gastropods live on the algae and feed on the encrusting organisms. Echinoids graze the algal turf.

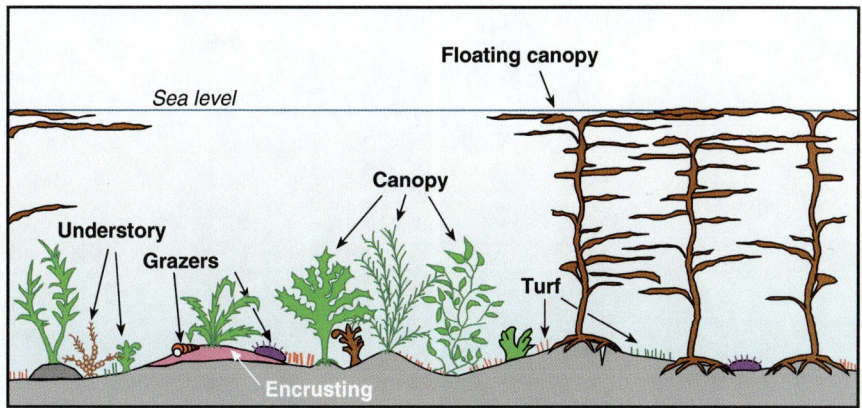

Fig. 4.7 Sketch of the principal algae and invertebrates on a rocky sub-tidal reef. Encrusting and articulated coralline algae form a soft non-calcareous algal turf that is grazed by echinoids and gastropods covers the hard substrate. The lower tier of erect bryozoans and sponges co-exist with an understory of green, brown and soft red macroalgae. The middle tier or canopy is composed of macrophytes, mostly phaeophytes (dominated by *Ecklonia* sp.) whereas the floating canopy or upper tier consists of large macrophytes. (*Macrocystis* sp. and *Durvillea* sp.)

The denser the canopy of macroalgae (*Ecklonia* or *Macrocystis*) the greater is the encrusting algal cover (e.g. *Lithophyllum*). If the canopy is removed, however, light intensity becomes too great and the encrusters undergo bleaching and reduced photosynthesis, resulting in substrate colonization by algal turf and geniculate corallines. This process results in increased sediment production in the form of coralline algal rods. Thus, in this setting, encrusting corallines equal low light and low sedimentation, whereas articulated corallines equal high light levels and high sedimentation.

Durvillea monopolizes space in high-energy lower intertidal and shallow subtidal environments (the coralline algal zone of the warm temperate coasts) to depths of 10 m on exposed coasts, especially along the west coast of Tasmania, but decreases in depth along moderately-exposed and sheltered shores. The brown laminarian kelp *Lessonia* and the fucoid *Phyllospora* form extensive beds below *Durvillea*. The huge sweeping fronds of these algae dislodge other invertebrates creating a monospecific habitat; numerous invertebrates are, nevertheless, associated with the holdfasts of *Lessonia*. The long stypes of *Macrocystis* in depths of 5–25 mwd can extend to the sea surface creating a canopy that buffers wave action and intercepts much solar radiation. Below the *Lessonia-Phyllospora* habitat, ranging from ~1 m on sheltered to ~20 m at wave exposed sites, this community gives way to a diverse assemblage of macrophytes of fucoid, red, and green algae. As depth increases and light levels decrease the diverse algal habitat merges into environments dominated by the kelp *Ecklonia radiata* to depths of 35 mwd (Crawford et al. 2000).

4.2.2 Calcareous Algae

All of the significant calcified algae are rhodophytes. Corallines (IMC and HMC) encrust hard substrates, seagrass blades, and macroalgal holdfasts, grow as rigid branching and articulated branching (geniculate) types, and form rhodoliths (coralline algal nodules). Peyssonnelids (aragonite) encrust and form sheets over loose sediment, but are not common.

4.2.2.1 Red Calcareous Algae

Encrusting and Rigid Branching: Encrusting types (Fig. 4.9) are epilithic and grow on hard rocky substrates, are epiphytic on seagrasses and macroalgae, and are epizooic on sponges, bryozoans, molluscs, crustaceans, and tunicates (Womersley 1981a). Our sampling shows that they live to 100 mwd on the Lacepede Shelf, to 110 mwd on the Lincoln Shelf and possibly to 150 mwd on the Baxter Sector of the Great Australian Bight. Non-geniculate somewhat rigid branching growth forms are also present across the region but are not important as sediment producers (Fig. 4.10).

Fig. 4.8 Calcareous epiphytes on macroalgae. (**a**) Calcareous spirorbid worm tubes on *Ecklonia*, Gulf St. Vincent, (**b**) encrusting coralline algae on *Ecklonia* sp., Gulf St. Vincent, (**c**) an intertidal platform eroded into Pleistocene aeolianite, fronted by a luxuriant growth of kelp, Robe, Bonney Coast; platform width–~10 m. (**d**) An intertidal platform eroded into mussel-encrusted Tertiary limestone, fronted by bull kelp, Marama, northwest coast of Tasmania. Kelp area ~3 m wide, (**e**) margin of a rocky intertidal platform (*left*) encrusted with coralline red algae and supporting prolific growth of the brown algae *Durvillea* (*large stipes*) and *Macrocyctis*; hammer scale, (*circled*), Warrnambool Otway Shelf, (**f**) holdfasts of kelp, partially obliterated by encrusting platy calcareous algae, Kingston, Lacepede Shelf

As stressed above, calcareous red algae do not usually encrust macroalgae except at the holdfast and on the blades of *Sargassum* and *Ecklonia*. *Lithophyllum*, and *Lithoporella* dominate in shallow waters whereas *Mesophyllum*, *Lithothamnion*, and *Melobesia* extend into deeper water. The most widespread types on seagrasses are *Neogoniolithon*, *Hydrolithon*, *Pneophyllum* (in shallow water), *Lithophyllum* (to ~20 mwd),

Fig. 4.9 Non-geniculate calcareous red algae. (**a**) Encrusting Pleistocene aeolianite, Robe, S.A. Depth 1 m, finger for scale, (**b**) fruticose rhodolith, Spencer Gulf, 2 mwd, (**c**) fruticose rhodolith, attached to *Amphibolis* stem (coin scale 2 cm in diameter, Spencer Gulf), (**d**) rhodoliths with a smooth nodular growth form; cm scale, Great Australian Bight, Baxter Sector, 46 mwd

and *Synarthrophyton* (to ~35 mwd) (James et al. 2009). Peyssonnelids grow from shaded intertidal locations across the shelf and extend into deep water. They are encrusting or loosely attached to the substrate.

There is a continuum in growth form between prone, encrusting types and shrub-like, fruticose algae. In some cases the holdfast is encrusting whereas most of the plant is fruticose. This variation is also common in rhodoliths (cf. Wray 1977). They can vary from fruticose nodules with discrete branches (*Sporolithon, Lithothamnion, Lithophyllum, Mesophyllum*) to smooth balls with no obvious protuberances (the foregoing as well as *Hydrolithon, Spongites, Neogoniolithon* and *Pymatolithon*).

There are few studies of living rhodoliths across this extensive region except at depths shallower than 30 mwd (Womersley 1984, 1987, 1996) but reconnaissance sedimentological examination (Ryan et al. 2008) reveals some trends. Most are an intergrowth of encrusting benthic foraminifers and calcareous algae,

particularly *Sporolithon* and *Lithothamnion*. Shapes vary from spherical to discoid to ellipsoid; sizes range from 1 to 4 cm with some up to 5 cm in diameter. They are generally isolated from one another in sparse to luxuriant seagrass environments. In deeper water, 10–35 mwd, they can form pavements of nodules in contact with one another, covering as much as 80% of the seafloor in layers up to 3 rhodoliths thick. The algae can then be encrusted by bryozoans, especially fenestrate growth forms.

Articulated Branching: The shallow rocky intertidal (Fig. 4.10) is dominated by *Corallina* whereas *Jania, Metagoniolithon*, and *Amphiroa* can grow to depths of ~50 m but are most abundant to depths of ~10 mwd (Shepherd and Womersley 1976). Epiphytes on seagrasses are mostly *Jania, Corallina*, and *Metagoniolithon*, all of which are abundant to depths of ~10 m, and decrease rapidly with increasing water depth (James et al. 2009).

Fig. 4.10 Articulated or non-geniculate calcareous red algae. **(a)** Articulated coralline algae growing in the intertidal zone on Pleistocene aeolianite, Investigator Strait; finger (under A) for scale, **(b)** red living and white dead coralline algae in the intertidal zone on Pleistocene aeolianite, 10 cm increments on scale bar; Robe, Bonney Coast, **(c)** sediment dominated by coralline algae fragments (rods) mixed with *Posidonia* fragments; finger (under C) for scale; Investigator Strait

4.2.2.2   Calcareous Green Algae

Calcareous green algae (Chlorophytes—Dasycladales and Udoteaceae) are ineffectual as carbonate sediment producers. The common tropical Udoteacean form *Halimeda cuncata* (Fig. 4.11) only occurs along the Albany coast to 6 mwd and is lightly calcified throughout. *Rhipiliopsis peltata* (Fig. 4.11) although occurring in relatively deep waters from 14 to 25 mwd, or in shaded shallower sites, is completely uncalcified. The dasyclad *Acetabularia* is present in shallow,

(3–20 mwd) settings, particularly in the gulfs, but is poorly calcified.

4.2.2.3   Sedimentology

Encrusting corallines are, in terms of sediment production, most important in seagrass environments where they break down into carbonate mud upon death of the grass blade. Articulated corallines disintegrate into innumerable sand-size rods. Both of the forgoing can be passively transported landward on seagrass blades and accumulate on beaches (Fig. 4.10). Nodular rhodoliths typically remain as cobble-size particles whereas fruticose and rigid branching forms can be broken down into granules and coarse sand. Corallines encrusting or growing within macroalgal holdfasts contribute large fragments to the sediment. Finally corallines encrusting rocky substrates remain attached or are broken down into sand size grains. Peyssonnelids generally break into sand size fragments.

4.2.3   Bryozoans

4.2.3.1   Introduction

Bryozoans are sessile, colonial organisms consisting of numerous minute (generally <1 mm in length) individuals (zooids). Zooids consist of (a) soft-parts of the animal (polypide) and include flexible proteinaceous support and/or connective tissue (referred to as kenozooidal), and (b) calcareous skeletal elements, mainly a protective exoskeleton, with shared skeletal walls common. Mineralogy of the calcareous marine forms (Classes Gymnolaemata and Stelolaemata: orders Cheilostomata—90% of living species and Cyclostomata—10% of living species, respectively) is LMC, IMC, or aragonite (Bone and James 1993; Smith et al. 1998, 2006). The opportunistic organisms can grow rapidly on a wide variety of seafloor habitats; rock, sediment, or other biota, living or dead, sessile or mobile, such as molluscs, hydroids, ascidian tunicates, sponges, soft worm tubes, octocorals, and other bryozoans (Hageman et al. 2000). They are themselves, however, often rapidly overgrown by sponges, colonial ascidians, corallines and other bryozoans, including other colonies of the same species. Individual colonies

Fig. 4.11 Calcareous green algae that are non-calcified. (**a**) *Halimeda cuncata*, ~2 mwd. Recherche Archipelago, Albany Shelf, (**b**) *Rhipiliopsis peltata* ~2 mwd. Recherche Archipelago, Albany shelf

live 1–10 years and produce carbonate at a rate of up to 3.6 cm ky^{-1} (Hageman et al. 2000; Yagunova and Ostrovsky 2008; Winston 2009). More details of the general organisms themselves can be found in McKinney and Jackson (1989), and Bock and Cook (1998; 2004).

There are relatively few sedimentological studies of bryozoans from the modern continental margin of southern Australia. Most of the bryozoan studies are from the Lacepede Shelf (Bone and James 1993; Hageman et al. 1995, 1998, 2000, 2003), the Lincoln Shelf (James et al. 1997), the Great Australian Bight and adjacent areas (Conroy et al. 2001). Bryozoans form at least 10% of the Holocene particles in all sediments in these vast areas. They are diverse, with over 140 species studied and a total diversity of well over 300 recognized species. The following classification generalities are based on the above studies.

4.2.3.2 Classification

The phylum is species-rich, with over 10,000 living species. Biological classifications, focused on the recurrence of growth forms through time and specific taxonomic relevance, have proven unwieldy from a geological perspective. Thus, early sedimentological and ecological studies utilized a morphological organization but added a sediment-production potential aspect. Nelson et al. (1988a) successfully reduced previous nomenclatural schemes for zooarial forms to universal utility. Bone and James (1993) focused on the potential sediment contribution of bryozoans when they modified this scheme, with further variations appearing over the following decade (see Hageman et al. 1998 for review). The new classification scheme herein (Fig. 4.12) is based on our earlier scheme, but with sedimentological refinement and recognition of significant morphological features. The significance of

substrate and its availability in a facies is introduced, e.g. the ability of free-living bryozoans to utilise shifting sands. The scheme, however, still does not refer specifically to such aspects as morphological plasticity, which is often substrate controlled, such as *Celleporaria fusca* (Fig. 4.13d) growing up from its substrate and *Celleporaria* sp. rejecting its primary roots and behaving like a free-living "biscuit" (Fig. 4.14b).

Colonies are initially divided into 3 simple overall groups reflecting their habit and position in the environment—*prone, erect* or *free-living*. There is neither taxonomic significance nor colony size implied in this first division. Most are encompassed within the first two groups—*prone* for those attaching themselves and growing on a substrate and *erect* for those growing above the seafloor into free space. The third group, the enigmatic *vagrant* or *free-living* bryozoans, use a sand-sized grain for initial attachment. The particle is encapsulated by the ancestrula early in colony growth, and thereby the animal becomes unattached for the remainder of its life and so can move about the seafloor (Winston 1988).

The next division is into 5 major growth style categories, each having post-mortem influence: (1) prone—*encrusting*, wherein the colony remains attached to the substrate, but typically breaking off into variably-sized fragments, (2) erect—*robust rigid* wherein the skeletons usually remain whole or break into moderately large fragments, (3) erect—*delicate rigid* wherein the colony disarticulates into medium to small fragments or grains, (4) erect—*delicate flexible* wherein the skeleton generally disarticulates into individual zooids, thus producing very fine grained or silt-sized particles, and (5) free living—*disc* wherein the colony is preserved as a single, completely flat or slightly-domed fragment. The final division is into morphology, within which different shapes are recognized. There is some overlap at this level, but this classification scheme provides reasonably

Fig. 4.12 Bryozoan colony classification

BRYOZOAN COLONY CLASSIFICATION

HABIT		GROWTH FORM NAME	MORPHOLOGY / SHAPE	SUBSTRATE / ATTACHMENT	SKETCH	EXAMPLES
PRONE	ENCRUSTING	UNI-LAMINAR	UNISERIAL MULTISERIAL / CRUSTOSE	R:S:O / C	3 mm	*Pyripora*
PRONE	ENCRUSTING	UNI-LAMINAR	MULTISERIAL / CRUSTOSE	R:S:O / C	1 mm	*Membranipora*
ERECT	ENCRUSTING	MULTI-LAMINAR	DOMAL / MOUND-LIKE	R:O / C	2 cm	*Celleporaria*
ERECT	ENCRUSTING	MULTI-LAMINAR	GLOBOSE / SPHERICAL	O / R	5 mm	*Sphaeropora*
ERECT	ENCRUSTING	MULTI-LAMINAR	ARBORESCENT / CYCLINDRICAL	O / C	1 cm	*Celloporaria*
ERECT	ROBUST RIGID	FOLIOSE	PALMATE / PLANAR	R:S:O / C/R	1 cm	*Parmularia*
ERECT	ROBUST RIGID	BRANCHING	RAMOSE / SHRUB - LIKE	R:S / C	5 mm	*Adeonellopsis*
ERECT	ROBUST RIGID	FENESTRATE	FLABELLATE / SHRUB - LIKE	R:S:O / C	A 1 cm 5 mm B	*Iodictyum* (A) & *Adeona* (B)
ERECT	DELICATE RIGID	FRONDOSE	PALMATE / TENTICULATE	O / C	1 cm	*Idmidronea*
ERECT	DELICATE RIGID	BRANCHING	BIFURCATING / DENDRITIC	S:O / C	5 mm	*Hornera*
ERECT	DELICATE FLEXIBLE	ARTICULATED BRANCHING	DIGITATE / SHRUB - LIKE	R:S:O / C/R	1 cm	*Cellaria*
ERECT	DELICATE FLEXIBLE	ARTICULATED ZOOIDAL	GENICULATE / SHRUB - LIKE	S:O / C/R	1 mm	*Othoscuticella*
ERECT	DELICATE FLEXIBLE	FENESTRATE	FOLIOSE / FLABELLATE	O / C/R	5 mm	*Flustra*
FREE	DISC	VAGRANT (Free Living)	DISCOID / CONE	S (< 1mm) / C	5 mm	*Lunularia*

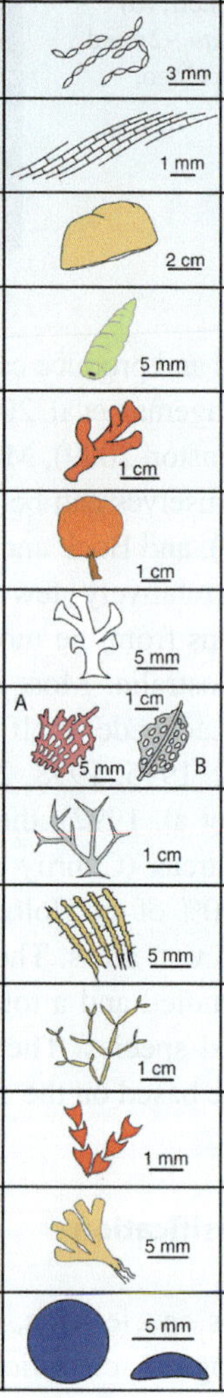

Substrate: **R** = Rock, **S** = Sediment grain(s), **O** = Organic (sessile, rarely mobile benthic biota).
Attachment: Initial larval settlement: **C** = Cemented, **R** = Pseudo roots/rootlets.

easy but specific allocation of bryozoans, while also allowing accuracy in communication and understanding between geologists, biologists and ecologists.

4.2.3.3 Encrusting Colonies

Unilaminar: The colony has only one layer of zooids. Unilaminar bryozoans may be *uniserial* (colony forms one row of zooids) or more commonly, *multiserial*, with the rows usually adjacent to another. Modern uniserial bryozoans are not important sediment producers and are mainly restricted to substrates such as the internal surface of mollusc shells (Fig. 4.13b). Multiserial bryozoans are, however, well adapted to survive in low nutrient environments as they may cover a large surface area, such as the cosmopolitan *Membranipora membranacea* on *Ecklonia radiata* (Fig. 4.13a). Their strength is substrate dependent, but if that substrate is dislodged, then the colony is passively moved to the new location. The best example of this is the common re-location of living colonies epiphytically attached to

Fig. 4.13 Encrusting colonies. (**a**) Multiserial crustose *Membranipora membranacea* on kelp, (**b**) numerous multiserial crustose colonies on the inside of a dead bivalve shell (cm scale), (**c**) multilaminar domal *Celleporaria* sp., (**d**) multilaminar *Celleporaria fusca*, (**e**) arborescent *Celleporaria* sp. on sponge (cm scale), (**f**) multilaminar (1) *Celleporaria crustata* and (2) *Densipora corrugata* on *Amphibolis* sp. stems

sea-grasses, particularly *Posidonia* spp. that are shed from the sea-grass meadow and carried on to the beach. This transport can be responsible, over time, for the deposition of large volumes of carbonate grains in the coastal environment (Brown 2005; James et al. 2009).

Multilaminar: These bryozoans exhibit self-overgrowth or the formation of thickened crusts, which may be a survival strategy where environmental factors are limiting, such as exposure of the colony during episodic, short term, water level changes. Such

Fig. 4.14 Encrusting colonies. (a) *Sphaeropora* sp., (b) Pseudovagrant form of *Celleporaria* sp

situations are common in estuaries, e.g. the Coorong Lagoon (Fig. 5.6e), where the time-span of individual *Conopeum aciculata* build-ups have been dated at many hundreds of years (Bone and Wass 1990; Sprigg and Bone 1993). It also allows morphological plasticity to develop when space becomes critical, particularly in high density, mixed biota ecosystems, so that the encrusting habit is exploited via specialized overgrowth behaviour. This leads to some bryozoan colonies invariably exhibiting specific morphologies—*domal* (Fig. 4.13c, d), sometimes looking like pliable colonies stacked on top of one another, *globose* (Fig. 4.14a) where the original substrate attachment to grains within the sediment is lost and the colony becomes a pseudovagrant, and the most common variant—*arborescent* (Fig. 4.13e) wherein the substrate is usually organic (Fig. 4.14b), such as cylindrical oscular sponges (Hageman et al. 1996). Conspicuous arborescent *Celleporaria* sp. forms are a major contributor to modern and Cenozoic sediments in southern Australia, growing locally today as thickets, between 120 and 250 mwd and especially on biogenic mounds (Bone and James 2002; James et al. 2004). All multilaminar bryozoan colonies are multiserial variants.

4.2.3.4 Robust Rigid Colonies

Foliose: These colonies (Fig. 4.15a) resemble palm fronds, including the occasional branching habit toward the apex of the frond. There can be minor colony flexibility at the attachment site. The number and style of fronds varies from a single frond (*Parmularia reniformis*), with its sturdy pedunculate root, to a gently convoluted frond (*Parasmittina* sp.), to numerous, cup-like fronds (*Tubilopora* sp.).

Branching: Such bryozoans are bifurcated (*Caleschara* sp.), usually sequentially and repeated, e.g. the ubiquitous *Adeonellopsis* sp. (Fig. 4.15b) so that the overall colony has the appearance of a tangled web. In this case, the branches are additionally flattened, allowing zooids on individual branches to equally access water with suspended particles. They are either cemented to the substrate or rooted, and commonly a combination of both for extra strength. These branching forms are most conspicuous in high-energy subaqueous dune fields. They break down into fragments that can be up to 1.5 cm long.

Fenestrate: This is the group that gave bryozoans their early common name of "lace corals". The conspicuous fenestrules (windows, e.g. *Iodictyum* sp.—Fig. 4.15c, e) enhance water movement around and through the colony, particularly in high-energy environments, without the likelihood of the rigid plates snapping off. The colonies are cemented to a hard substrate or to seagrass (Fig. 4.13f) and grow across the depositional spectrum.

One distinctive atypically large form, *Adeona* sp. (Fig. 4.15d, f) is a colony that grows as a rigid structure composed of juxtaposed plates, but which is atypically flexible at its sea-floor attachment site. This part of the organism is very distinctive, with calcareous kenozooidal branches and trunk, attached below the sea-floor to a gelatinous lower trunk and roots that may extend as much as a meter into the sediment. Alternatively, the base of the trunk may be cemented directly to a hard substrate. It is a major pioneer species, growing in mobile sand substrates where it becomes a substrate for other biota. Its large fenestrules allow high volumes of water to flow through the colony, which has the ability to orientate itself to enhance this passage of water. It breaks up into gravel-size pieces of sediment upon death.

Fig. 4.15 Robust rigid colonies. **(a)** *Parmularia reniformis*, **(b)** *Adeonellopsis sulcata*, **(c)** *Iodictium phonecium*, **(d)** *Adeona* sp. **(e)** *Iodictium phonecium* growing on *Amphibolis* sp. stems and encrusted by *Celleporaria* sp., **(f)** Close view of *Adeona* sp. with attached *Cellaria* sp. and other delicate branching types (cm scale)

4.2.3.5 Delicate Rigid Colonies

Frondose: Bryozoans of this type are usually small to medium sized colonies (0.3–5 mm), vase-shaped and with branches stemming from or close to the attachment site. They usually grow close together resulting in multiple coarse—sized sediment grains consisting of entire colonies (Fig. 4.16a). Colonies are typically found weakly attached to organic substrates, especially sea-grasses and algae in the coastal environment. They are rare in habitats where larger forms of bryozoans thrive.

Branching: These relatively small colonies (1 cm-2 cm) consist of so many bifurcating branches that they appear to be fenestrate (*Hornera* sp.), but the gaps between the branches are fortuitous, not by design (Fig. 4.16b). They survive best, and are ubiquitous, in high diversity ecosystems where their rigidity is protected by the baffling affect of other biota.

4.2.3.6 Delicate Flexible Colonies

Articulated Branching: These bryozoan colonies are highly successful, being the most commonly seen and recognizable forms in fine- to medium-grained sediment in the modern seafloor and in southern Australian Cenozoic limestones. They consist of small rods/cylinders of about 5 zooids (3–12 range,

Fig. 4.16 Delicate rigid branching colonies. (a) *Idmidronea* sp., (b) *Hornera foliacea*

Fig. 4.17) interconnected by joints with proteinaceous tissue, particularly in the *Cellaria* genus (Figs. 4.15f, 4.17a). Such flexibility enables their survival in high-energy environments, so that they are frequently pioneer species, particularly using *Adeona* sp. and sponge as attachment sites (Fig. 4.17b). Articulation enables these bryozoans to grow in relatively high-energy situations, but most of them are found below swell wave

Fig. 4.17 Articulated branching colonies. (a) *Cellaria* sp., (b) *Calpidium* sp. growing on a tubular sponge, Articulated zooidal colonies. (c) *Orthoscuticella ventricosa*, (d) close view of zooids (*O. ventricosa*), (e) *Flustra* sp. (f) Vagrant (free living) colonies 1 *Lunularia capulus* (*blue* living); 2 pseudovagrants *Celleporaria* sp. & *Sphaeropora* sp. Scale in mm

base. Colony density increases over time to a cover that has the appearance of low-growing shrubbery.

Articulated Zooidal: Bryozoan colonies of this type are structured such that individual zooids (colloquially known as cats, singlets, or doublets) are joined to one another by proteinaceous tissue (Fig. 4.17c, d). These opportunistic and/or pioneer forms, either cemented or rooted, can rapidly colonize all available substrates in numerous environments not subject to terrigenous sediment input. Colony density in deeper water gives the appearance of a turf (Fig. 7.19). These bryozoans disintegrate upon death, producing hundreds of delicate, fine sand and silt-size particles that are easily broken down even further into mud. The grains are a common component in aeolianites along the southern coast of Australia.

Fenestrate: The bryozoans in this group are only common locally and because of their lightly calcified zooids are rare in sediments. They (e.g. *Flustra*—Fig. 4.17e) are often mistaken for algae, as algae are their most frequently observed substrate.

4.2.3.7 Disc—Shaped Colonies

Vagrant: These **free-living**, as disc-shaped, aragonitic colonies (Fig. 4.17f) are between 0.25 and 5.0 mm in diameter (locally up to 3.0 cm). Such cheilostomes e.g. *Selenaria* sp. and *Otionella* sp.—(Cook and Chimonides 1978) grow in a wide range of environments, including energetic situations where rooted bryozoans cannot survive. This is because the colony has the ability to exhume itself after being swamped by sediment via a concerted movement of its setae. Some forms can move about the sea floor at rates of up to 1 m h^{-1}. If the colony is broken each segment has the ability to regenerate into a new colony.

4.2.3.8 Importance of Substrate

Substrate Type: Substrate has a basic and fundamental control on the overall distribution of bryozoans; (1) determining the relationship between anchoring of the colony to a site, (2) influencing the manner in which the individual zooids grow, and (3) controlling the rate at which the sediment grains are produced (Fig. 4.12).

Substrates can be divided, simply, into (1) lithic, (2) sediment, and (3) organic, with the manner of attach-

ment of the larva, the growth of the first 1–5 zooids (ancestrula) and the expansion and stabilization of the adult colony, each necessitating different physical and/or organic techniques.

1. *Lithic:* Bedrock, hardgrounds or calcareous shells. Encrustation on such a substrate is generally via cementation and so the skeleton either remains attached after death or, if upright, must be broken off to form sediment particles.
2. *Sediment:* Particulate sediment in which the bryozoan is held in place via a series of cuticular roots. This is similar to those on organic substrates, once the roots rot, the calcareous skeleton becomes part of the sediment.
3. *Organic:* Generally attached to another organism (algae, grass, or invertebrate). Many organic hosts lack external hard parts so that when the host dies and decays, the bryozoan skeletons falls to the seafloor and forms sediment. The nature of this substrate, however, often results in the bryozoan grains being transported and deposited elsewhere.

Animal-Substrate Relationship: The animal—substrate couplet can be summarized into five simplified ecologic groups:

1. *Epiphytic:* Bryozoans that live attached to algae and kelp (sea grass, green and brown algal leaves, roots, and stypes–e.g. *Discoporella* sp.) are most important on the inner shelf (<40 mwd).
2. *Rigid, rooted and vagrant:* Bryozoans that are adapted to life in loose, shifting sand and include colonies with robust skeletons anchored by kenozoidal roots (*Adeona* sp.) and free living forms (e.g. *Otionella* sp.). Such bryozoans are found scattered across the sea floor as isolated colonies. They are pioneers and typically, once established, act as substrates for other forms. Outer shelf and upper slope bryozoans live in isolated, complex communities of intergrown bryozoans, sponges, hydroids, tunicates and polychaete worms (Cook and Chimonides 1981). Bryozoans are present in these communities in a variety of forms.
3. *Flexible rooted:* These bryozoans are characterized by articulated forms (zooidal—*Orthoscuticella* sp.; branching—*Cellaria* sp.), that have chitinous kenozooids that permit colony articulation and provide root systems that allow the animals to stabilize loose sediment.

4. *Epizoic:* Bryozoans attached (rooted or cemented) to invertebrate substrates such as rigid rooted and flexible rooted bryozoans, sponges, hydroids, ascidians, and polychaete worm tubes, are highly diverse and abundant on the outer shelf and upper slope (120–220 m). Such communities are also present on regions of the inner and middle shelf that are not dominated by shifting sand.

5. *Cemented:* A diverse, but sedimentological insignificant group of unilaminar and multilaminar as well as uniserial and multiserial (e.g. *Lichenopora* sp.) bryozoans that cement directly to a hard substrate.

4.2.3.9 Depth Limits of Living Bryozoans

Living bryozoans can, with some overlap, be divided into five depth-related assemblages;

1. *Coastal (0–30 mwd):* The wide variation in environment controls their distribution and density, such that encrusting epiphytic forms are common in sea-grass meadows, free-living forms are restricted to open shifting sands, articulated forms are found attached to algae and other biota in rocky reef environments, along with fenestrate and other erect forms,

2. *Shelf (30–110 mwd):* Small numbers of the large, conspicuous fenestrate *Adeona* sp. along with large numbers of articulated zooidal, encrusting, articulated branching, robust rigid branching, fenestrate, and vagrant types,

3. *Shelf Edge (110–200 mwd):* The zone of most diverse growth comprising ubiquitous delicate branching forms, particularly cyclostomes, and articulated zooidal cheilostomes together with large numbers of articulated branching, fenestrate, robust fenestrate, and fewer encrusting, vagrant, and nodular-arborescent cheilostomes,

4. *Top of Slope (200–220 mwd):* Articulated branching, articulated zooidal and flat robust branching cheilostomes, delicate, branching cyclostomes, and lesser nodular-arborescent cheilostomes of variable morphology,

5. *Upper Slope (>220 mwd):* Delicate branching cyclostomes and articulated cheilostomes with variable nodular-arborescent-globose cheilostomes. The components in sediments generally correspond to these assemblages, with particles displaced by gravity downwards as much as 50 m in deep neritic and upper slope environments.

4.2.3.10 Sedimentology

It is clear from the foregoing that bryozoans form carbonate particles across the grain-size spectrum in all marine environments under discussion (Fig. 4.18) (Bone and James 1993). Mud is produced mainly by articulated zooidal animals, fine sand is generated by articulated zooidal and delicate branching bryozoans, medium to coarse sand is formed by delicate branching, articulated zooidal, types and fragments of larger colonies, granules are mainly fragments and whole skeletons of fenestrate, encrusting robust rigid branching and vagrant morphotypes. The largest biofragments are from the few bryozoans that often grow into spectacularly large colonies, such as the fenestrate *Adeona* sp., *Tubilopora* sp. and *Celleporaria fusca* in shallow water, and nodular-arborescent forms at the shelf edge and on the upper slope. Encrusting types, once liberated from their substrate (e.g. grasses), form particles across the grain-size spectrum.

4.2.3.11 Mineralogy

There is also a mineralogical partitioning in the phylum (Fig. 4.18) (Bone and James 1997; James et al. 2005; Smith et al. 1998, 2006; Taylor et al. 2009). Approximately 50% of bryozoans produce skeletons that are IMC, 30% use aragonite, and about 20% LMC or are biminerallic. As a result most bryozoan particles are calcite-dominated. In depth-related terms, sediments in water depths to 130 mwd are mostly IMC and lesser aragonite, and at water depths >130 m they are IMC and LMC. In terms of grain-size, sand and mud is mostly LMC and IMC whereas granules and cobbles are both IMC and aragonite, never LMC.

4.2.4 Molluscs

4.2.4.1 Gastropods

Gastropod diversity is moderate with 24 genera identified. *Coxiella* sp. is most numerous in estuarine settings. The small *Bembecium* sp. and other cerithids are abundant on muddy tidal flats. Rocky peritidal surfaces are covered with neritids and limpets, whereas subtidal shallow rock reefs host the robust top shell (*Turbo* sp.) and abalone

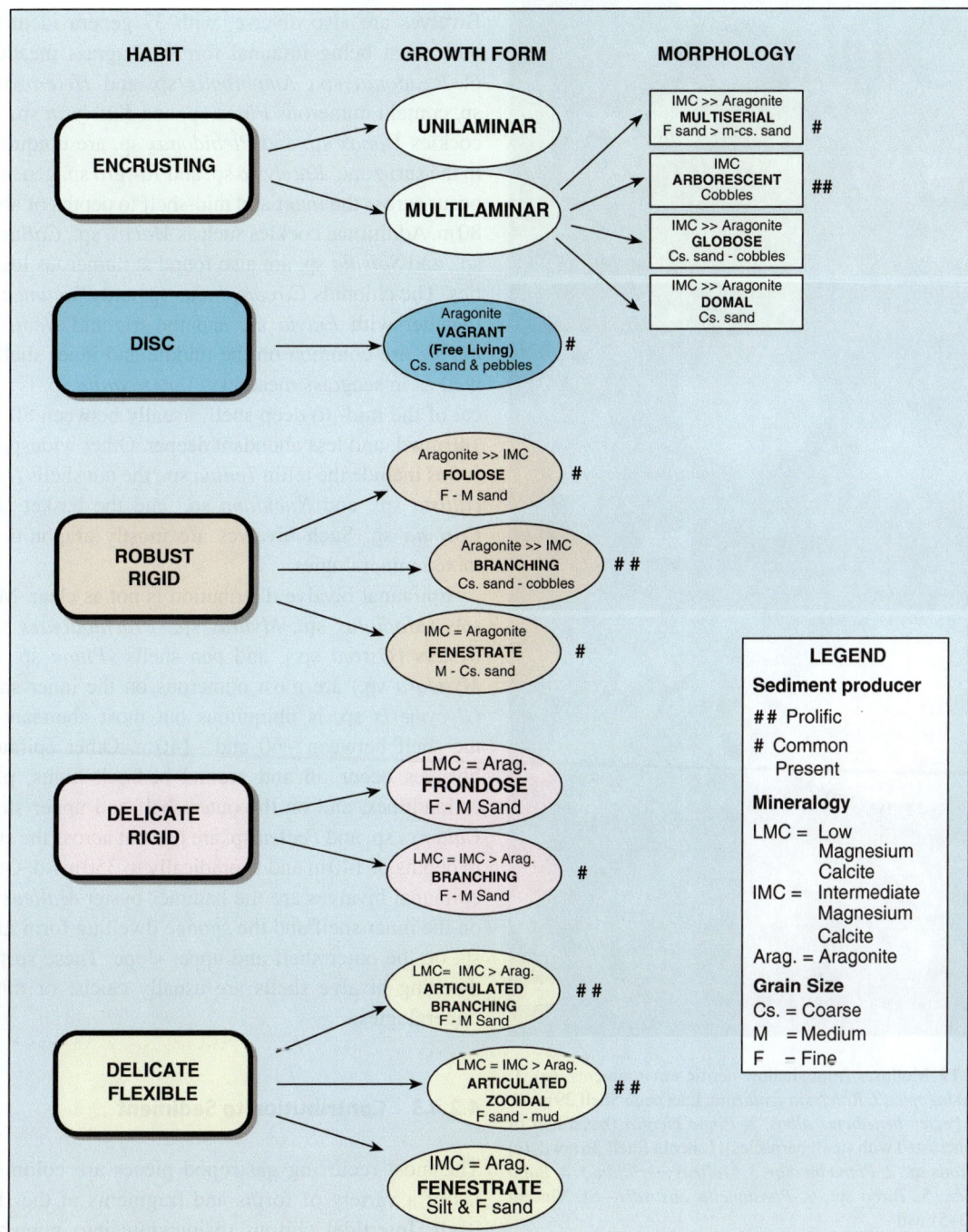

**BRYOZOAN GROWTH FORM
SEDIMENT GRAIN SIZE & MINERALOGY**

Fig. 4.18 Grain size and mineralogy of different bryozoan growth forms

(*Haliotis* sp.). Seagrass meadows are populated by numerous *Bulla* sp., *Batillaria* sp. and *Phasianotrochus* sp. Open sand plains contain Nautiidae (*Polinces* sp.) as well as *Cassis* sp. Otherwise the most common snails are the suspension and detritus feeders *Calyp-*

traea sp. (which in life is mostly attached to dead bivalves), the periwinkle *Clanculus* sp., the pheasant shell *Phasianella* sp. various turitellids, olives, mitres, sponge-dwelling siliquarids (*Tenagodus* sp.), and cowries (*Cyprea* sp.). Semi-infaunal forms include a

Fig. 4.19 Molluscs from shallow neritic environments. **(a)**. 1. *Tawera lagopus*, 2. *Katelysia scalarina*, Lacepede Shelf 29 mwd, **(b)** 1. *Pecten benedictus albus*, 2. *Pinna bicolor* (basal half of shell encrusted with small barnacles), Lincoln Shelf 36 mwd, **(c)** 1. *Haliotis* sp., 2. *Pinna bicolor*, 3. *Malleus meridianus*, 4. *Bulla botanica*, 5. *Turbo* sp., 6. *Phasianella australis*—St. Vincent Gulf, 3–5 mwd

variety of turitellids and scaphopods (*Dentalium* sp.). Predatory types comprise moon shells (*Polinices* sp.), the large gastropod *Campanile* sp., and rock shells (*Murex* sp.) (Figs. 4.19, 4.20, 4.21).

4.2.4.2   Bivalves

Bivalves are also diverse, with 37 genera identified with most being infaunal forms. Seagrass meadows of *Posidonia* sp., *Amphibolis* sp. and *Heterozostra* sp. contain numerous *Pinna* sp. and *Katelysia* sp. The cockles *Donax* sp. and *Plebidonax* sp. are ubiquitous in the surf zone. *Katelysia* sp. and *Tawera* sp. generally occur across the inner and mid-shelf to depths of about 80 m. Additional cockles such as *Mactra* sp., *Callucina* sp., and *Spisula* sp. are also found at numerous localities. The chionids *Circomphalus* sp. and *Placamen* sp. together with *Fulvia* sp. and the trigonid *Neotrigonia* sp. are common on the middle and inner shelf as well as in seagrass meadows. *Venericardia* sp. is typical of the mid- to deep shelf, usually between 80 and 140 mwd, and less abundant deeper. Other widespread forms include the tellin *Tellina* sp., the nut shells *Notocallista* sp., and *Nuculana* sp., and the basket shell *Corbula* sp. Such bivalves are mostly aragonitic or mixed mineralogies.

Epifaunal bivalve distribution is not as clear. Mussels (*Modiolus* sp., *Mytilus* sp., *Brachiodontes* sp.), oysters (*Ostrea* sp.), and pen shells (*Pinna* sp. and *Myadora* sp.) are most numerous on the inner shelf. *Glycymeris* sp. is ubiquitous but most abundant on the shelf between −60 and −140 m. Other epifaunal bivalves occur on and around bedrock highs, paleostrandlines, and on the outer shelf and upper slope. *Chlamys* sp. and *Pecten* sp. are present across the shelf to depths of 140 m and sporadically to 350 mwd. Other epifaunal bivalves are the hammer oyster *Malleus* sp. on the inner shelf and the sponge dwelling form *Lima* sp. on the outer shelf and upper slope. These surface dwelling bivalve shells are usually calcite or mixed mineralogies.

4.2.4.3   Contribution to Sediment

The most recurring gastropod pieces are columnellas of a variety of forms and fragments of the shell itself. Intertidal chitons disintegrate into numerous granule-size plates. Bivalves are numerous in both the sand-size and gravel-size fractions. Fragmented bivalves comprise more than 50% of the skeletal coarse and fine fraction in most sediment shallower than 90 mwd.

Fig. 4.20 Molluscs from middle neritic environments. **(a)** 1.
Katelysia sp., 2. *Nuculana* sp., 3. *Placamen* sp., 4. *Batillaria* sp.,
5. *Neotrigonia* sp. Lacepede Shelf 59 mwd, **(b)** 1, 2, 8. *Katelysia*
sp., 3. *Circomphalus* sp., 4. *Katelysia* fragment, 5. *Bittium grana-
rium*, 6. A columella, 7. *Batillaria* sp., remainder are fragments.
Lacepede Shelf 55 mwd, **(c)** 1. *Placamen flindersi*, 2. *Pecten benedictus albus*, 3. *Katelysia scalarina*, 4 *Fulvia* sp., 5. *Bachidon-*
tes erosus, Great Australian Bight, Baxter sector, 64 mwd, **(d)**
Chlamys bifrons (worn), 2, 3, 5. *Katelysia scalarina*, 4. *Mytilus
edulis*, 5. *Fulvia* sp., 6. *Glycymeris radians*, 7. *Fulvia* sp., 8.
Calyptraea sp., Great Australian Bight, Ceduna sector, 75 mwd,
(e) 1, 2 *Glycymeris radians*, 3 *Katelysia scalarina*, 4, *K. scalarina* with cemented calcareous sand in concavity, Lincoln Shelf,
80 mwd, **(f)** *Tenagodus australis*, Lincoln Shelf 64 mwd

Fig. 4.21 Molluscs from the deep neritic environment. **(a)** 1. *Conus* sp., 2. *Oliva* sp., 3. *Dentalium sp.*, 4. *Lima lima*, Albany Shelf, 110 mwd; **(b)** *Gazameda iredalei*, Great Australian Bight, Baxter Sector, 180 mwd

4.2.5   Foraminifers

Foraminifers, both planktic and benthic are a comparatively minor, 10–30%, but recurring component of sediments across all sectors. The mid-latitude fauna is relatively diverse comprising free and encrusting, large and small individuals with a few endemic forms. The assemblages have been studied across the southwestern continental margin (Li et al. 1996a, 1998, 1999), in the gulfs (Cann et al. 1993, 2000, 2002) and as far eastward as the Lacepede Shelf (Li et al. 1996b). Somewhat less is known about them off southwestern Victoria and Tasmania. As with other particles, there is a clear distinction between living and fresh tests as opposed to brown-stained relict forms; stranded types have not been distinguished as readily (Fig. 4.22).

4.2.5.1   Planktonic Foraminifers

There are generally between 15 and 25 species. The most distinctive taxa are *Globorotalia menardii, Neogloboquadrina dutertrei, Globigerinoides trilobus, Globigerinoides ruber*, and *Globorotalia inflata* (Li et al. 1996b, 1999).

4.2.5.2   Benthic Foraminifers

In most areas there are more than 200 species, generally 8–10 times more diverse than the planktics. In general terms, they are mostly miliolids, discorbids, and elphidiids (Fig. 4.22) with increasing numbers of cibicidids, agglutinated and other trochospiral rotalid tests, especially infaunal bolivinid-uvigerinids and cassulinids, offshore. The symbiont-bearing forms *Marginopora*

and *Peneropilis* are found in shallow, euphotic, relatively warm-water environments. The large milioline *Marginopora vertebralis* has been identified regionally but it is, in most cases likely *Amphisoris hemperchii* (Li et al. 1996a). The two forms are difficult to separate with certainty when relict and so these types are generally referred to as the *Marginopora-Sorites* group. They generally signify low mesotrophic to near oligotrophic nutrient levels. Another distinctive form is the rotaliid *Ammonia beccarii*, a euryhaline form that typifies marine environments of reduced salinity. The pink encrusting *Miniacina miniacea* is common on the outer shelf in the bryozoan-sponge communities (Fig. 4.22).

Marine grass beds have a distinctive suite of benthic foraminifers that are especially well known from Spencer Gulf and Gulf St. Vincent (Belperio et al. 1984; Cann et al. 1993, 2002, 2006). Amongst these are *Nubecularia*, miliolids (*Quinqueloculina, Triloculina, Milionella, Spirolculina*), rotaliids (*Discorbis, Elphidium*), and large symbiont-bearing sortids (*Peneropolis*).

4.2.5.3   General Distribution

There are several recurring spatial trends. There are relatively few planktics nearshore but they increase and are locally equal in number to benthic tests at >100 mwd. This is paralleled by an increase in the number of benthic species from inboard to outboard; the outer shelf and upper slope is rich and diverse. In general terms the numbers of benthics is moderate nearshore, drops off and is lowest on the mid-shelf, and then rises rapidly to the most diverse assemblages on the upper shelf and upper slope. The mid-shelf is highest in brown-stained relict taxa.

Fig. 4.22 Benthic foraminifers. Large, photosymbiont-bearing protists, **(a)** *Marginopora vertebralis*, *S* stranded, *R* relict; **(b)** *Peneropolis planatus*; Small benthic protists **(c)** *Nubecularia lucifuga* (planar ventral attachment view); **(d)** *Elphidium fichtellianum*; **(e)** *Spiroloculina antillarium*; **(f)** *Discorbis dimidiatus*. All images except (A) are scanning electron photomicrographs. (Modified from James et al. 2009)

Benthic assemblages are depth dependent and remarkably consistent across large segments of the shelf. The inner shelf <60 mwd, typically contains numerous relict forms (see below) and distinctive shallow-water types such as *Elphidium* spp. and *Discorbis dimidiatus* and a low planktic:benthic ratio. The midshelf with a low planktic:benthic ratio generally has a conspicuously high relict component (>50%), many extant shallow-water benthics, and especially numerous *Cibicides* spp. The outer shelf has conspicuously fewer relict individuals, but deep water and calcareous species are common to dominant. The upper slope contains many deep-water species including many agglutinated forms. The biota is 10–20 times more abundant on the outer shelf and upper slope compared to that on the inner and mid-shelf.

Variations on this theme are present locally. There are, for example, comparatively few relict benthics on the mid shelf in the western Otway, Bonney Sector region. This is a region of strong summer upwelling and likely represents high modern productivity.

There is a conspicuous west-to-east, warm-to-cold trend in both planktonic and benthic taxa. A typical warm-water, oligotrophic group of planktics including *Globigerinoides trilobis*, and *Globorotalia menardii* is present in the far west Albany Shelf, in the Great Australian Bight the planktics are dominated by the temperate form *Globigerinoides inflata*, in the Lincoln and Lacepede shelves still cooler water species dominate (*Globorotali inflata*, *Globorotalia ruber*, *Globigerinoides bulloides*, and in deeper sites *Globorotalia truncatulinoides*). This is ascribed to the decreasing influence of the warm-water Leeuwin Current eastwards (Li et al. 1998).

The Western Australia shelf facing the Indian Ocean contains large, warm-water, benthic forms such as *Heteristegina, Amphistegina,* and *Planorbulinella,* but only *Marginopora* (*Amphisoris*) occurs east of Cape Leeuwin. It is common in shallow waters together with non-calcified green calcareous algae and scattered zooxanthellate corals along the Albany Shelf to about Esperance (Li et al. 1999). These observations are interpreted to reflect the somewhat warmer and lower trophic resources (low mesotrophic to near oligotrophic) conditions of the southwestern continental margin as opposed to the colder and mesotrophic conditions on the Lincoln Shelf and further east (Li et al. 1998). They further suggested that Kangaroo Island acts today as a barrier to the movement of warm currents eastward across the southern continental margin.

4.2.6   Echinoderms

Particles from echinoids, plates, pieces of plates, and spines, consistently form 5–10% of sediment grains. Most are from infaunal spatangoids and clypeasters with the larger coarse sand and granule-sizes coming from epifaunal tests and spines, especially cidaroids. Infaunal echinoid tests are rarely recovered whole. Crinoids, (free-living comatulids), asteroids, and ophiroids although locally abundant, are never significant as sediment producers, except very locally. Holothurians contribute small numbers of stellate IMC spicules, ~100 μm in size.

4.2.7   Barnacles

Barnacles are not an important part of the carbonate sediment producing biota in this region but are locally numerous in peritidal environments. Acorn barnacles (*Pedunculata* sp.) attach themselves to the substrate whereas goose barnacles (*Sessilia* sp.) attach by means of a stalk. Their shells are composed of several plates that spontaneously separate from one another upon death forming coarse sand to gravel-size fragments. These animals need relatively high nutrient levels (mesotrophic) and depend upon active phytoplankton growth. Their mineralogy is either IMC or HMC.

4.2.8   Calcareous Worms

Irregular to coiled serpulids, typically attached to a hard substrate, have calcareous tubes 2–5 cm long and form sand in the coarse fraction. Spirorbid tubes are tightly coiled, ammonite-like, and their small 1–2 mm diameter tubes are common on rocky substrates, seagrass blades, and locally attached to macroalgae stipes. Their skeletons are either aragonite and/or Mg-calcite.

4.2.9   Ostracods

These small arthropods have a calcitic carapace of two shells composed of Mg-calcite joined by chitin. They crawl or swim above the seafloor, burrow into sediment, and are epiphytic on seagrass and macroalgae. Most shells are 0.5–3.0 mm long and upon death the chitinous ligament springs apart freeing the valves which remain whole in the sediment. Whereas benthic forms are well calcified, calcification in pelagic types is weak and so they do not contribute significantly to the sediment. They molt as instars (young forms before maturity) at least 6 times, and so contribute significant material to the sediment, all of which is in the finer fraction. They range in mineralogy from LMC to IMC.

4.2.10   Coccoliths

Much of the mud-size fraction on the outer shelf and upper slope is nannoplankton remains, especially coccoliths. The numerous calcitic discs that comprise the coccosphere are 2–6 μm in size, with individual scales even smaller. Their mineralogy is LMC.

4.2.11   Sponges

Sponges are a significant part of the seafloor biota. Most are attached to hard substrates and many harbour a multitude of commensual invertebrates such as gastropods, tunicates, bryozoans, worms, and echinoderms. Others have holdfasts that allow them to grow on sediment substrates. Calcareous sponges are present but they are not numerous, comprising about 10% of the taxa; their spicules are triaxons. The much more numerous demosponges, those with siliceous spicules, are especially abundant, with about 200 genera in southern Australia waters. Many that live in shallow, high-energy environments are encrusting with brightly coloured, upright forms increasing in waters deeper than 20 mwd. They can be large, reaching 20 cm in height and over 40 cm in diameter. Spicules are either rod-shaped megascleres from 100 µm to several millimeter in length or smaller microscleres in a variety of shapes ranging from triaxons to variably straight to curved monaxons. These spicules are conspicuous throughout the fine-grained mud and fine-sand fraction.

4.2.12 Ascidians

These delicate, filter-feeding, sessile tunicates are a common part of the subtidal rocky substrate biota, in many locations being as abundant as sponges and bryozoans. They are solitary and colonial and provide a significant substrate for bryozoans, sponges and brachiopods. The animals grow from the intertidal zone to the shelf edge. Although not possessing a calcareous skeleton they contain distinctive fine sand-size to silt-size mace-like aragonite spicules. They are not voluminous sediment producers but the spicules form an insignificant but recurring part of the fine-grained sediment fraction.

4.3 Other Components

Corals and brachiopods are large and conspicuous, but not volumetrically significant components in all sediments.

4.3.1 Corals

There are four aragonitic zooxanthellate corals in the neritic zone. The largest, *Plesiastrea versipora*, is a massive to platy form whose skeleton can be as large as 3 m in diameter and is found mainly in the gulfs to 25 mwd. The platy form *Coscinaraea* ssp. (Fig. 4.23a) grows to depths of 30 m whereas the smaller, generally prone, branching *Culicia* sp. is found in sheltered caves on open coasts and vertical rocks walls and as deep as 100 mwd on shells and bryozoans. The small cup-coral *Scolemia* sp. (2–3 cm diameter) also lives in cryptic habitats to depths of ~30 m.

There are, in contrast, more than 50 species of aragonitic azooxanthellate corals. Especially prominent are *Platytrochous laevigatus, Caryophyllia planilamellata*, and *Flabellum pavonium.*

Octocorals (soft corals), especially sea fans (alcyonarioans), and sea pens (gorgonians), contain a variety of Mg-calcite spicules embedded in their tissue. These calcareous sclerites are typically straight with

Fig. 4.23 Corals. (**a**) *Pleisiastrea* sp., (10 cm wide) an orange zooxanthallate colonial coral (*arrow*) that usually occurs in shallow, cryptic environments, Investigator Strait, 2 mwd, (**b**) close up of *Pleisiastrea* sp., Gulf St. Vincent, (**c**) numerous dead and living azooxanthellate corals (*Caryophyllia planilamellata*) from 230 mwd, Lincoln Shelf, (**d**) other azooxanthellate corals, *Flabellum pavonium* (*left*) and *Stephanocyathus* sp. (*right*) from 500 mwd, Ceduna Sector, Great Australian Bight

Fig. 4.24 Brachiopods.
(a) *Anakinetica cummingi*:
Lacepede Shelf, 126 mwd,
(b) *Magellania flavescens*:
Spencer Gulf, 3 mwd

numerous tubercules and of fine to medium sand size. Curved straight orange stylasterine spicules (*Distichopora* sp.) are especially conspicuous.

4.3.2 Brachiopods

All brachiopods are terebratulids, four are Terebratallidae *Magellania flavescens, Anakinetica (Magadina) cumingi, Jaffaia jaffensis,* and *Magadinella mineuri,* and two are Cancellothyridae, *Cancellothyris hedleyi* and *Terebratulina cf. cavata.* All are pedically attached except *A. cumingi* which can live in sand, attaches to sediment grains (Richardson 1987) and uses its pedicle to lever itself out of the shifting sediment (Fig. 4.24).

They fall into four groups. The relatively small somewhat thickened *A. cumingi* occurs in all areas sampled and is orders of magnitude the most numerous form. Usually found as a few valves per sample, there are localities where several hundred individuals were present. The somewhat larger *M. flavescens* is the dominant form inboard, especially in the gulfs and the northwest Great Australian Bight, Baxter Sector but is also found scattered across the shelf at a few localities to depths of ~500 m. *C. hedleyi, J. jaffensis* and *T. cavata* occur inboard <50 mwd and outboard >80 mwd, but are never numerous. The delicate forms *M. mineuri* and *M. hurleyi* are only found in waters >100 m. All brachiopods have LMC shells.

4.4 Authigenic Mineral Particles

4.4.1 Glauconite

Glauconite, the green, K-poor, Fe-rich smectite clay that matures into a non-expandible glauconitic mica

(Amorosi 2003), is not present above trace amounts in any facies. Repeated analyses of black particles using standard chemical and X-Ray diffraction techniques have failed to find any significant amounts of these authigenic minerals.

4.5 Detrital Grains

Detrital grains, comprising a varied suite of siliciclastic particles, mineral oxides, dolomite, and older Cenozoic carbonate clasts, are of varying abundance.

4.5.1 Siliciclastic Particles

The most significant are terrigenous clastics, of which there are four types with differing provenances. The first is a suite of poorly sorted, compositionally and texturally immature, sands and gravels composed of rock fragments and mineral grains eroded from nearby igneous and metamorphic crystalline and hard sedimentary rocks. The second is well-sorted, fine to medium grained, texturally and compositionally mature, almost orthoquartzitic sand. Such sediment is multicycle and derived from erosion of largely unlithified, already mature Permian or Cenozoic sands. The third is heavy minerals that are prominent as beach lags and locally numerous in shelf sediment, especially oxides such as hematite, magnetite, rutile, and ilmenite. The fourth rare but clearly visible component comprises other heavy mineral grains (especially garnet, staurolite, andalusite, and sillamanite) from crystalline rocks and are locally sorted into placers (Fig. 4.25).

Fig. 4.25 Detrital Grains. (**a**) A beach placer of heavy minerals (mostly ilmenite), northwest side of Investigator Strait (m-scale circled), (**b**) extensive heavy minerals on a 80 m wide beach along the northern side of the Lacepede Shelf near Kingston, (**c**) dolomite rhombs picked from sediment sample at 42 mwd adjacent to Sanders Bank, Lacepede Shelf, where the dolomite constitutes 25% of the grains, photographed under plain light

4.5.2 Dolomite

Dolomite, in the form of colourless to orange to dark red particles and crystals, either single rhombs or rhomb clusters (Fig. 4.25), is variably present in surficial sediments across the area. Where studied in detail on the Lacepede Shelf (Bone et al. 1992) they can form up to 25% of the sediment, but are usually <5%. The rhombs (Fig. 3.32) range from rounded to sharp-edged, have a cloudy core and a well-zoned cortex, are typically medium- to coarse-grained sand size and chemi-

cally Ca-rich (av. 56.3 mol% $CaCO_3$). Rhomb clusters are either numerous crystals or individual crystals cemented by white microcrystalline HMC (av.12 mol% $MgCO_3$).

Originally interpreted as entirely modern, the dolomite has been shown to comprise cores of Tertiary dolomite (cf. James et al. 1991; Kyser et al. 2002) that are overgrown with a final stage dolomite precipitated in sediments just below the modern seafloor (Bone et al. 1992). Similar multicycle detrital dolomite is common in the Pliocene Roe Calcarenite (James et al. 2006; James and Bone 2007) and Bridgewater Formation aeolianites (Wilson 1991).

4.5.3 Older Cenozoic Carbonates

Many of the Cenozoic carbonates are soft and easily eroded. These sediments form conspicuous outcrops on the shelf edge and locally the upper slope. They do not, as confirmed by Li et al. (1996a, 1998, 1999), contribute a significant amount of material to the neritic sediments. There are, nevertheless, local minor additions of material as demonstrated by the presence of Cenozoic benthic foraminifers (Milnes and Ludbrook 1986) and planktic foraminifers (Li et al. 1999).

4.6 Synopsis

The marine carbonate factory in this temperate environment is producing a heterozoan assemblage of carbonate particles today and has done so throughout the Pleistocene. Sediments are palimpsest, a mixture of recent, stranded, and relict biofragments. The grains come from a spectrum of algae and invertebrates and are dominated by coralline algae, bryozoans, molluscs, and benthic foraminifers. These and other relatively minor components produce sediment composed of LMC, IMC, and aragonite.

1. Seagrasses and macroalgae are not sediment producers as such but host a variety of epiphytes, mainly coralline algae, benthic foraminifers and molluscs whose remains collect to form significant deposits. Macroalgae are largely confined to rocky seafloor outcrops whereas seagrasses grow mostly in

sediment substrates. Shallow photic environments west of Cape Jaffa sustain seagrass and macroalgal growth whereas cooler marine environments to the east support only sparse seagrass but prolific kelps. Seagrasses, dominated by *Posidonia* and *Amphibolis*, thrive in warm waters <30 mwd with active epiphyte production mostly in <10 mwd. Macroalgae, mostly reds and browns, grow in waters <40 mwd. The calcareous epiphyte growth on these algae appears to be much less than on seagrasses and is concentrated on the holdfast, except for the brown alga *Ecklonia* whose blades can be encrusted by bryozoans and spirorbid worms.

2. Corallines (IMC and HMC) are the only important carbonate algae, green calcareous algae, although present on shallow parts of the Albany Shelf, are not calcified. Encrusting corallines grow on rocky substrates but do not contribute much to the sediment. They do, however, form rhodoliths (coralline algal nodules) that are numerous in shallow water and grow, because of the water clarity, to ~120 mwd in the Great Australian Bight. Articulated coralline algae are prolific sediment producers in <10 mwd, especially on intertidal rocky platforms and in seagrass meadows.

3. Major invertebrates

   - Bryozoans: These ubiquitous sediment producers are found in all parts of the depositional system and produce particles from silt to cobble size. Volumetrically, about 30% of the particles are LMC, 60% are IMC, and 10% are aragonite. They are herein classified under a revised system that integrates skeletal habit, morphology-shape, and substrate. The sediment, lithic, and biotic substrates upon and in which they grow are further related to water depth.
   - Molluscs: Gastropods (aragonite) are of moderate diversity and although ubiquitous, are particularly numerous in seagrass environments. Bivalves, both infaunal (generally aragonitic) and epifaunal (usually calcitic) are again of modest diversity.
   - Foraminifers: These predominantly LMC and IMC protists comprise from 10 to 30% of the sediment in all environments. It is a classic temperate water protist biota of small benthics, with large benthics (*Peneropilis* and *Marginopora*) present only in warm temperate, near-

shore settings. Seagrass beds have a distinctive benthic foraminifer fauna. Benthic numbers are relatively high inboard, decrease on the middle shelf, and rise again on the outer shelf. There is a west-to-east, warm-to-cool water trend in the biota.

4. Minor invertebrates

   - Infaunal echinoid spines and plates (IMC—HMC) comprise a steady 5–10% of the sediment; epifaunal echinoid particles are least numerous.
   - Calcareous worm tubes (especially spirorbids—aragonite and Mg-calcite) form a similar proportion of most sediment.
   - Barnacles, common heterozoan fragments elsewhere, are not important.
   - Ostracods (LMC—IMC) are ubiquitous in the fine sand fraction, but do not contribute volumetrically to most sediment.
   - Sponges, although a conspicuous component of the epibenthic biota, only contribute siliceous spicules to the very fine sand and mud fraction.
   - Coccoliths (LMC) are present everywhere in the mud fraction, and especially in upper slope sediments.
   - Zooxanthellate (reef building) corals (aragonite) are rare and of low diversity, typically growing as isolated colonies, and localized to very shallow waters. Azooxanthellate (solitary) corals (aragonite) are diverse and usually preserved as whole colonies. Octocoral spicules (IMC) are rare.
   - Brachiopods (LMC—terebratulids) are not significant sediment producers. They are present in the sediment as a few single shells or large monospecific aggregations.

5. Authigenic and detrital grains.

   - Glauconite, a common constituent in other cool temperate carbonate deposits, is rare.
   - Siliciclastic grains are dominated by quartz and are either transported into the environment by fluvial processes today, (especially in Tasmania) or are relict from when terrigenous particles were delivered to the shelf during the LGM lowstand.
   - Dolomite particles, in the form of single rhombs or rhomb clusters, are a minor but recurring particle type. They principally come from erosion of Cenozoic sediments.

Chapter 5
Marginal Marine Deposystems

5.1 Introduction

Southern Australia, with its spectrum of carbonate depositional environments within reach of major universities and government surveys, is famous for detailed studies on marginal marine environments and facies (Lowry 1970; Gostin et al. 1984, 1988; Thom 1984a, b; Belperio 1995; Sanderson et al. 2000) and their applicability to the rock record (Muir et al. 1980). Of particular significance are the seminal works on muddy tidal flats (Ferguson et al. 1983; Belperio et al. 1988; Belperio 1995; Edgar 2001), evaporative lacustrine systems (Warren 1991), and syndepositional dolomite (von der Borch 1976; De Deckker et al. 1982; Gostin et al. 1988; Dutkiewicz and von der Borch 1995; Wright and Camoin 1999; Wright 2000). Most of these studies have taken place in South Australia, much less is known about similar environments in the rest of southern Australia. The purpose of this short chapter is to provide an overview of these marginal marine environments so that they can be later integrated into the overall sedimentary system. The work is mainly that of others with our findings integrated where applicable.

5.2 Setting

The marginal marine setting in southern Australia can be separated into two end member situations, erosional coasts and depositional coasts (Figs. 5.1, 5.2). This separation comes with the caveat that intermediate situations are also present because of the vagaries of time and environmental change. Erosional coasts are easily separated into hard bedrock and soft sediment types. Resistant lithologies are mostly Precambrian crystalline and well-indurated Neoproterozoic-early Paleozoic rocks. The more easily eroded generally cliffed coasts are composed of Permian, Cretaceous, and Cenozoic siliciclastic and carbonate sediments and sedimentary rocks. Depositional coasts are also divisible, in this case, into exposed beach-dune–aeolianite and protected mudflat settings. Whereas there is little geological record of erosional coasts, there is an excellent late Quaternary record of depositional coasts, as summarized by Gostin et al. (1988) and Belperio (1995).

The shallow offshore subtidal and adjacent peritidal environments are always inexorably linked in carbonate depositional systems; in general, the former is the sediment source and the latter is the sediment sink. All these environments are in the euphotic zone and so organisms such as phototrophic bacteria, macroalgae, and seagrasses are of inordinate importance.

Sediments are principally the remains of numerous calcareous benthic invertebrates and calcareous algae that are variably mixed with siliciclastic sediment or older carbonate grains. Holocene skeletons are mostly molluscs (particularly bivalves), benthic foraminifers, and calcareous algae. Bryozoans, although present, are less numerous than the other components. Seagrass meadows and macroalgae (including kelp) forests are important sediment factory sites as are calcareous phototrophs. There are low numbers of stranded particles because this is the highest sea level stand since MIS 5. Relict particles, although present, are only locally abundant.

5.3 Rocky Peritidal

Open ocean coasts in southern Australia are battered year-round by waves and pounded by swells from the southwest. Hard crystalline bedrock localities are

N. P. James, Y. Bone, *Neritic Carbonate Sediments in a Temperate Realm*,
DOI 10.1007/978-90-481-9289-2_5, © Springer Science+Business Media B.V. 2011

Fig. 5.1 Sketches of the end members of coasts in southern Australia (**a**) An erosional coast developed on bedrock of either hard Precambrian crystalline or Paleozoic sedimentary rocks; many such bedrock localities are capped by Pleistocene aeolianites, (**b**) an erosional coast cut into stacked, multigeneration Pleistocene aeolianites, (**c**) a depositional coast formed by a series of prograding aeolianites with modern and Pleistocene interdune corridors occupied by marine lagoons or lacustrine depositional systems, (**d**) a depositional coast in a location largely protected from open ocean waves and swells with an extensive muddy tidal flat system

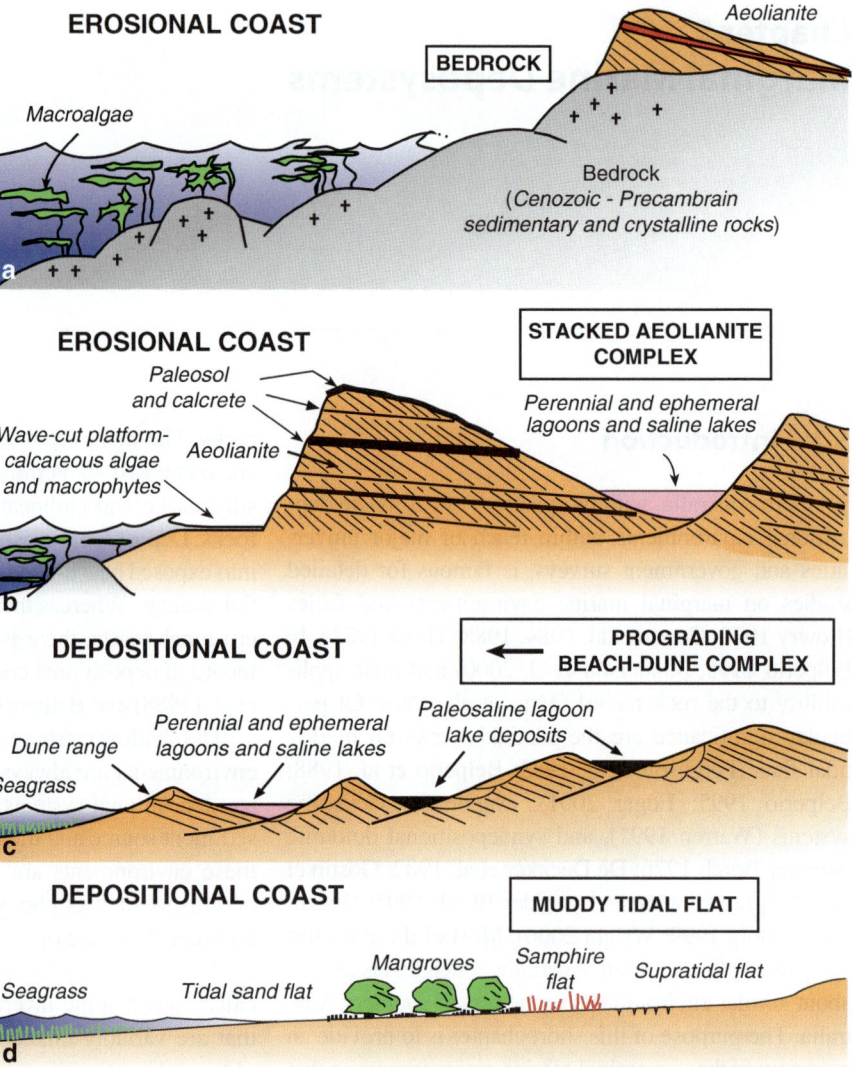

typically smooth, well-rounded outcrops (locally called turtlebacks, Fig. 5.3b) whereas indurated layered sedimentary rocks, depending on their structural attitude, are a series of several projecting ridges. Softer, relatively flat-lying strata typically form upstanding cliffs fronted by an erosional low-tide platform wherein the cliff base can be locally obscured by fallen blocks.

The biota on rocky shores is, as elsewhere, strongly zoned relative to tidal range and wave intensity. Outcrops and boulders in the supratidal zone are veneered with cyanobacteria, calcareous algae, and lichens that are locally grazed by gastropods. Barnacles, limpets, mussels, and calcareous polychaete worm tubes colonize intertidal rock surfaces. Sheltered habitats have fewer species of barnacles and more grazing gastropods (e.g. *Turbo, Austrocochlea* sp., *Bembecium* sp.,

and *Nerita* sp.). The intertidal zone is characterized by abundant macroalgae, especially where constantly swept by strong waves (Shepherd and Womersley 1976; Womersley 1981b; Shepherd 1983).

These environments can be prolific sediment sources but not settings of much accumulation. The calcareous debris is typically swept into adjacent shallow shoreface environments (see below) or accumulates in ephemeral rock pools. Macroalgal fronds are not generally encrusted by calcareous epiphytes until death and so, apart from *Ecklonia* sp., are not prolific carbonate producers. The holdfasts do, however, support a community of coralline algae, bryozoans, and polychaete worms. The main carbonate production comes from corallines and a few barnacles that grow on the rock surface together with molluscs and

Fig. 5.2 Coastal systems. **(a)** A granite coastline, Woody Island, Recherche Archipelago, Albany Shelf, cliffs are ~20 m high, **(b)** a rapidly eroding cliffed coastline cut into Cenozoic limestone, east of Portland, Otway Shelf, **(c)** a calcrete-capped cliff of stacked aeolianite dunes, Innes National Park, north coast of Investigator Strait, ~25 m high, **(d)** a depositional coast, Long Bay Beach, north of Robe, Bonney Coast

echinoids that graze the algal turf. Although there are no quantitative studies of the amount of carbonate produced here it appears that it is a less productive carbonate factory compared to broad-leaf seagrass meadows, and the sediment is coarser grained.

5.4   Barrier Island—Aeolianite—Lagoon—Saline Lake Complexes

High-energy depositional coasts across southern Australia are classic barrier island complexes. These systems can form in front of older Pleistocene aeolianite systems with a lagoon lying between the Holocene and Pleistocene deposits (Fig. 5.1). The lagoon can with time, however, become completely cut off and barred from direct marine influence leading to the formation of saline lake complexes, especially under

a semi-arid climate. Depositional systems comprise a suite of beach, dune, lagoon, and saline lake environments.

5.4.1   Beaches

Most sandy beaches are subject to 0.5–2 m high waves that break 20–200 m offshore with a shallow shoreface characterized by numerous shallow water bars, rips, and beach cusps. Sediment is moved onshore, winnowed, and sorted with the finer fraction swept into back-beach dunes that are either active or rapidly vegetated (Fig. 5.4).

Beach and upper foreshore sediments are seaward-dipping, cross-bedded, generally well-sorted, medium to coarse grained quartz sand or a mixture of finely fragmented biofragments, generally molluscs, benthic

Fig. 5.3 Erosional coasts. (**a**) White orthoquartzite sand beach and an offshore island of Proterozoic granite, Duc of Orleans Bay, Albany Shelf, (circled automobile for scale), (**b**) rounded granite (G) outcrop ~8 m high (a turtleback) overlain by Pleistocene aeolianite (P) north side of Investigator Strait, (**c**) rounded volcanics forming a cliff ~8 m high, Wardang Island, Spencer Gulf, (**d**) cliffed coast and low tide platform (~20 m wide) cut into Pleistocene aeolianite, Robe, Bonney Coast, (**e**) numerous small intertidal mussels on granite, Bicheno NE Tasmania; pen scale, (**f**) bivalves on Tertiary limestone covered with intertidal acorn barnacles, Maramah, west Tasmania near Cape Grim

foraminifers, coralline algae and minor relict grains, with whole or abraded molluscs, bivalves, or a mixture of both (Shepherd and Womersley 1976; Cann et al. 1999; Murray-Wallace et al. 1999). High-energy sand beaches contain large numbers of the surf clam *Donax*.

Moderate energy beaches have a more diverse biota with the molluscs mostly *Venerupis, Katelysia, Notospisula, Anapella,* and *Polinices. Anadara,* the relict Pleistocene warm-water bivalve, is locally mixed with the Holocene forms.

Fig. 5.4 Beach environments. (**a**) Areal view of the southeastern margin of the Lacepede Shelf and the town of Kingston illustrating the extensive prograding beach system, (**b**) beach-dune system near Victor Harbor, Lacepede Shelf, (**c**) modern beach-dune system along the northeastern shore of the Lacepede Shelf at 'The Granites' a series of rounded Cambrian outcrops at right center; vehicle for scale (circled), (**d**) beach cusps along the northern shore of the Lacepede shelf with numerous beach bivalve (*Donax deltoides*) shells; geologist for scale, (**e**) eroded back beach at same locality as D with concave beach cusp laminations (50 cm scale circled), (**f**) close view of E illustrating laminations composed entirely of the surf clam (*D. deltoides*) shells (10 cm scale)

These sediments cannot be simply categorized in terms of composition because they vary so widely. They are a product of both the hinterland source and the offshore marine environment. Beach sediments along the Albany coast eastward to the Roe Plain are orthoquartzitic such that they squeak when walked upon. Those beaches in bays along western Eyre Peninsula, in the gulfs, and along the sweep of the Coorong are variably siliciclastic and carbonate rich. The beaches along the Otway and western Tasmania coast, because of somewhat higher rainfall, are more siliciclastic.

5.4.2 Aeolianite Dunes

On erosional coasts, the variably lithified Pleistocene Bridgewater Formation aeolianite complexes are cliffed along their leading, seaward margins with former interdune corridors behind them now sites of saline lake deposition similar to those behind Holocene beach-dune complexes. Alternatively, the Pleistocene aeolianite ridges are breached by marine erosion resulting in elongate leeward lagoons and bays like those behind some Holocene barrier islands (Fig. 5.5, 5.6).

Dune sands range in composition from pure carbonate to pure quartz and all compositions in between. The deposits are typified in outcrop by spectacular high-angle planar and trough cross-bedding with dune geometries quite obvious in Pleistocene sections. Modern dunes (Fig. 5.5) are typically vegetated by low growing prostrate shrubs, both halophytic varieties and salt-tolerant angiosperms (e.g. *Tetragonia*) (Reeckmann and Gill 1981; Schwebel 1983; Short 1988).

Their heterogeneous composition defies strict categorization but the fine-grained carbonate particles are mostly subangular mollusc grains, pieces of coralline algae, benthic foraminifer tests (particularly grass-dwelling species), and bryozoan zooids. Other grains include echinoid pieces, intraclasts, and locally, abraded dolomite rhombs.

Rhizoconcretions are common in most Pleistocene examples and are also partially developed in modern dunes. Complex, multigeneration calcrete horizons with local pisoliths comprise bounding horizons in most Pleistocene aeolianite successions. Modern dunes are variably encrusted by a case-hardened crust (Warren 1983).

Fig. 5.5 Aeolian systems. (**a**) Areal view of Coffin Bay in the southeastern part of the Great Australian Bight illustrating an extensive dune complex that is currently migrating northward across the peninsula, the Bay is 10 km wide, (**b**) carbonate dunes ~25 m high along the coast at Cactus Beach, western Eyre Peninsula, (**c**) extensive wind-blown sands and dunes along the north shore of Investigator Strait, Innes National Park, (**d**) a vast dune system inland from the coast at Eucla along the northern margin of the Great Australian Bight

5.4.3 Lagoons

The character of lagoons is highly variable and depends to a large extent on the degree of water circulation. They range from large embayments with classic shallow marine facies, to semi-restricted schizohaline environments, the most well-known of which is the Coorong lagoon (Fig. 5.6).

The sedimentary facies are mostly shelly and muddy carbonate sands with numerous gastropods,

Fig. 5.6 Lagoons. (**a**) Areal view of the mouth of the River Murray looking east along the Coorong Lagoon, Lacepede Shelf to the right, (**b**) vertical areal photograph near the southern end of the Coorong Lagoon at Salt Creek illustrating the Southern Ocean beach-dune complex, the Lagoon, and adjacent saline lakes; the 3 arrowed lakes precipitate carbonates, with the first magnesite (M) and the other 2 dolomite (D), whereas the small lake at the right hand side (H) is halite, (**c**) view looking south across the Lagoon to the landward-prograding coastal dune system, (**d**) an accumulation of the gastropod *Coxiella* along the northern margin of the Coorong Lagoon; individual gastropods are 2 mm long, (**e**) numerous encrustations of the bryozoan *Conopeum aciculata* along the northern shore of the Coorong lagoon

and large intertidal to subtidal bivalves (*Anadara* [Pleistocene], *Katelysia, Ostrea*). Interbedded calcitic and dolomitic lacustrine muds with greenish clay and clayey quartz sand contain a freshwater to brackish water biota, especially ostracods and the small gastropod *Coxiella*. The Coorong in particular is subject to spasmodic blooms of the single species of the bryozoan *Conopeum aciculata* which encrusts all suitable substrates including serpulid worm bioherms (Bone and Wass 1990; Bone 1991; Sprigg and Bone 1993). Stromatolites, thrombolites, teepee structures, desiccation polygons, subtidal gypsum and ephemeral halite are typical (von der Borch 1965; von der Borch and Lock 1979 for detailed descriptions; see Gostin et al. 1988). These facies likewise characterize many of the late Pleistocene interdune corridors that are also sites of deflation lags, carbonate sands and muds, ephemeral saline lakes, and evaporites.

5.4.4 Lakes

The ephemeral lakes, because of their easy accessibility and variable character, have been intensively documented; as reviewed by Gostin et al. (1988) and Belperio (1995). They range from brackish to saline and can be as small as a few hundred meters across to many kilometers wide with commercially exploited evaporites. The smaller ones, although geographically close to one another, can have very different mineral assemblages. The deposits are mostly muddy carbonates, gypsum and minor halite. They are of particular scientific interest because of the suite of carbonate mineral precipitates, including dolomite (Alderman and Skinner 1957; von der Borch et al. 1977; von der Borch and Lock 1979; Warren 1982a, b, 1988; Bowler and Teller 1986; Rosen et al. 1988, 1989; Henderson 1997). Many contain sedimentary structures that

Fig. 5.7 Saline lakes. (**a**) Areal view of an extensive salt lake system behind a beach-dune complex that stretches 22 km between the Point Dempster (bottom) and Point Malcolm (top) at Israelite Bay, Great Australian Bight, (**b**) flooded Marion Lake in winter, Innes National Park, southern Yorke Peninsula, (~1 km across), (**c**) desiccated saline lake near the southern end of the Coorong Lagoon; people for scale, (**d**) dolomitic mud in salt lake near the southern end of the Coorong Lagoon that is black and reducing just below the surface (C.C. von der Borch feet for scale)

are also found in muddy tidal flat environments (see below) namely microbial mats, stromatolites, desiccation cracks, teepees, flat clast conglomerates, and a low diversity, high abundance mollusc (mostly gastropods) biota (Figs. 5.7, 5.8).

5.5 Muddy Tidal Flats

Muddy intertidal and supratidal sediments occur in protected locations such as intrashelf basins or along the margins of lagoons between beach-dune complexes (Fig. 5.9).

Fig. 5.8 Saline lakes. (**a**) Desiccation cracks in a lake near southern end of the Coorong lagoon, 50 cm scale bar (circled), (**b**) extensive teepees, Deep Lake, Innes National Park, southern Yorke Peninsula, scale (circled) has 10 cm bars, (**c**) domal stromatolite at Lake MacDonnell, west coast of Eyre Peninsula, hammer handle is 20 cm long, (**d**) exposed lacustrine gypsum crystals, Marion Lake, Innes National Park, southern Yorke Peninsula. Muddy Tidal flats, (**e**) areal view of northern Spencer Gulf and the town of Port Pirie. The shallow neritic zone is a series of seagrass beds, the black zone comprises intertidal microbial mats and mangroves ~1 km wide, the white areas inland comprise the supratidal zone, (**f**) areal view of the extensive mangrove shoreline north of Port Pirie, Spencer Gulf, tidal creeks are ~30 m wide

Fig. 5.9 Muddy Tidal flats. (**a**) Intertidal mangrove woodland a few kilometers north of Adelaide, trees are ~3 m high, (**b**) the intertidal zone composed of numerous split and partially desiccated microbial mats; foreground ~20 m wide, (**c**) close view of a series of split and rolled mats covering carbonate mud; scale divisions are 2 cm, (**d**) areal view 100 m across of a zone of groundwater resurgence and extensive large teepees, (**e**) ground level view of the teepees illustrated in D, (**f**) the supratidal zone with extensive growth of the halophytic shrubs including *Arthrocnemum*; northern Spencer Gulf; foreground ~20 m wide

They are not present facing the open ocean. Although somewhat different in setting from saline lake deposits the sediments are similar in composition, a mixture of marine, terrestrial, and evaporite materials. Such mud flats have many of the attributes displayed by tropical peritidal environments such as the Bahamas and the Persian Gulf (Pratt 2010). Although somewhat modified by local Holocene uplift (Belperio 1993) they are characterized by the classic suite of intertidal features such as shelly beaches, tidal creeks, mangrove woodlands, small ephemeral saline ponds, desiccation cracks, teepees, split and blistered microbial mats, stromatolites, and crab burrows. The microbial mats are also locally veneered with halophytic groundcovers that can withstand diurnal salt water (e.g. *Salicornia* spp.). Supratidal samphire flats, exposed to prolonged desiccation, also host growth of dolomite and gypsum in the mixed terrestrial and marine muds (Fig. 5.9).

More specifically, wide intertidal and supratidal mudflats are characterized by sand flats, woodlands populated by the white mangrove (*Avicennia marina*) that grows to 5–10 m high and has prolific pneumatophores, and clay-rich mud flats containing gypsum (Cann et al. 2009). The muddy sediment is, in part, derived from land, in part swept onto the flats from adjacent grass beds during storms and in part from epiphytes on mangroves. Whereas much of the molluscan biota also comes from adjacent shallow neritic grass beds, some is indigenous, as are some of the benthic foraminifers. These *in situ* components come from the roots of the mangroves, the sediment surface, and the shallow subsurface. The upper, predominantly exposed supratidal part of the flat is mostly allochthonous mud and sand with scattered dolomite, layers and nodules of gypsum, and microbial mats, with teepees and megapolygons in areas of groundwater resurgence.

5.6 Synopsis

The coastline of southern Australia has erosional and accretionary segments. Erosional coasts are cut into hard crystalline and sedimentary rocks or relatively soft Permian, Cretaceous, and Cenozoic sediments and sedimentary rocks. Depositional coasts are formed by sediment derived from the adjacent marine environment or via fluvial input. The semi-arid climate west of Cape Jaffa produces carbonate–evaporite marginal marine deposits whereas the relatively humid climate along the coasts of Victoria and Tasmania results in siliciclastic-dominated sand bodies. The well-studied coasts in the west are either exposed beach-dune–aeolianite complexes or protected peritidal mudflats.

1. Calcareous algae, gastropods, limpets, and barnacles dominate the calcareous biota on high-energy erosional shores; articulated corallines and gastropods are the most important sediment producers. Macroalgal holdfasts support a community of calcareous algae, bryozoans, and polychaete worms. Whereas this environment is a source of sediment there does not appear to be much local accumulation.

2. The high-energy depositional coasts are classic barrier island complexes. The system comprises a suite of beach, dune, lagoon and saline lake environments.

   • Beaches range in composition from pure carbonate to pure quartzose sand. The dominant calcareous biota comprises a suite of robust beach and shallow offshore bivalves.

   • Aeolianite dunes are likewise compositionally variable with modern examples vegetated by halophytic shrubs. The carbonate particles usually come from adjacent offshore seagrass beds.

   • Lagoons range from wholly marine to semi-restricted and schizohaline. Sediments are generally shelly and muddy carbonate sand with bivalves and prolific gastropods. Stromatolites, thrombolites, teepees, desiccation polygons, subtidal gypsum, and ephemeral halite are common.

   • Ephemeral lakes range from brackish to saline, but the evaporitic ones have been most studied because of synsedimentary dolomite formation. Sedimentary structures are the same as those in desiccated parts of lagoons together with a low diversity, high abundance gastropod biota.

3. The low-energy peritidal mudflats are similar to many semi-arid tropical systems worldwide and exhibit the same suite of sedimentary textures and fabrics with a mixture of marine, terrestrial and evaporite materials. The intertidal zone is commonly vegetated by mangroves and the supratidal flats are locally disturbed by teepees and megapolygons associated with groundwater resurgence.

Chapter 6
Neritic Sedimentary Facies

6.1 Introduction

Understanding this vast depositional realm has been a protracted study. As each new sector of the continental margin was documented, the new findings necessitated re-evaluation of previous interpretations. This chapter is the result of a total re-study of all bottom samples acquired since 1989 and their placement in a facies scheme that is applicable to the entire continental margin.

Sedimentary facies (23) are herein defined in terms of composition, and so are lithofacies in the traditional sense (Boggs 1995, p. 290). The sediments, as outlined in Chaps. 3 and 4, are composed of carbonate particles of various ages, with locally important siliciclastic grains. The principal components are recent biofragments, relict carbonate particles, and siliciclastic grains (mainly quartz). Recent particles include both latest Pleistocene (mostly stranded) and Holocene grains. This is because they were all produced during the latest eustatic sea level rise, and more practically, the nature of stranded particles has not yet been studied on large parts of the shelf. Relict particles are separated because they formed during a completely separate sea level cycle prior to the LGM.

Megafacies (5) are defined when the sediment is composed of 50% or more of any one of the above major components (e.g. Relict Megafacies have more than 50% relict grains). Each megafacies contains from one to several individual facies (Fig. 6.1). Each facies description should be read in conjunction with Table 6.1 that summarizes all pertinent attributes. Also see the facies maps in Chaps. 7, 8, and 9.

6.2 Megafacies C—Recent Carbonate

Sediments that are almost wholly recent carbonate (>50%) are mainly located in nearshore or on outer–upper slope environments. They range in texture from poorly sorted deposits with a complete spectrum of grain sizes to clean carbonate sands. These variations reflect both the architecture of the skeletons and the hydrology of the depositional environment. Grainy carbonates, those with little or no mud, are mostly on the open shelf, with molluscs and corallines dominant inboard and bryozoans most conspicuous outboard. Muddy carbonates are either in grass beds, in deeper parts of intrashelf basins, (the gulfs and Bass Strait), or in sub-wavebase settings. Two facies, articulated coralline and intraclast sands (facies C5) and coral, arborescent bryozoan gravel and mud (facies C12) stand out. Both are largely composed of stranded particles; the former is a shallow-water grass bed accumulation and the latter, a moribund bryozoan mound deposit (Table 6.1).

6.2.1 Grainy Carbonate Facies

6.2.1.1 Facies C1—Bryozoan Sand and Gravel

Deeper parts of the mid-shelf and much of the outer shelf to ~200 mwd, across the entire southern margin, is covered with Holocene bryozoan sand and gravel. This sediment, less than 25% of which is relict material and in which bryozoans are far more numerous

N. P. James, Y. Bone, *Neritic Carbonate Sediments in a Temperate Realm*,
DOI 10.1007/978-90-481-9289-2_6, © Springer Science+Business Media B.V. 2011

SEDIMENTARY FACIES

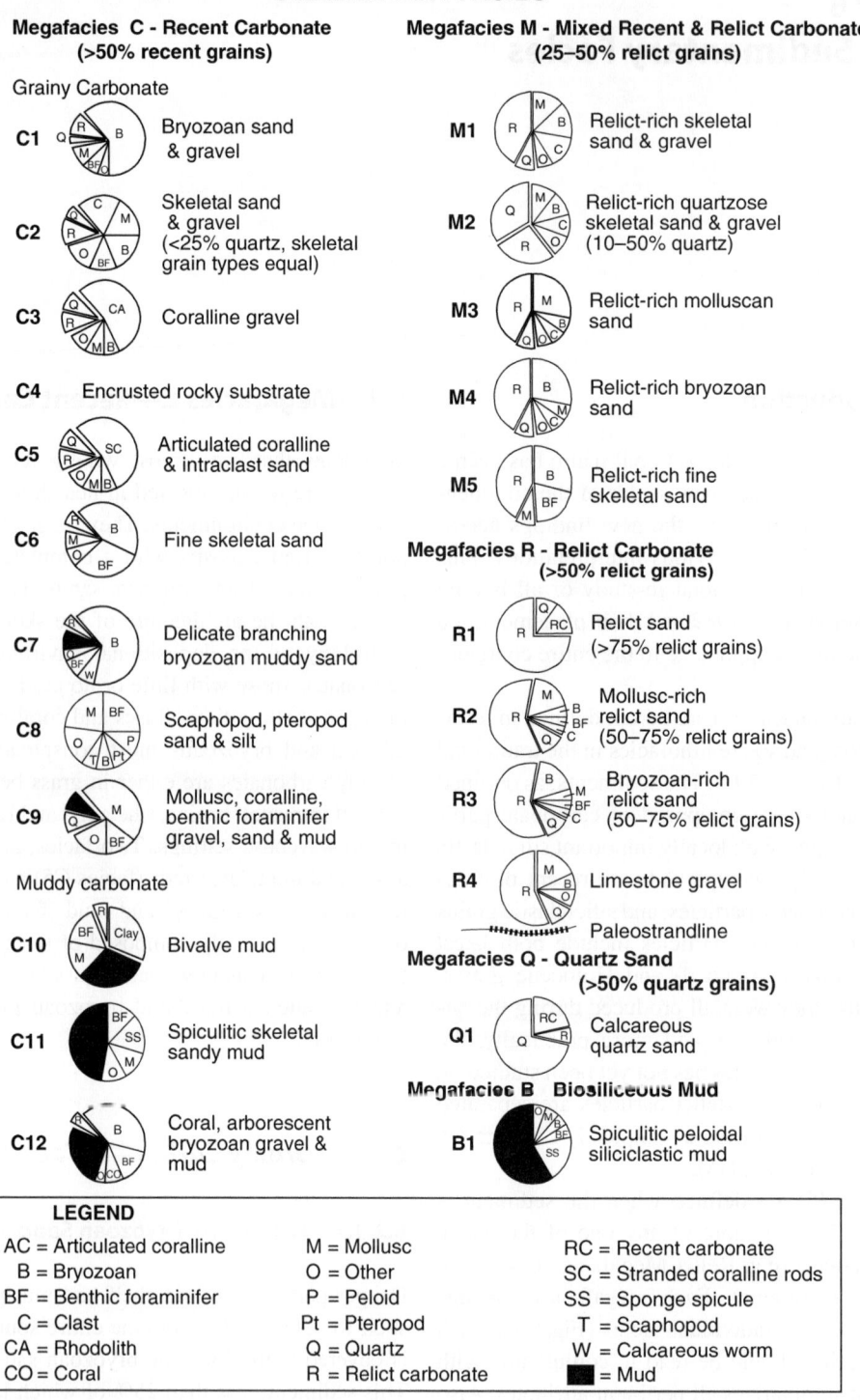

Fig. 6.1 Pie diagrams (average compositional plots) of the major megafacies and their composite facies that comprise the surficial sediments

Table 6.1 Sedimentary facies attributes

Facies	Location	Depth mwd	Grains %	Recent Carbonate	Relict	Notes
Megafacies C—Recent Carbonate						
Grainy Carbonate						
C1 **BRYOZOAN SAND & GRAVEL** (Bryozoan Rudstone–Grainstone Packstone)	**Albany** **Great Australian Bight, all Sectors** **Lacepede Shelf** south of **Kangaroo Island** & locally in **Spencer Gulf** **Otway Shelf** **Tasmania** all sectors	100–325 60–210 80–140 30–50 70–130 50–200	Siliciclastic=0–10 Recent carbonate=75–100 Relict carbonate=10–25	*Coarse* (5–50%) = bryozoan dominated, moderate to extremely diverse, fenestrate (especially *Adeona*), flat robust branching, foliose, arborescent, encrusting, vagrant, articulated zooidal; *Siliquaria*; turitellid gastropods when muddy; diverse bivalves (*Glycymeris, Katalysia, Chlamys*); benthic foraminifers (*Textularia, Elphidium, Cibicides, Maniacina*); *Heterestegina* inboard in GAB; cemented clasts. *Sand* (60–95%) = bryozoans >> bivalves > benthic foraminifers = coralline algal rods; numerous nodosariid foraminifers; planktic foraminifers common locally; many stranded coralline algal rods and other grains locally in GAB and Neptune Sector. *Mud* = (0–20).	Coarse (70%) = lithoclasts > molluscs Sand (30%) = benthic foraminifers >> coralline algal rods.	Extensive mid–deep neritic bryozoan sands.
C2 **SKELETAL SAND & GRAVEL** (Biogenic Grainstone & Rudstone)	**Albany** very local **Roe Terrace & Baxter sector** **Investigator Strait** Eastern **Spencer Gulf** **Bass Strait near islands & Tasmania** south of Hobart	35 30–50 20–50 <50 30–50	Siliciclastic=0–35 Recent carbonate=50–100 Relict carbonate=10–15	*Coarse* (0–50%, most <10%) = Dominated by bivalves and bryozoans. Diverse bivalves; infaunal and epifaunal, whole and fragmented. Most conspicuous = pectens & infaunal forms. Bryozoans = fenestrate (particularly *Adeona*), flat robust branching & vagrant types, sediments locally rich in articulated branching fragments. Pink foraminifer *Miniacina* & corallines encrust larger skeletons. Other grains include branching corallines, *Dentalium* shells & serpulid clusters. *Sand* = (50–100%, most >90%) = Molluscs = bryozoans = coralline rods = all other skeletons. Bivalves = delicate infaunal fragments, gastropods = small & whole or large columellas. Echinoids = infaunal as small spines and plate pieces. Most bryozoan fragments = fenestrate, flat robust branching, & vagrant. *Miniacina* are conspicuous. Serpulid clusters and barnacle plates locally numerous. Particles = fresh & angular to abraded and rounded. *Mud* = 0.	10–15	Facies with roughly equal ratios of the main skeletal elements.

Table 6.1 (continued)

Facies	Location	Depth mwd	Grains %	Recent Carbonate	Relict	Notes
C3 **CORALLINE GRAVEL** (Rhodolith Rudstone)	**Albany** north of Cape Leeuwin **Great Australian Bight** **Roe Terrace (west)** **Western Spencer Gulf** **Gulf St-Vincent** locally **Lacepede Shelf** on shallow banks	0–100 35–50 20–40 <30	Siliciclastic=0–40 Recent carbonate=60–100 Relict carbonate=0–30	*Coarse* (70–100%)=rhodoliths or coralline branches, numerous pectens, & epifaunal bivalves; *Adeona*-dominated bryozoans. *Sand* (0–20%)=Local quartzose & rock fragments, <1/3 relict; up to 15% blackened grains, traces of dolomite. *Mud* (0–10%)	Coarse (100%)=calcarenite clasts, mollusc fragments.	A classic Maerl facies
C4 **ENCRUSTED ROCKY SUBSTRATE** (Hardground)	**Great Australian Bight & Ceduna sector** **Investigator Strait** **Otway Shelf** **Lacepede Shelf topographic highs (banks)**	25 <70 <50	No sediment, just local boring & encrustation.	Prolific algal & in vertebrate growth on rocky substrates; encrusting and basally-attached demosponges, branching and encrusting coralline algae (*Metagoniolithon, Lithophyllum*), encrusting peyssonnelids red algae, molluscs, and bryozoans; bryozoans extremely diverse (large foliaceous encrusting, articulated zooidal, flat robust branching, and fenestrate (*Adeona* sp.); bivalves *Glycymeris, Katalysia, Neotrigonia, Chlamys* sp., fragments *Malleus*, rare small limpets, diverse gastropods; large serpulid worms, thick-walled terebratulid brachiopods, small crustaceans, benthic foraminifers (variable miliolids, & *Miniacina*), and echinoids. Fleshy macrophytes to 45 mwd.	Minor	Rocky seafloor or hardground bored & encrusted; carbonate sediment factory; little deposition but much contribution to surrounding sediment.
C5 **ARTICULATED CORALLINE & INTRACLAST SAND** (Coralline Algal, Intraclast Grainstone & Rudstone)	**Albany** **Great Australian Bight & Ceduna Sectors** **Baxter & Eyre Sectors** **Lincoln Shelf** **Bass Strait near Islands**	90–210 120–250 60–100 90–150 30–45	Siliciclastic=0–10 Recent carbonate=100 Relict carbonate=0–10	*Coarse* (20%);=arborescent, globose & fenestrate bryozoans, serpulid worms, *Siliquaria*, large solitary corals. *Sand* (40–80%)=coralline rods=intraclasts >bryozoans=bivalves; rods are predominantly stranded. *Mud*=(0–10%)	Calcarenite clasts.	A stranded & reworked sea grass facies.

Table 6.1 (continued)

Facies	Location	Depth mwd	Grains %	Recent Carbonate	Relict	Notes
C6 **FINE SKELETAL SAND** (Fine Skeletal Grainstone)	**Albany** **Great Australian Bight Baxter sector** Eyre sector Ceduna sector **Lincoln Shelf** **–Spencer Gulf Entrance, Gulf St-Vincent** Lacepede Shelf Otway Shelf Bass Strait Tasmania very local	410–475 60–70 115–125 135–155 80–200 95–200 130–180 40–60 120	Siliciclastic=0–20 Recent carbonate=100 Relict carbonate=0–20	*Coarse* (0–20%)= bryozoans, *Siliquaria*, azooxanthellate corals, gorgonian spicules, calcarenite clasts encrusted with *Miniacina*, serpulid worms, infaunal & epifaunal echinoid pieces. *Sand* (80–100%)=homogeneous, fine; well-sorted, bryozoans (articulated zooidal) ≥ benthic foraminifers, infaunal echinoid fragments, planktonic foraminifers, sponge spicules >> serpulids, bryozoan fragments, gastropods, bivalves, pteropods. *Mud* (0–10) benthic biofragments	Coralline rods.	Skeletal elements winnowed from other facies.
C7 **DELICATE BRANCHING BRYOZOAN MUDDY SAND** (Delicate Branching Bryozoan Rudstone–Floatstone)	**Great Australian Bight, Ceduna &Eyre Sector** Lincoln & Lacepede Shelf Otway Shelf Bass Strait Tasmania south & east	150–360 60–350 180–350 70–205	Siliciclastic=tr Recent carbonate=90–100 Relict carbonate=0–10	*Coarse* (10%)= Delicate branching bryozoans >>> azooxanthellate corals, serpulid worms tubes, infaunal echinoid fragments; rare *Adeona, Celleporaria*; scaphopods and pteropods; molluscs (*Chlamys, Venericardia, Siliquaria*), solitary corals, stranded rhodoliths in shallow water. *Sand* (50–75%)=Delicate branching bryozoans and benthic foraminifers (*Textularia, Pyrgo, Uvigerina*); siliceous sponge spicules, pteropods, serpulid tubes. *Mud*=(0–40%).	*Coarse* (70%)=lithic intraclasts. *Sand* (30%)=Benthic foraminifers > coralline rods.	Deep neritic bryozoan, substorm wavebase sands & muds
C8 **SCAPHOPOD PTEROPOD SAND & SILT** (Scaphopod–Pteropod wackestone–mudstone)	**Great Australian Bight, Eyre & Ceduna Sectors** Lacepede Shelf Bass Strait Tasmania east	350–550 425 75 100	Siliciclastic=0 Recent carbonate=100 Relict carbonate=0	*Coarse* (10–30%)=*Dentalium*, corals, pteropods, bryozoans. *Sand* (40–80%)=1/3 of the sediment=*Dentalium* (whole & fragments), pteropods & benthic foraminifers (infaunal biserial agglutinants, and epifaunal porcellaneous types, e.g. *Pyrgo*). Roughly equal fecal pellets, echinoids (infaunal spines and plates), bivalves (pieces of delicate infaunal shells), gastropods (whole small and columella), serpulids (clusters). *Silt* (10–60%) = Formed by unidentifiable skeletal fragments, small plankitc foraminifers, siliceous sponge spicules, and echinoid pieces.	None	Upper slope deposits that may be locally winnowed.

Table 6.1 (continued)

Facies	Location	Depth mwd	Grains %	Recent Carbonate	Relict	Notes
C9 MOLLUSC, CORALLINE, BENTHIC FORAMINIFER GRAVEL, SAND & MUD (Molluscan Coralline Foraminiferal Floatstone)	**Gulf St Vincent, Spencer Gulf & Small Embayments**	5–25	Siliciclastic=20–60 Recent Carbonate=40–80 Relict Carbonate=trace	*Coarse* (10–50%)=bivalves (*Spisula, Dosinia, Pinna, Ostrea, Katalysia, Circe*; gastropods (*Phasianella, Canthardus*) *Sand* (20–80%)=Mollusc fragments, benthic foraminifers (*Nubecularia, Peneropilis, Discorbis*), branching coralline algae (*Metagoniolithon*), articulated coralline algae (*Amphira*). *Mud* (10–30%)=mixed carbonate and siliciclastic	trace	*Posidonia* - dominated grass beds; highly variable texture, burrowed, plant roots.
Muddy carbonate						
C10 BIVALVE MUD (Molluscan Floatstone, Wackestone)	**Gulf St Vincent, Spencer Gulf & Small Embayments** **Bass Strait**	20–50 30–80	Siliciclastic=20–50 Recent Carbonate=40–50 Relict Carbonate=tr–40	*Coarse* (10–30%)=Bivalves (*Malleus, Pinna, Pecten*), minor bryozoans (*Adeona*) *Sand* (20–50%)=Molluscs, benthic foraminifers planktic foraminifers, echinoids *Mud* (20–70%)=mg-calcite dominated.	Lithoclasts, bivalves.	Open basin floor; intensively burrowed.
C11 SPICULITIC SKELETAL SANDY MUD (Spiculitic Wackestone-Mudstone)	**Great Australian Bight All Sectors** **Tasmania east, Bass Strait** **Slope all sectors**	100–500 27–176 210–500–	Siliciclastic=tr Recent carbonate=80–100 Relict carbonate=0–20	*Coarse* (0–10%)=Delicate branching & vagrant bryozoans; rare flat robust branching, fenestrates, and tubular *Celleporaria*; bivalves=*Chlamys & Venericardia*, gastropods=*Siliquaria*, turitellids, olives, and whelks; scaphopods (*Dentalium*); azooxanthellate corals, infaunal echinoids, crustaceans, solitary serpulids, pteropods. *Sand* (25–65%)=Delicate branching bryozoans, benthic foraminifers, (especially infaunal forms)=molluscs=micromolluscs and ostracods=sponge and gorgonian spicules. *Mud* (40–70%)=-mixture ~2/3 benthic components and ~1/3 planktic components	*Coarse* (65%)=lithic intraclasts >> molluscs; relict rhodoliths & coralline rods in shallow water. *Sand* (35%)=Benthic foraminifers > coralline rods.	Deep neritic sponge-bryozoan, sub storm wave-base muds Fine sands and hemipelagic muds

Table 6.1 (continued)

Facies	Location	Depth mwd	Grains %	Recent Carbonate	Relict	Notes
C12 CORAL—ARBO-RESCENT BRYOZOAN GRAVEL & MUD (*Celleporaria*–Coral Rudstone–Packstone)	**Albany** **Great Australian Bight Baxter & Ceduna Sector** **Lincoln Shelf** **Lacepede Shelf** **Tasmania**	290–430 160–500 (most extensive at 250–360) 200–360 165 400	Siliciclastic = 0 Recent carbonate = 90–100 Relict carbonate = 0–10	*Coarse* (15–30%) = Recent – conspicuous large dead *Celleporaria* (hollow tubular, sheet-like), variable numbers of living & dead azooxanthellate corals - diverse bryozoans (encrusting, fenestrate, flat robust branching, delicate branching, articulated zooidal), large gastropod fragments, *Siliquaria*, pectens, oysters, turitellids, serpulids, stylasterines, calcarenite clasts. *Sand* (20–55%) = Recent – delicate branching and vagrant bryozoans, benthic foraminifers, planktic foraminifers, rare gorgonian spicules. *Mud* (30–50%) = articulated zooidal bryozoan singlets, siliceous sponge spicules.	Coarse = tr	Exposed bryozoan mound facies & modern coral growth. Bimodal

Megafacies M—Relict Rich Carbonate

Facies	Location	Depth mwd	Grains %	Recent Carbonate	Relict	Notes
M1 RELICT-RICH SKELETAL SAND & GRAVEL (Intraclast–Skeletal Grainstone—Rudstone)	**Albany** **Head of Bight & Baxter Sector** **Lincoln Shelf** **Lacepede Shelf,** east of Kangaroo Is. **Spencer Gulf** **Bonney Shelf** **Bass Strait** **Tasmania** north-west Otway Shelf	65–100 40–10 30–80 30–115 35–55 160–220 45–70 110–130 50–70	Siliciclastic = 0–20 Recent carbonate = 50–75 Relict carbonate = 25–50	*Coarse* (0–60%, most 5–20%) = Similar to Facies C2 and dominated by a diverse suite of whole and fragmented bivalves and bryozoans. *Sand* (40–100%) = Similar to the sand in Facies C2 except that the particles are generally more rounded and abraded. Relict grains are locally particularly well rounded and polished. *Mud* = 0	*Coarse* (0–20%) = lithoclasts, generally rounded. Sand (80–100%) = Benthic foraminifers (including local *Marginopora*), lithoclasts and bivalve fragments (with local cemented carbonate sand attached).	A mixture of skeletal sand and relict particles.

Table 6.1 (continued)

Facies	Location	Depth mwd	Grains %	Recent Carbonate	Relict	Notes
M2 RELICT-RICH QUARTZOSE SKELETAL SAND & GRAVEL (Quartzose, fossiliferous Grainstone – Packstone)	**Albany** **Great Australian Bight, Ceduna Sector** (inshore) **Lacepede Shelf** **Bonney Shelf** **Otway Shelf** **Bass Strait** **Tasmania** south	55–85 25–65 35–75 175–285 <70 20–55 20–105	Siliciclastic = 10–50 Recent carbonate = 25–55 Relict carbonate = 35–50	*Coarse* (5–80%)=Molluscs (small mussels, *Lima*, cones, periwinkles, limpets, oysters, *Pinna*, broken *Pecten*, barnacles) –robust gastropods, (e.g. Turbo); large benthic foraminifers; diverse bryozoans (broken *Adeona*, *Parmularia*, flat robust branching, arborescent, articulated zooidal forms); infaunal echinoids; cemented clasts; coal. *Sand* (20–95%) = Quartzose (including crystalline rock fragments); bryozoans ≤ molluscs. *Mud*=none.	*Coarse* (40–80%)=calcareous lithoclasts, coralline rods, *Marginopora*; *Sand* (20–60%)=benthic foraminifers=coralline rods; abraded and rounded intraclasts	Mollusc-dominated seagrass assemblage that has been reworked. Reworked, very coarse sand & gravel; grains extensively fragmented, abraded, polished and bioeroded on Otway shelf.
M3 RELICT-RICH MOLLUSCAN SAND (Mollusc–intraclast Rudstone–Grainstone)	**Investigator Strait** **Lacepede Shelf** **Bass Strait**	50–70 30 35–50	Siliciclastic = 5–20 Recent carbonate = 30–65 Relict carbonate = 25–50	*Coarse* (10%) = robust bivalves (*Glycymeris*, *Katelysia*), and minor bryozoans (fenestrate, vagrant & *Adeona*), local *Chlamys*. *Sand* (70–90%)=molluscs > bryozoans; abundant *Marginopora* (5–30%) and coralline algal rods in GAB-diverse bryozoans. *Mud* = (0–20%)	*Coarse* (65%) = brown gravel and pebble intraclasts > worn bivalve shells, (*Marginopora* in GAB). *Sand* (35%)=benthic foraminifers > coralline rods.	Mollusc-dominated carbonate sand rich in relict grains and containing minor quartz.
M4 RELICT-RICH BRYOZOAN SAND (Bryozoan, Intra-clast Rudstone – Grainstone-Packstone)	**Albany** **Lacepede, Lincoln & Bonney Shelf** **Tasmania**	85–110 30–100 (up to 200) 40–175	Siliciclastic=0–20 Recent carbonate=30–75 Relict carbonate=25–50	*Coarse* (5%) = dominated by bryozoans (fenestrates, *Adeona*, *Celleporaria*, flat robust branching, vagrant, articulated zooidal forms); a few bivalves (*Glycymeris*, *Chlamys*). turitellid gastropods, *Siliquaria*, epifaunal echinoid spines, cemented clasts. *Sand* (80%)=bryozoans >> molluscs=coralline algal rods=benthic foraminifers; locally numerous stranded coralline rods. *Mud* (0–15%)= Not present eastward of Albany	*Coarse* (60%)=numerous brown lithoclast cobbles >> bivalves. (minor *Marginopora* in GAB; *Sand* (40%)=coralline rods=benthic foraminifers	Bryozoan-dominated carbonate sand rich in relict grains.

Table 6.1 (continued)

Facies	Location	Depth mwd	Grains %	Recent Carbonate	Relict	Notes
M5 RELICT-RICH FINE SKELETAL SAND (Intraclast-fine skeletal grainstone)	**Bass Strait Tasmania**	45–75 50–170	Siliciclastic=0–20 Recent carbonate=45–75 Relict carbonate=25–55	*Coarse* (0–20%)=bryozoans, *Siliquaria*, azooxanthellate corals, gorgonian spicules, calcarenite clasts encrusted with *Miniacina*, serpulid worms, infaunal & epifaunal echinoid pieces. *Sand* (80–100%)=homogeneous, fine; well-sorted, bryozoans (articulated zooidal)≥benthic foraminifers, infaunal echinoid fragments, planktonic foraminifers, sponge spicules >> serpulids, bryozoan fragments, gastropods, bivalves, pteropods. *Mud* (0–10) benthic biofragments	Coralline rods, benthic foraminifers & catenicellenids	
Megafacies R—Relict Carbonate						
R1 RELICT SAND (Intraclast Grainstone)	**Albany** **Great Australian Bight, Ceduna Sector** **Baxter Sector** inshore off Roe Terrace **Lacepede Shelf** South of Kangaroo Is. **Bass Strait**	75–90 50–95 50–80 30–55 65–85 45	Siliciclastic=tr Recent carbonate=10–25 Relict carbonate≥75	*Coarse* (10%)=mostly epifaunal (pectens and glycymerids) and thin-shelled infaunal bivalves; minor bryozoans (living and dead vagrants, fenestrate, fewer flat robust branching and *Adeona*) increasing outboard; rhodoliths - alive at 60–65 mwd, dead at 80–100 mwd; conspicuous brachiopods (*Anakinetica*); *Marginopora* throughout. *Sand* (90%)=intraclasts >> molluscs > bryozoans (except off Smoky Bay); bryozoans fragmented (vagrants and *Parmularia*), increasing diversity outboard (fenestrate and articulated branching); numerous coralline rods; conspicuous *Marginopora* fragments inboard. *Mud*=none.	*Coarse* (55%)=intraclasts >>> bivalves; *Marginopora*, with coralline rods and rhodoliths locally; *Sand* (45%)=Benthic foraminifers=coralline rods=intraclasts.	Widespread relict carbonate sands with only a minor recent carbonate component.
R2 MOLLUSC-RICH RELICT SAND (Intraclast–Mollusc Grainstone)	**Great Australian Bight, Baxter & Ceduna Sectors** **Spencer Gulf** entrance, Investigator Strait **Lacepede Shelf** **Bonney Shelf** **Bass Strait**	65 50–90 35–45 40–80 270–280 75–85	Siliciclastic≤10 Recent carbonate=15–40 Relict carbonate=50–75	*Coarse* (10–60%)=robust and delicate bivalves (*Tawera, Chione, Glycymeris, Ostrea, Pecten*); bryozoans (fenestrate, vagrant & minor pseudovagrant, nodular, encrusting forms, *Adeona*); gastropods; brachiopods. *Sand* (40–90%)=molluscs > benthic foraminifers=coralline algal rods > bryozoans; local *Marginopora*; bryozoans most diverse in Baxter Sector; blackened grains increase inshore. *Mud*=none	*Coarse* (50%)=pebbles, including lithoclasts >>bivalves rhodoliths locally; *Marginopora* in GAB Baxter Sector, coralline rods. *Sand* (50%)=Coralline rods=benthic foraminifers	Relict sands with <50% mollusc-dominated recent carbonate.

Table 6.1 (continued)

Facies	Location	Depth mwd	Grains %	Recent Carbonate	Relict	Notes
R3 BRYOZOAN-RICH REL-ICT SAND (Intraclast–Bryozoan Grainstone)	**Great Australian Bight (all sectors)** **Lacepede Shelf** **Bass Strait** **Tasmania** (western & southern areas)	50–120 50–120 75–130 30–90	Siliciclastic ≤20 Recent carbonate = 25–50 Relict carbonate = 50–75	*Coarse* (20–50%) = dominated by bryozoans, fenestrates (*Adeona, Conoscharalina*), flat robust branching, vagrant, local *Celleporaria* & bivalves (*Glycymeris, Chlamys, Malleus, Katalysia*), *Siliquaria*, conspicuous lithoclasts; living rhodoliths at ~70mwd, dead stranded rhodoliths at ~80mwd. *Sand* (45–75%) = bryozoans > benthic foraminifers = molluscs = coralline algal rods with minor pelagic foraminifers. *Mud* (0–10%).	*Coarse* (55%) = lithoclast pebbles >>> bivalves >>> bivalves *Marginopora*. *Sand* (45%) = benthic foraminifers > coralline rods, all abraded.	Relict sands with <50% bryozoan-dominated recent carbonate.
R4 LIMESTONE GRAVEL (Abraded Mollusc Rudstone)	**Baxter Sector** **Lacepede Shelf** **Bonney Shelf** **Bass Strait &** **Tasmania**	180 35–60 60 15–60	Siliciclastic = 10–45 Recent carbonate = 35–50 Relict carbonate ≥50	*Coarse* (70%) = robust bryozoans, molluscs, encrusting coralline algae in shallow water, numerous azooxanthellate corals (*Flabellum*), in deep water. *Sand* (30%) Quartz; molluscs > bryozoans in shallow water; sponge spicules, pteropods, bryozoan fragments in deep water. *Mud* = 0–10%.	*Coarse* (80–100%) = brown calcarenite cobbles & boulders; bored & encrusted by living peyssonnelids & coralline algae (shallow water), abraded, oblate; stained, bored and abraded bivalves (*Chlamys, Glycymeris, Katelysia, Cleidothaerus, Ostrea, Nuculana, Barbatia, Malleus*) gastropods = turitellids & tulips; bryozoans = *Adeona*, articulated zooidal, articulated branching, vagrants. *Sand* (20%) = stained lithoclasts.	Shore-parallel linear gravel bands = paleostrands; deep water gravels = lowstand paleostrands.

Tabel 6.1 (continued)

Facies	Location	Depth mwd	Grains %	Recent Carbonate	Relict	Notes
Megafacies Q—Quartz Sand						
Q1 CALCAREOUS QUARTZ SAND (Skeletal Quartz Arenite)	**Albany**, **Lacepede Shelf**, **Gulf St-Vincent**, **Bass Strait** around islands, **Tasmania**	60–70 5–70 <20 15–85	Siliciclastic ≥50 Recent carbonate = 10–50 Relict carbonate = 20–50	*Coarse* (15%) = mollusc dominated *Donax* < 40mwd, *Katelysia* > 40 mwd. *Sand* (85%) = quartz > molluscs > bryozoans, benthic foraminifers, echinoids, worm tubes. *Mud* = none	*Sand* = 25% lithoclasts.	Quartz sand with up to 40% biogenic content.
Megafacies B—Biosiliceous						
B1 SPICULITIC, PELOIDAL, SILICICLAS-TIC MUD (Biosiliceous Cherty Mudstone)	**Western Tasmania Estuary**	5–10	Siliciclastic = 60% (clays) Recent spicules & carbonate = 40% Relict carbonate = 0	*Coarse* (0) *Sand* (40%) = spicules & peloids–a few molluscs, echinoid spines, benthic foraminifers, bryozoans. *Mud* = (0)	None	Inner estuary, sub-photic, low pH environment

than molluscs, is one of the most widespread facies. Deposits range from poorly sorted to well sorted with very fine sand- to cobble-size particles (Fig. 6.2). Poorly sorted sediment is always made up of angular fragments whereas better sorted sediment, more common in <120 mwd, is somewhat abraded. Cores of this facies on the Otway Shelf are cross-stratified sand (Boreen et al. 1993).

These vast spreads of uniform sand have minor coarse material, generally bryozoans, ubiquitous slit shells (*Tenagodus*–implying abundant sponges), scattered calcarenite clasts encrusted by the red foraminifer *Miniacina miniacea* and serpulid worms, epifaunal echinoid spines, and irregular rhodolites (clumps of corallines and peyssonnelids), clusters of serpulids, and crab claws. Grains and granules range from fresh to abraded to encrusted by bryozoans, corallines, and foraminifers. Bryozoans are numerous and

diverse. Although there are generally less than 10% Recent mollusc particles by volume, large molluscs are diverse and conspicuous. Small ahermatypic corals such as *Trematotrochus* and *Caryophyllia* are attached to larger bryozoans and rhodolites. Sediments in the Ceduna Sector on the eastern Great Australian Bight Shelf are distinguished by 20–40% particles of stranded articulated coralline algae.

6.2.1.2   Facies C2—Skeletal Sand and Gravel

Biofragmental sands in which no single particle type dominates are generally inner neritic deposits (20–50 mwd in the Gulfs (Spencer Gulf and Gulf St. Vincent) that have the general composition of a sea grass facies but with little mud, implying winnowing after deposition. Sediments usually contain some quartz that

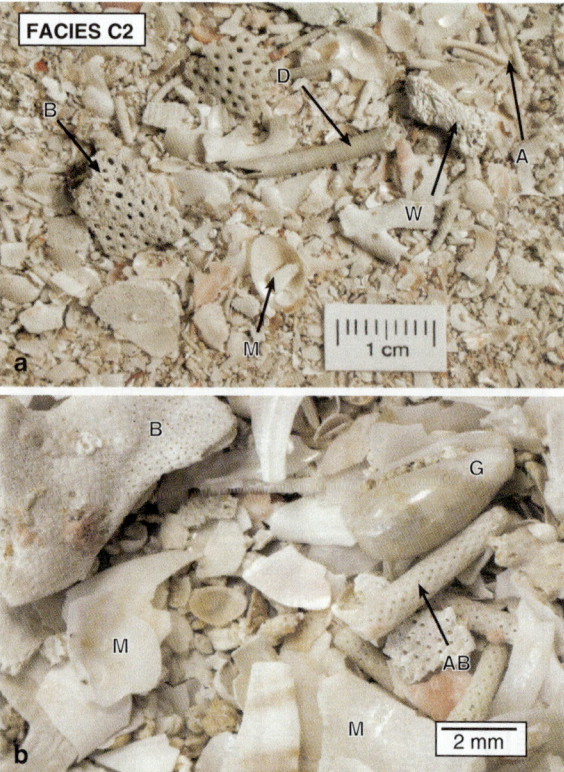

Fig. 6.2 Facies C1—Bryozoan Sand and Gravel. (**a**) A diverse suite of coarse sand to granule size bryozoan particles with conspicuous large fragments of robust rigid branching (B) and fenestrate (F) growth forms, (**b**) a close view of these dominant growth forms together with smaller fragments; the pink encrustation (*arrow*) is the foraminifer *Miniacina miniacea*. (Lacepede Shelf, VH89 01–171 mwd)

Fig. 6.3 Facies C2—Skeletal Sand and Gravel. (**a**) A diverse suite of biofragments including serpulid worm tube clusters (W), bivalves (M), coralline algal rods (A), bryozoans (B), and scaphopods (D), (**b**) a close view of robust rigid branching bryozoans (B), articulated branching bryozoans (AB), gastropods (G), bivalve fragments (M), and many smaller skeletal remains. (Great Australian Bight, ACM106–51 mwd)

locally may be as high as 20%. The most conspicuous whole bivalves are pectens and infaunal types whereas bryozoans are dominated by fenestrate, robust and rigid branching, and free-living types (Fig. 6.3). The sand fraction is similar in composition to these coarse particles but with the addition of benthic foraminifers. Relict sediment is generally minor.

6.2.1.3   Facies C3—Coralline Gravel

Coralline gravels are either rhodoliths (cf. Bosellini and Ginsburg 1971; Ginsburg and Bosellini 1973) or rigid branching fruticose nodules (Wray 1977) and their fragmented branches. Such sediments outboard of grass banks in 30–50 mwd (<30 mwd in the gulfs) are like the maerl facies at similar depths in the sub-tropical Mediterranean (Carannante et al. 1988). Pectens and other epifaunal bivalves together with conspicuous *Adeona* are locally numerous. The sand can be locally

quartzose (up to 40%). Relict sediment is generally minor (Fig. 6.4).

6.2.1.4   Facies C4—Encrusted Rocky Substrate

This part of the seafloor is one of encrustation and carbonate production rather than sediment accumulation. Numerous rocky substrates, a widespread hard-bottom biota, and a thin, patchy, sediment veneer characterize the environment. The prolific growth on subaqueous outcrops, on all shelves in <70 mwd, comprises both macroalgae and benthic invertebrates, especially sponges, coralline algae, numerous different free and byssally attached bivalves, diverse gastropods, and assorted bryozoans, pedically attached brachiopods, a spectrum of benthic foraminifers, and grazing regular echinoids. Although the biota is prolific, most components are shed onto the surrounding seafloor, only to be redistributed by waves and currents. The preserved facies is that of

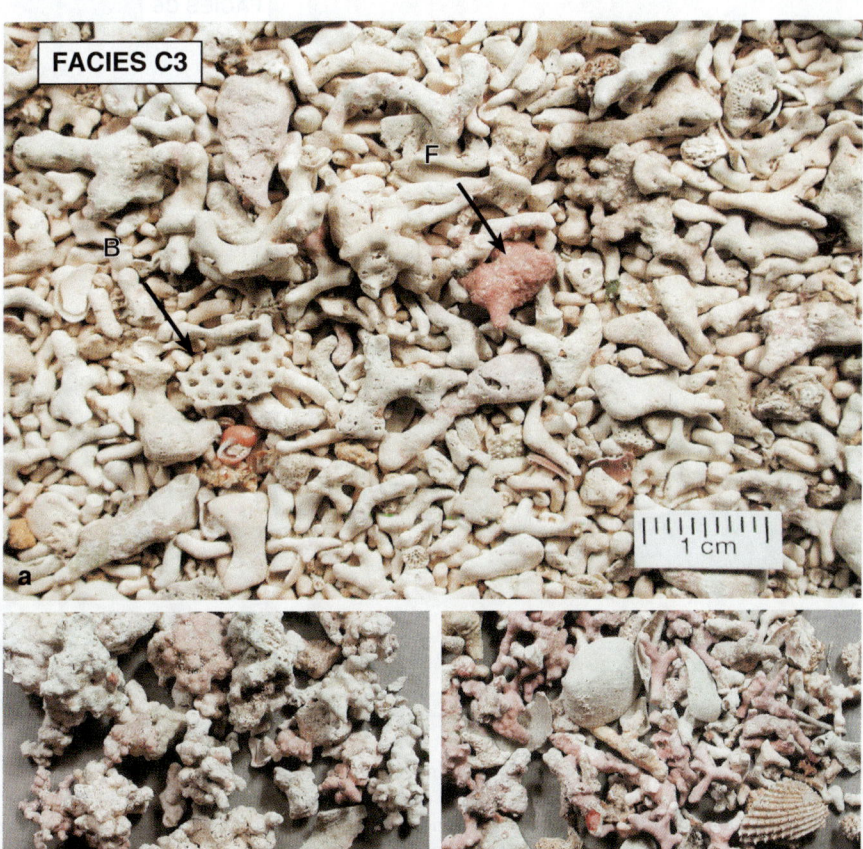

Fig. 6.4 Facies C3—Coralline Gravel. (**a**) Coarse fragmented branches of fruticose coralline algae with a few bryozoans (B) and encrusting foraminifer fragments (F) (*Miniacina miniacea*) (Great Australian Bight, GAB 064–42 mwd), (**b**) living (pink) and dead (white) fruticose rhodoliths (Great Australian Bight, GAB 061–46 mwd), (**c**) numerous robust branching coralline algal sticks and scattered bivalves. (Lacepede Shelf, VH89 15B–37 mwd)

a hardground (or stoneground) excavated by a suite of boring invertebrates, especially echinoids, bivalves, and sponges, that can be encrusted by coralline algae.

6.2.1.5   Facies C5—Articulated Coralline and Intraclast Sand

This is largely a stranded facies. Coralline rods from articulated red algae together with irregular intraclasts have a striking distribution; they are conspicuous components in all Great Australian Bight sediments between 60 and 250 mwd (most numerous between 90 and 150 mwd) but are virtually absent or at a low level in Lacepede Shelf and Investigator Strait sediments. Bryozoans, serpulid worms, siliquarids, and corals dominate the minor coarse fraction. Intraclasts are mostly skeletal particles with minor synsedimentary cement (Fig. 6.5).

6.2.1.6   Facies C6—Fine Skeletal Sand

This homogenous, well sorted, fine to very fine-grained sand consists of fragmented, fresh particles (Fig. 6.6). It generally contains <20% relict material and an infinitesimal (0–20%) coarse fraction. The material can, however, be locally bimodal with up to 20% coarse to very coarse-grained sand particles of bryozoans with lesser soft coral spicules, *Tenagodus*, and serpulid worm tubes. The fine sediment is mostly erect flexible articulated zooidal bryozoan zooids, planktic foraminifers, benthic foraminifers, broken ostracod shells, infaunal echinoid plates and spines, and siliceous sponge spicules. The minor mud fraction is generally silt–sized skeletal biofragments, many of which are unidentifiable. The deposits are generally in deeper water on the outer shelf and upper slope (80–200 mwd) but can range from 40 to 475 mwd.

Fig. 6.5 Facies C5—Articulated Coralline and Intraclast Sand. (a) Sand composed almost entirely of geniculate coralline algal rod-like segments; most of these sands are stranded, (b) close view of this sand that also contains some intraclasts (*arrows*) and a few other skeletal fragments. (Great Australian Bight, GAB 087–90 mwd)

Fig. 6.6 Facies C6—Fine skeletal sand. (a) Fine and very fine biofragments with a few bryozoans (B) and a demosponge holdfast (S), (b) close view of the numerous fragmented particles together with articulated zooidal bryozoan zooids (*arrows*). (Great Australian Bight, GAB 056–73 mwd)

6.2.1.7 Facies C7—Delicate Branching Bryozoan Muddy Sand

Erect flexible articulated branching cyclostomes (plus some cheilostomes) dominate the sediment, forming an almost mud-free sediment (Fig. 6.7). Sands also contain abundant infaunal benthic foraminifers, siliceous sponge spicules, pteropod cones, and serpulids. The bryozoan component is distinct from shallower shelf assemblages in the relatively greater abundance of delicate cyclostomes, and absence of robust shelf forms such as *Adeona* sp. Although fragmented, most bryozoans are not significantly worn, and preservation of extremely delicate structures and whole fragile bryozoans is common. Mud is distinctively silty and composed of biofragments. Deposits are generally confined to the outer part of the shelf and upper slope between 60 and 350 mwd.

6.2.1.8 Facies C8—Scaphopod, Pteropod Sand and Mud

The material is a fine sandy and silty carbonate with conspicuous scaphopods (*Dentalium*), corals, pteropod cones, small delicate *pectens*, and bryozoans (Fig. 6.8). The facies is usually deep water (300–500 mwd), but is as shallow as 75 mwd in Bass Strait and on the eastern Tasmania shelf. The sand is composed of infaunal and epifaunal benthic foraminifers, fecal pellets, infaunal echinoid segments, gastropods, and serpulids. The silt is mainly skeletal grains such as planktic foraminifers, siliceous sponge spicules, echinoid pieces, and unidentifiable grains. The sediment is surprisingly coarse-grained for such depths and thus could be somewhat winnowed.

6.2.1.9 Facies C9—Mollusc, Coralline, Benthic Foraminifer Gravel, Sand and Mud

Sediments in *Posidonia*- and *Amphibolis*-dominated seagrass beds (<25 mwd) are composed, for the most part, of poorly sorted muds and sandy muds with conspicuous whole molluscs and benthic foraminifers (Fig. 6.9). This facies occurs as shore-attached banks along the margins of the gulfs and as shoals in small embayments, especially along the western coast of Eyre Peninsula.

The Quaternary sediment in and beneath grass meadows and banks has been carefully documented (Hails et al. 1984a; Barnett et al. 1997; Cann et al. 2000) and

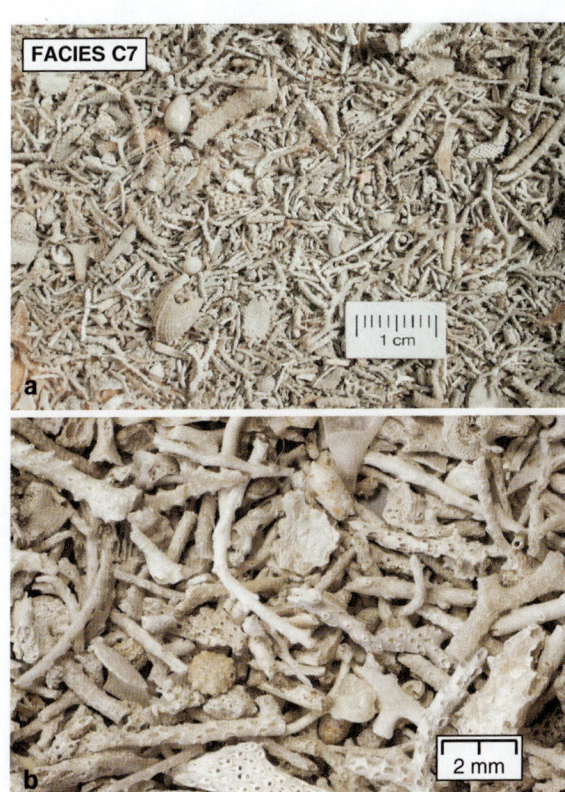

Fig. 6.7 Facies C7—Delicate Branching Bryozoan Muddy Sand. (**a**) Numerous small branches of dominantly cyclostome bryozoans forming an almost mud-free sediment, (**b**) a close view of the bryozoans wherein retention of delicate zooidal architecture indicates little movement or abrasion. (Great Australian Bight, ACM 085–196 mwd)

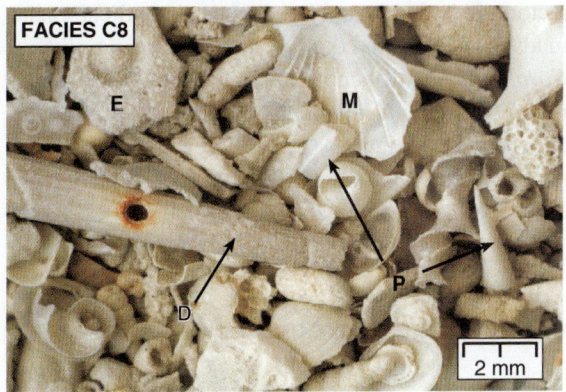

Fig. 6.8 Facies C8—Scaphopod, Pteropod Muddy Sand. A close view of the coarse fraction composed of delicate bivalves (M) echinoid plates (E), scaphopods (D) and pteropods (P). (Great Australian Bight, ACM 100–495 mwd)

Fig. 6.9 Facies C9—Mollusc, Coralline, Benthic Foraminifer Gravel, Sand and Mud. (a) Cores from beach sands (*1,2*) and adjacent offshore seagrass facies (*3,4,5*), Spencer Gulf (courtesy R.V. Burne). Grassbed cores contain numerous roots (G) and bivalves (M), (b) poorly sorted sediment from a shallow grassbed containing large bivalves (M) gastropods (G) and numerous *Posidonia* blades (P). (Investigator Strait near Kangaroo Island, PL94 71–15 mwd)

its poorly-sorted character reflects the nature of this factory. It is dominated by siliciclastic and carbonate muds, sands rich in coralline algae (*Metagoniolithon, Corallina, Jania*), benthic foraminifera (*Nubecularia*, miliolids (*Quinqueloculina, Triloculina, Milionella, Spirolculina*), rotaliids (*Discorbis, Elphidium*), large soritids (*Peneropolis*)), and mollusc fragments (Gostin et al. 1988; Cann et al. 2002).

6.2.2 Muddy Carbonate Facies

6.2.2.1 Facies C10—Bivalve Muds

This facies floors the large gulfs and the center of Bass Strait from 25 to 80 mwd. Mollusc fragments, planktic foraminifers, and infaunal echinoid pieces dominate the muds and sands. The bivalves are mostly epifaunal or semi-epifaunal forms such as *Pinna*, *Maleus*, and *Chlamys*.

6.2.2.2 Facies C11—Spiculitic Skeletal Sandy Mud

The sediment is a mixture of fine neritic benthic particles swept seaward, pelagic calcareous foraminifer and nannoplankton fallout, skeletal sand with local small segments of erect flexible delicate branching bryozoans floating in the mud and *in situ* production (Fig. 6.10). It mostly extends from depths as shallow as 100 mwd down to ~500 mwd across all shelf sectors. Carbonate content ranges from 23 to 80%, the remainder being clay minerals. Sediment is variably rich in catenecellid singlets, with the only larger bryozoan of importance

Fig. 6.10 Facies C11—Spiculitic, Skeletal Sandy Mud. (a) Fine sediment with scattered pteropods (P) and small gastropods (G) (Lacepede Shelf, VH91 115–582 mwd), (b) close view of the sandy nature of the sediment composed, of fine biofragments together with siliceous sponge spicules (*arrows*). (Great Australian Bight, GAB 022–374 mwd)

being small free living (vagrant) lunuliform bryozoans. The sediment is locally winnowed into muddy sands and gravels rich in *Dentalium,* together with pteropods, oblate and small high-spired gastropods, small infaunal echinoid plates, bryozoans (spherical and vagrant forms), epifaunal and infaunal benthic foraminifers, ostracods, micromolluscs, and angular limestone clasts. Other locally abundant skeletons include solitary corals and small pectens. Agglutinated benthic foraminifera are conspicuous. The small number of molluscs are tiny cerithid-like gastropods and pteropods. Shelf edge skeletal debris decreases progressively downslope. Large thin-walled brachiopods, high spired gastropods, scaphopods, and solitary corals are rare but conspicuous. Sediment is thoroughly bioturbated and homogenous, with local vague burrow mottling in cores from on the Otway Shelf (Boreen et al. 1993).

6.2.2.3 Facies C12—Coral–Arborescent Bryozoan Gravel and Mud

This facies comprises conspicuous, large, stranded arborescent bryozoans (*Celleporaria*), and an otherwise diverse assemblage of rigid robust branching, fenestrate, and free-living bryozoans (Fig. 6.11). Growing on this coarse substrate are variably abundant solitary corals as isolated colonies or as candelabra-like associations. The facies begins at a depth of 160 m where it is transitional into shallower shelf facies but is

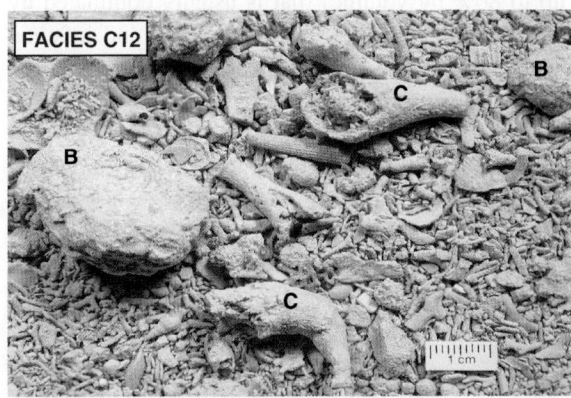

Fig. 6.11 Facies C12—Coral, Arborescent Bryozoan Gravel and Mud. A poorly sorted mixture of bryozoan-coral sand and gravel with conspicuous *Celleporaria* colonies (B) and azooxanthellate corals (C). (Lincoln Shelf, PL94 22–320 to 350 mwd)

best developed between 250 and 300 mwd, and is gradational downwards into slope muds between 350 and 400 mwd. Five skeletons of *Celleporaria* sp. from four localities in the Ceduna sector and Lincoln Shelf have ^{14}C ages ranging from 21.2 ka to 13.7 ka and from cores on the Otway Shelf have ^{14}C ages of 21.0–10.3 ka.

A similar bryozoan assemblage (James et al. 1997) is associated with large topographically-positive features on the Beachport Terrace off the Otway Shelf. The structures occur are present in 200–400 mwd, and have mound-like geometries with up to 120 m of relief. The intimate association of dense *Celleporaria* sp. bryozoan-sponge accumulations and the mound features suggests that they may be exhumed parts of biological buildups similar to the late Pleistocene bryozoan reef mounds off the Eyre Sector (James et al. 2004). Delicate cyclostome bryozoan sands occur on the mound flanks. Similar large bryozoan mounds have been sampled along the eastern margin of Bass Strait, in an area of elevated nutrient and chlorophyll concentrations (Gibbs et al. 1986; Heggie and O'Brien 1988).

In summary, the facies is essentially a Holocene deep-water biota of corals (including stylasterines), bryozoans, gastropods, and bivalves growing on the coarse exhumed mainly bryozoan remains of stranded, moribund bryozoan-sponge mounds.

6.3 Megafacies M—Relict-Rich Carbonate

Deposits that are a mixture recent and relict carbonate particles cover large expanses of all shelves. Such relict-rich facies comprise a background of relict sediment (25 to 50%) that is being added to by the production of younger carbonate. This megafacies is spatially transitional between recent carbonate deposits megafacies C outlined above and relict sediments of megafacies R described below (Table 2.1).

6.3.1 Facies M1—Relict-Rich Skeletal Sand and Gravel

No one single skeletal particle type dominates these sediments. They mostly occur in the mid-neritic zone

Fig. 6.12 Facies M1—Relict-rich Skeletal Sand and Gravel. **(a)** A mixture of relict foraminifer (F) (particularly *Marginopora*) and other brown skeletal fragments together with light coloured recent particles such as bivalves (M) and gastropods (G), **(b)** a close view of recent bryozoans (B) and bivalves (M), together with relict large benthic foraminifers (*Marginopora*) (F) and bivalves (MR), and numerous unidentified relict grains (R). (Great Australian Bight, ACM 080–60 mwd)

Fig. 6.13 Facies M2—Relict-rich Quartzose Skeletal Sand and Gravel. **(a)** Sand composed of a mixture of abraded brown relict grains and light recent grains, **(b)** a close view of the rounded brown relict particles (R), Recent bryozoan fragments (B), and quartz grains (Q). (Lacepede Shelf, VH89 053–47 mwd)

but range between 30 and 120 mwd. Recent fragments are generally a diverse suite of bryozoans and bivalves. Particles are conspicuously rounded and abraded. The relict components are diverse and comprise lithoclasts, benthic foraminifers (including *Marginopora*), and bivalves (Fig. 6.12).

6.3.2   Facies M2—Relict-Rich Quartzose Skeletal Sand and Gravel

Sediment is generally quartz-rich, bivalve-bryozoan sandy gravel, a heterogeneous, very poorly sorted mixture of roughly equal relict and recent bioclasts, and quartz (Fig. 6.13). It is found at 20–100 mwd on most shelf sectors but is conspicuously deeper

(175–285 mwd) on the Bonney Shelf. The deposits can have a substantial coarse (>2 mm) component (up to 80% by volume) that is usually dominated by an extremely diverse suite of epifaunal and infaunal bivalves, robust as well as well as diminutive gastropods, diverse bryozoans, and infaunal echinoids. Molluscs dominate the sand–sized fraction. The relict components are a suite of coralline rods, lithoclasts, large benthic foraminifers (e.g. *Marginopora*) and worn bivalves. The deposits are generally mud-free. The overall biota resembles a seagrass assemblage.

6.3.3   Facies M3—Relict-Rich Molluscan Sand

The sediment is quartzose molluscan sand with up to 50% relict grains. The sediments are shallow water

Fig. 6.14 Facies M3—Relict-rich Molluscan Sand. (**a**) Sand that is a mixture of brown relict grains and white recent particles together with numerous whole bivalve shells (*Katelysia*), (**b**) close view of the sand fraction composed of well rounded, abraded, and polished brown relict grains and white recent biofragments. (Lacepede Shelf, VH89, LB 2–28 mwd)

Fig. 6.15 Facies M4—Relict-rich Bryozoan Sand. (**a**) A sand of Recent bryozoan pieces against a background of dark brown relict grains, (**b**) close view of the dark brown, worn relict grains and the white recent bryozoan fragments. (Lacepede Shelf, VH89, 23–58 mwd)

generally between 30 and 70 mwd. The material is typified by shells of the robust bivalves *Katelysia*, and locally, *Chlamys*. Sands are also dominated by mollusc fragments but with lesser bryozoans. Relict components comprise brown and grey-black pebble lithoclasts and worn bivalves (Fig. 6.14).

6.3.4   Facies M4—Relict-Rich Bryozoan Sand

A mixture of Holocene bryozoan sand and gravel and obvious brown relict particles gives this sediment a speckled brown and cream or 'salt and pepper' appearance (Fig. 6.15). The sediment is present on most shelf sectors between 30 and 175 mwd with most shallower than 100 mwd. Quartz comprises up to 20% of the sed-

iment on the Lacepede and Otway shelves. The coarse fraction (>2 mm) is mostly a diverse suite of bryozoans and a few bivalves, mainly *Glycymeris*, *Katelysia*, and local *Chlamys*, gastropods and echinoid spines, together with some cemented calcarenite clasts. The sands are overwhelmingly bryozoan particles. Relict grains are dominantly lithoclast cobbles, with a few bivalves, and large benthic foraminifers in the Great Australian Bight.

6.3.5   Facies M5—Relict-Rich Fine Skeletal Sand

This sediment is similar in composition to that of facies C6 and can contain up to 50% relict grains. The sediment is homogenous, well sorted, fine to very fine-grained sand and consists of fragmented, fresh

Fig. 6.16 Facies M5—Relict-rich Fine Skeletal Sand. (**a**) Fine and very fine biofragments with a few bryozoans, (**b**) close view of the numerous fragmented particles together with erect, flexible articulated zooidal bryozoan zooids. (Western Tasmania Shelf, GA 2062–148 mwd)

particles, with an infinitesimal (0–20%) coarse fraction (Fig. 6.16). The material is mostly catenecellid bryozoan zooids, planktic foraminifers, benthic foraminifers, broken ostracod shells, infaunal echinoid plates and spines, and siliceous sponge spicules. The minor mud fraction is generally silt–sized skeletal biofragments, many of which are unidentifiable. This sediment is found east of Cape Otway, in Bass Strait, and covers large areas off the Tasmanian coast between 45 and 175 mwd.

6.4   Megafacies R—Relict Carbonate

Relict carbonate sediments, wherein such grains form >50% of the sediment, cover wide areas of the middle shelf, especially in the Great Australian Bight. Whereas some sediments are almost entirely relict, others have substantial mollusc and bryozoan components. These

sediments are typically bimodal with relict sand-sized grains and much larger recent skeletons. One particular deposit, limestone gravel (facies R4), is disposed in linear ridges on the shelf and interpreted to be a series of paleostrandlines (Table 2.1).

6.4.1   Facies R1—Relict Sand

Distinguished by little recent material, these sediments are almost all brown-stained relict intraclasts (Fig. 6.17). They are found between 30 and 95 mwd on most shelf sectors with the majority at 50–90 mwd. The obvious shells are mostly pectens and thin-shelled infaunal bivalves, a small variety of bryozoans, rhodoliths in mid to deep neritic environments, and *Marginopora*. The sand is almost wholly relict intraclasts, small benthic foraminifers, large benthic foraminifers

Fig. 6.17 Facies R1—Relict Sand. (**a**) Bimodal sediment composed of well rounded and well sorted relict sand dominated by brown relict particles and a few larger white recent biofragments, (**b**) close view of the varied brown intraclasts that comprise the sediment (Great Australian Bight, ACM 076–91 mwd)

(*Marginopora*) with accessory recent mollusc and bryozoan fragments and coralline rods.

6.4.2 Facies R2—Mollusc-Rich Relict Sand

This well-sorted, rounded, medium sand to coarse gravel is composed of relict intraclasts, minor quartz, and numerous large recent molluscs (more abundant than bryozoans) (Fig. 6.18). It occurs between 30 and 90 mwd but also lies in deep water (270–280 mwd) on the Bonney Shelf. The whole to slightly fragmented bivalves are mainly *Tawera, Chione, Glycymeris, Ostrea,* and *Pecten,* the sands are overwhelmingly composed of mollusc fragments, together with numerous benthic foraminifers. *Marginopora* occurs in all locations in the Great Australian Bight but there are no live

forms east of Esperance Bay. Sediments in the far west contain a higher percentage of intraclasts (~65%), fewer coralline algal rods, and up to 40% benthic foraminifers in the sand size fraction. The numerous relict components comprise mostly brown lithoclast pebbles.

6.4.3 Facies R3—Bryozoan-Rich Relict Sand

Relict-rich sands with numerous and diverse recent bryozoans are present between 50 and 120 mwd. The abundant recent coarse fraction (>2 mm) is dominated by fenestrates, robust rigid branching, vagrant and locally arborescent growth forms; bivalves are less numerous and rhodoliths are present locally (Fig. 6.19). The sands are mostly bryozoans even though the sediments contain numerous benthic foraminifers, molluscs, and coralline rods. The relict particles are lithoclast pebbles

Fig. 6.18 Facies R2—Mollusc-rich Relict Sand. (**a**) Sand with a salt-and-pepper appearance due to the mixture of brown relict and white recent particles with whole and broken bivalve shells, (**b**) close view of the numerous rounded brown relict grains and the angular Holocene bivalve fragments. (Investigator Strait, PL94 59–40 mwd)

Fig. 6.19 Facies R3—Bryozoan-rich Relict Sand. (**a**) Sand with a salt-and-pepper appearance due to the mixture of brown relict intraclasts and white recent particles most of which are bryozoans, (**b**) close view of brown relict particles and white recent bryozoan fragments. (Lacepede Shelf south of Kangaroo Island PL94 50–114 mwd)

with a few bivalves, large benthic foraminifers and coralline rods, all of which are conspicuously abraded.

6.4.4 Facies R4—Limestone Gravel

These sediments are difficult to summarize because of the wide depth range over which they occur (15–180 mwd), but in general the deposits are coarse in shallow water and become finer grained with more planktic elements in deeper water. The facies is distinguished by chocolate brown limestone pebbles and cobbles with numerous brown-stained and abraded bivalves, gastropods, and lesser bryozoans (Fig. 6.20). They are essentially quartz sands and abundant worn lithoclasts resembling beach cobbles and abraded shallow-water bivalves, together with numerous Holocene bivalves in shallow water, and bryozoans, living azooxanthellate corals, and planktic foraminifers in deep water. The sed-

iments are interpreted to be a drowned late Pleistocene shallow-water beach facies that is now serving as a substrate for Holocene epibenthic growth.

6.5 Megafacies Q—Quartz Sand

Whereas many facies have a quartz sand component this megafacies is almost all quartz with only a minor carbonate component (Table 2.1).

6.5.1 Facies Q1—Calcareous Quartz Sand

Fine-to medium grained and well sorted grained quartz sand (>50%), these yellow, brown, or orange sands on the Lacepede Shelf, and almost pure quartz sands on the Albany Shelf, containing variable proportions of relict and recent carbonate, particularly infaunal bivalves and limestone cobbles (Fig. 6.21). They cover

Fig. 6.20 Facies R4—Limestone Gravel. (**a**) Brown, mostly rounded relict limestone pebbles, cobbles, and relict sand grains mixed with white recent biofragments most of which are bivalves. (Lacepede Shelf, VH89 085–56 mwd), (**b**) a variety of selected pebbles illustrating the diversity of clast types. (Lacepede Shelf, VH91 152D–62 mwd)

Fig. 6.21 Facies Q1—Calcareous Quartz Sand. (**a**) Medium to coarse-grained quartz sand with a few angular white bivalve fragments, (**b**) close view of the subangular to rounded clear, yellow, and orange quartz. (Lacepede Shelf, VH89 011–42 mwd)

much of the central Lacepede Shelf where most of the quartzose sand is relict or stranded and came from the ancient River Murray when it flowed across the shelf during the LGM. They also lie along the inner shelf of western Tasmania shelf, a region of relatively high rainfall, significant runoff, and fluvial sediment input.

6.6 Megafacies B—Biosiliceous Mud

Sponge spicules are a recurring component of many outer neritic—upper slope deposits but these sediments, localized to the estuarine environments in western Tasmania, are dominated by siliceous spicules (Table 2.1).

6.6.1 Facies B1—Spiculitic Peloidal Siliciclastic Mud

The shallow floor of Bathurst Harbour estuary is completely dark and populated by a marine biota. Somewhat acidic waters, however, dissolve most carbonate grains and leave an organic-rich peloidal terrigenous mud rich in siliceous sponge spicules. Spicules, constituting up to 20% of the sediment, are predominantly siliceous monaxons and triaxenes with calcareous triactines. The complete siliceous sponge spicules occur together with scattered, partially dissolved fine sand sized particles of foraminifers, bryozoans, echinoderms, molluscs and calcareous sponge spicules. Peloids are the fecal remains of numerous burrowing heart urchins (*Echinocardium* sp.) (Fig. 6.22).

6.7 Synopsis

The sediments are divided into five megafacies within which there are 23 facies. The different deposits have been classified largely on the basis of the relative proportions of recent carbonate, relict carbonate, and siliciclastic particles.

1. Megafacies C—Recent Carbonate, in which all sediment contains >50% recent carbonate biofragments is present across the environmental spectrum from <20 to >350 mwd. Deposits comprise 12

Fig. 6.22 Facies B1—Spiculitic Siliciclastic Mud. (**a**) Pelleted clay-mineral-rich mud and siliceous sponge spicules, (**b**) a washed sample from which the mud has been removed illustrating the numerous siliceous sponge spicules and angular carbonate skeletal particles. (Images courtesy of C. M. Reid) (Bathurst Harbour, Tasmania—6 mwd)

facies, eight of which are grainy and 4 of which are muddy.

2. *Grainy carbonate facies* on the inner and middle parts of the shelf are dominated by bryozoans, molluscs, and benthic foraminifers (facies C1, C2). Shallow (<50 mwd) relatively warm parts of the shelf, in the Albany sector, in the western Great Australian Bight, and in Spencer Gulf in particular, have spreads of rhodoliths (facies C3). The mollusc, coralline, benthic foraminifer gravel, sand and mud (facies C9) defines grass beds. The seafloor across the shelf edge, between ~80 and 200 mwd, is usually covered with fine skeletal sand (facies C6) or delicate branching bryozoan muddy sands (facies C7) that alternate spatially along strike. The upper slope is locally covered with scaphopod, pteropod sand and silt (facies C8). This facies, although mostly pelagic, contains relatively little mud.

3. *Muddy carbonate facies* are localized to small shoreline embayments, the gulfs, or the deep-water slope. The shallow parts of the gulfs <50 mwd are covered with bivalve muds (facies C10). The seafloor across the shelf edge to at least 500 mwd can be covered by spiculitic skeletal sandy mud (facies C11). Like facies C8, the sediment is mainly pelagic and spiculitic with a few bryozoans but locally numerous *Dentalium*. The coral—arborescent bryozoan gravel and mud (facies C12—200–400 mwd) is clearly the top of an exposed and marooned bryozoan mound that now forms a substrate for coral growth that is also being blanketed by pelagic carbonate mud.

4. Megafacies R—Relict carbonate wherein >50% of the particles are relict and generally lies between 20 and 120 mwd with most deeper than ~50 mwd. The sediments are largely confined to the open shelf and comprise four facies. These facies are either dominated by relict particles (facies R1), are mollusc-rich (facies R2), are bryozoan-rich (facies R3), or contain abundant fine skeletal sand particles (facies M5). Facies R4 (limestone gravel) is a series of linear ridges that are interpreted to represent paleostrandlines. Like facies C12, these hard substrates are now sites of epibenthic growth.

5. Megafacies M—Relict-rich carbonate, wherein the sediment is a mixture of recent grains and relict particles comprises a consistent 25–50% of the sediment. The deposits lie spatially between the previous two megafacies. The relict-rich sediments usually cover the seafloor between 60 and 100 mwd and laterally grade into megafacies 1 or 3. The four facies are relict–rich examples of many recent carbonate megafacies; skeletal sand and gravel (facies M1), quartzose skeletal sand and gravel (facies M2), molluscan sand (facies M3), bryozoan sand (facies M4), and fine skeletal sand (facies M5).

6. Megafacies Q—Quartzose sand comprises only facies Q1 (calcareous quartz sand). The quartzose component can be relict, left over from when siliciclastic sand was delivered to the shelf during the LGM lowstand, or modern if there is significant fluvial modern input. The calcareous component (especially molluscs) is from the modern calcareous infaunal and epifaunal community.

7. Megafacies B—Biosiliceous mud has only one facies (B1—spiculitic peloidal siliciclastic mud) and is confined to the Bathurst Inlet estuary floor. Much of the carbonate has been removed from the seafloor via dissolution by organic acids. The seafloor here is also completely dark because of the high tannin content of the water.

Chapter 7
Neritic Depositional Environments

7.1 Introduction

The southern continental margin of Australia today has been much the same throughout the late Cenozoic and so the Holocene depositional system is a continuation of one that has been in place for ~2 my. It has been dominated by strong SW winds that drive high-energy waves and swells directly onto exposed coasts. The climate, however, has become progressively drier since the mid-Pliocene such that much of southern Australia is today semi-arid. Exceptions are the Albany coast and the somewhat mountainous coasts of southern Victoria and Tasmania. This chapter is a documentation of the suite of modern marine depositional environments that are present across the continental margin.

The carbonate depositional system extends from the shoreline across the shelf onto the upper slope (Fig. 7.1). The wide continental margin is best categorized as an open shelf (cf. Ginsburg and James 1974). The shelf itself ranges from 20 to 230 km wide, extends to ~200 mwd, and passes gradually seaward into a progradational or erosional continental slope. It lacks an elevated rim, is relatively deep, and many depositional environments are disturbed by storm waves and swells. It is a swell-dominated, hydrodynamically partitioned, high-energy system (Fig. 7.1) that resembles a siliciclastic shelf except that sediment production takes place throughout (Fig. 7.2).

The seafloor of the continental shelf cannot be easily partitioned. Different parts are bathymetrically alike but the surface is also profoundly influenced by hydrodynamics. In bathymetric terms, the shelf is most simply divided into a relatively narrow inner shelf to ~60 mwd, a middle shelf that extends outboard to a small cliff that was formed by strandline erosion during the LGM and the base of which today lies at ~120 mwd (Fig. 7.1). The commonly degraded crest of this escarpment is at ~90–100 mwd. The outer shelf in this scheme extends from 120 to the shelf edge at ~220 mwd.

Such a division, although useful, bears little relationship to the present nature of shelf deposition or the distribution of modern seafloor biological communities both of which are in part controlled by hydrodynamics. Thus, the seafloor is better partitioned in terms of modern oceanography. The shelf is herein defined (Fig. 7.1) as a neritic setting (between the intertidal and 100 fathoms [~200 mwd] or the shelf edge; Bates and Jackson 1984). The two important hydrodynamic interfaces are normal storm wave base at ~60 mwd and deeper swell wave base at ~140 mwd. For practical purposes there are, therefore, two active hydrodynamic zones; (1) *the zone of wave abrasion*, wherein grainy sediment is moved more or less constantly by a combination of year-round swells and seasonal storms and the base of which lies at ~60 mwd, (2) the *zone of swell and episodic storm wave disturbance*, wherein grainy sediment is moved and winnowed episodically, mostly by swells but also by exceptionally intense storms. Numerous seafloor images, especially those from the Lincoln Shelf, Lacepede Shelf, and the West Tasmanian Shelf, indicate that physical sedimentary structures, especially subaqueous dunes and rippled coarse-grained sands and gravels, produced by these waves and swells, extend to ~140 mwd but rarely deeper.

On the basis of the foregoing, *Shallow Neritic Environments* are classified as those found from the strandline to ~60 mwd, most of which on the open shelf lie within the zone of wave abrasion. These environments also include the floors of most intrashelf basins such as Spencer Gulf, Gulf St. Vincent and Bass Strait. That

N. P. James, Y. Bone, *Neritic Carbonate Sediments in a Temperate Realm*,
DOI 10.1007/978-90-481-9289-2_7, © Springer Science+Business Media B.V. 2011

Fig. 7.1 A sketch illustrating the different morphological and neritic environments on the shelf

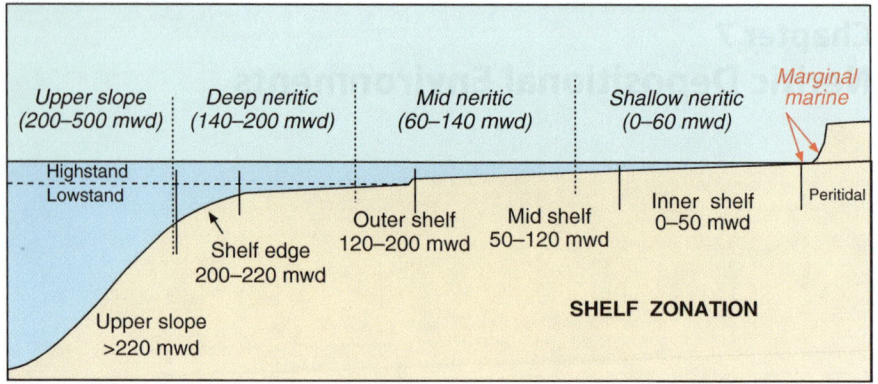

part of the shelf between ~60 and ~140 mwd, the zone of swell and episodic storm wave disturbance, is herein defined as the *Middle Neritic Environment*. The *Deep Neritic Environment* in the case of southern Australia extends from ~140 mwd to the shelf edge, which lies at 200–220 mwd. Sediments in this zone are characteristically muddy. Beyond this depth the seafloor lies within the *Upper Slope Environment*, again a tranquil environment. The shallow upper slope to ~400 mwd is

the site of progressively decreasing epibenthic growth and below 400 mwd the seafloor is just sediment.

These defined zones do not strictly correspond to carbonate ramp environments as defined by Burchette and Wright (1992) because, although the physiographic setting is similar, hydrodynamics are different. Normal storm wave base is not the deepest high-energy element swell as wave base is generally deeper. Nevertheless, shallow, middle and deep neritic environments

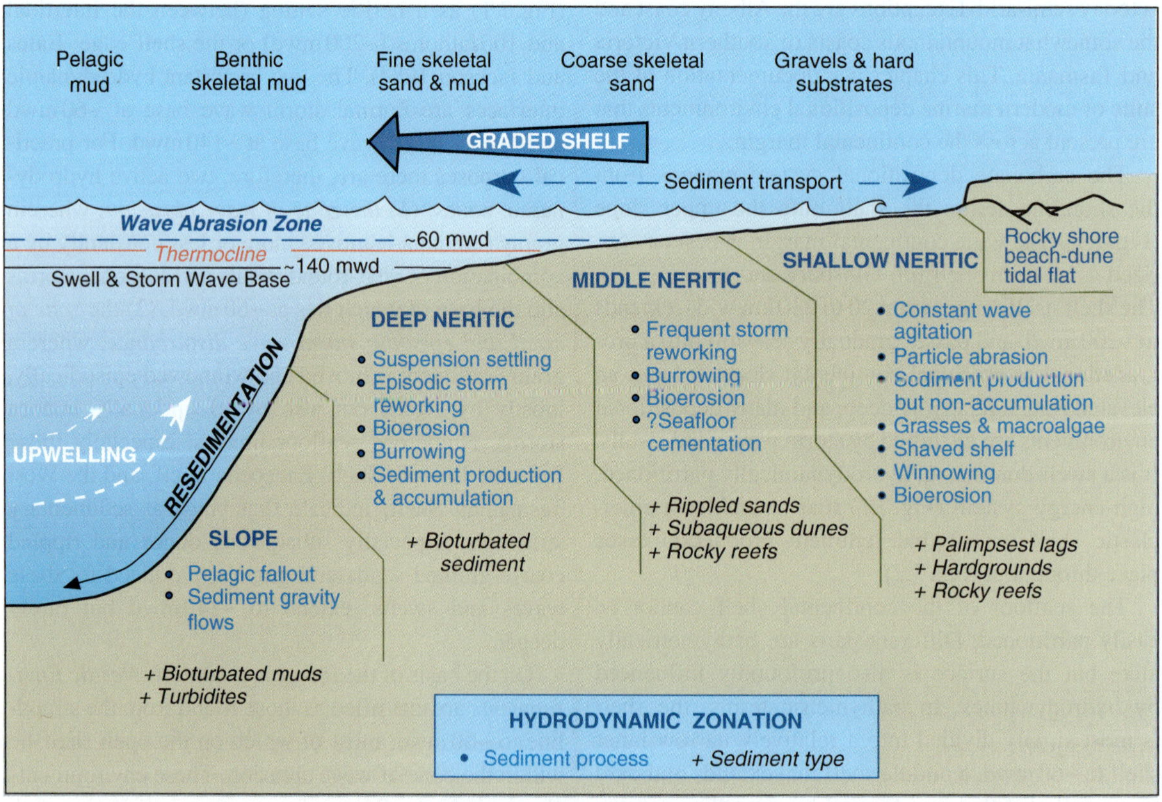

Fig. 7.2 A sketch summarizing the hydrodynamics, the neritic environments, sedimentary processes, resultant deposits, and character of the graded shelf

can, in a general way, be encompassed within inner, mid, and outer ramp settings respectively.

Depth of the photic zone, as expected, varies greatly across the spectrum of environments. It is <10 mwd over large areas of Spencer Gulf and Gulf St. Vincent where the water contains abundant suspended matter. By contrast, most of the open shelf has a deep photic zone, generally between 65 and 102 mwd, (as measured by growing coralline algae). When taking video images in the Great Australian Bight, the camera could be operated without lights to depths of 120 mwd. This illuminated seafloor wherein light is able to penetrate deeply into the water column is likely due to the low trophic resource levels and thus low planktic biomass.

A total of 13 neritic depositional environments are recognized (Figs. 7.3, 7.4, 7.5, 7.6). Names for some of these marine environments have, as is common practice, been borrowed from the terrestrial biological literature, e.g. turf, coppice, and scrub.

7.2 Shallow Neritic Environments

7.2.1 Setting

The shallow neritic environment (Fig. 7.1) extends to ~60 mwd and is wholly within the photic zone. The seafloor lies entirely above fairweather wave base, and so the open shelf is incessantly swept by waves and swells, particularly from the southwest.

The zone, because of the complex nearshore geography and geology of southern Australia, includes a wide variety of settings such as estuaries, small embayments, lagoons, and intrashelf basins (Figs. 7.3, 7.4, 7.5). Spencer Gulf, Gulf St. Vincent, and most of Bass Strait are intrashelf basins. These areas are, because of geography, somewhat protected from the open ocean swells and waves that sweep across the shelf. As a consequence, sediments can be muddy, influenced by

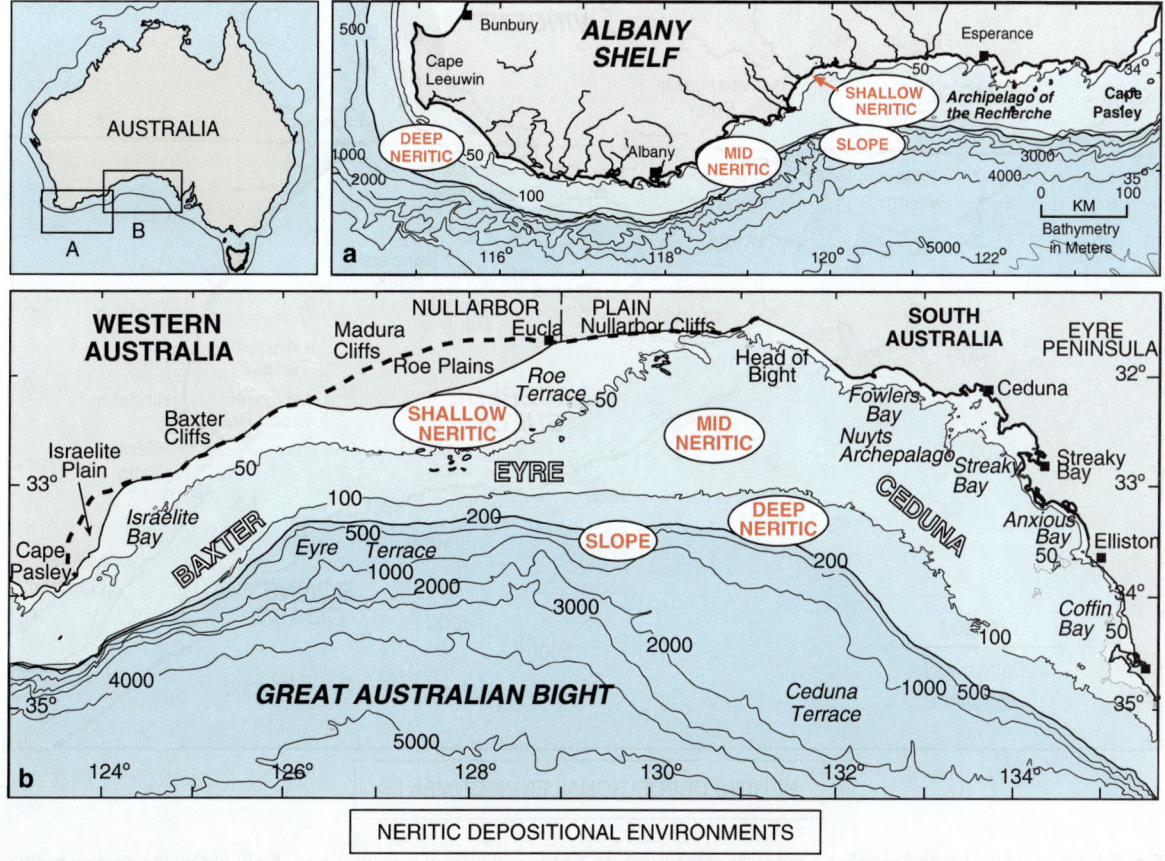

Fig. 7.3 Charts of the Albany and Great Australian Bight sectors of the southern continental margin (see inset map) highlighting the modern neritic zones in each sector

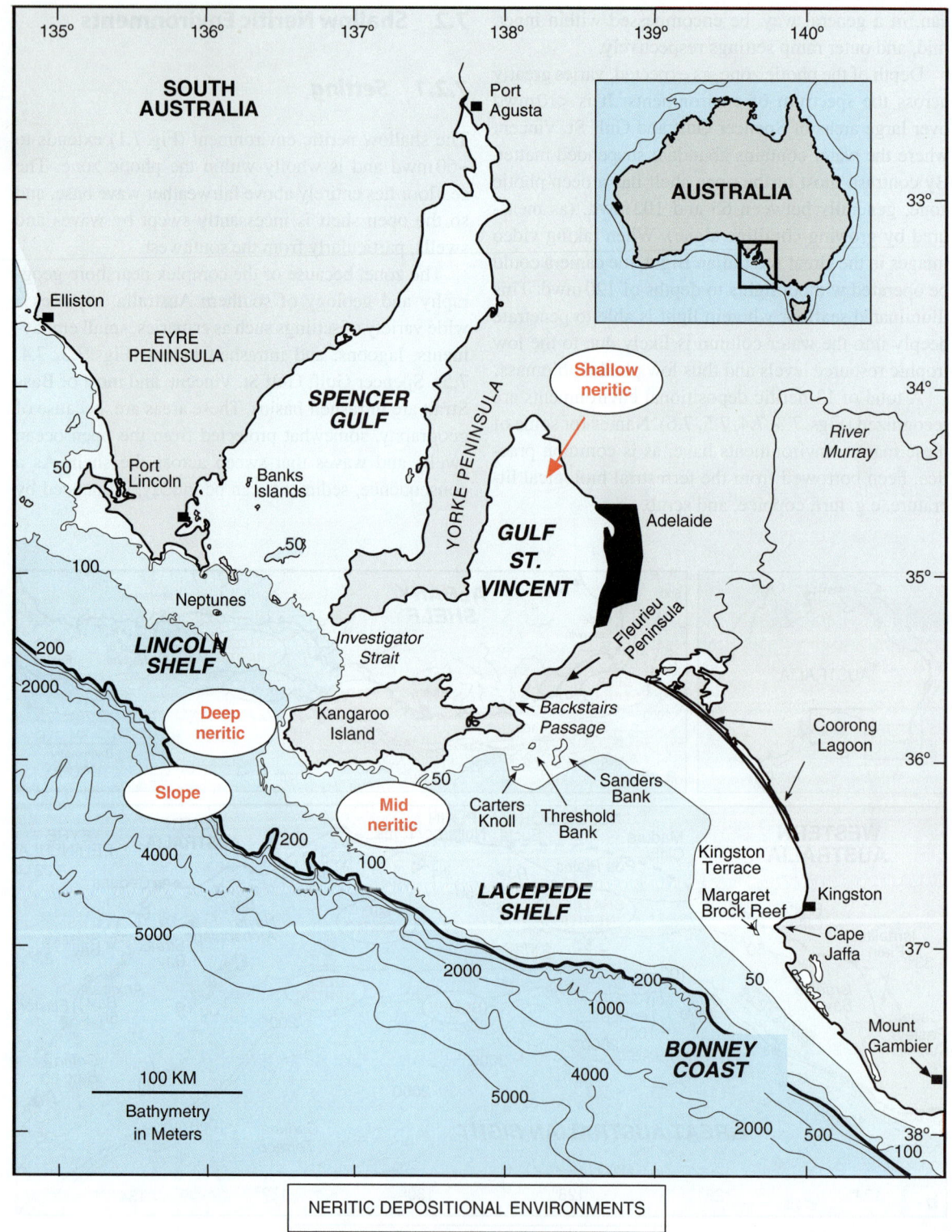

NERITIC DEPOSITIONAL ENVIRONMENTS

Fig. 7.4 Charts of the South Australian Sea sector of the southern continental margin (see inset map) highlighting the modern neritic zones

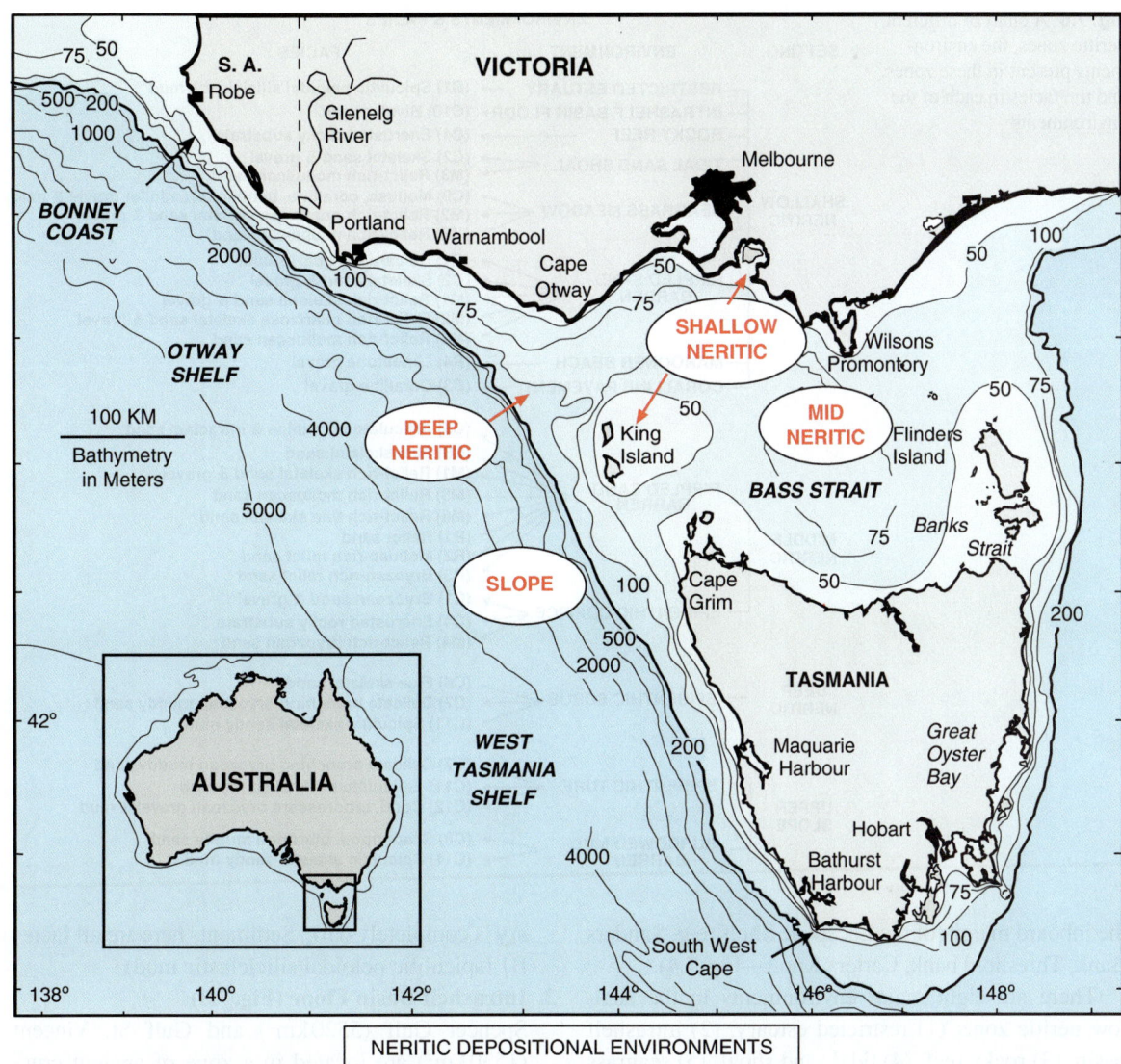

Fig. 7.5 Charts of the Southeastern Continental Shelf sector of the southern continental margin (see inset map) highlighting the modern neritic zones

terrestrial processes, and strongly affected by climate. Water clarity and thus seafloor illumination is reduced in many of these environments by suspended sediment or floating macroalgae.

The environment on the open shelf typically descends rapidly to ~40 mwd over distances of <5 km and then gradually flattens out to ~50 mwd. The shallow neritic environment is particularly narrow on the Albany Shelf, the Ceduna sector of the Great Australian Bight, and the Otway and western Tasmanian shelves. By contrast, in other areas it is a wide irregular to flat bedrock surface partially veneered with sediment,

such as the Roe Terrace in the Great Australian Bight (30–60 mwd–70 km wide), or the Kingston Terrace and Margaret Brock Reef on the Lacepede Shelf (12 mwd) (Figs. 7.3, 7.4). Where the shelf is adjacent to Precambrian crystalline rock cratons, there are numerous small submarine outcrops and groups of offshore islands e.g. Recherche Archipelago islands in the western Great Australian Bight, the Nuyts Archipelago in the eastern Great Australian Bight, the Althorpes and Banks islands in the Neptune Sector, and islands in Bass Strait. On the other hand, numerous highs adjacent to the Delamerian fold belt have created shallow subaqueous banks along

Fig. 7.6 A chart of different neritic zones, the environments present in these zones, and the facies in each of the environments

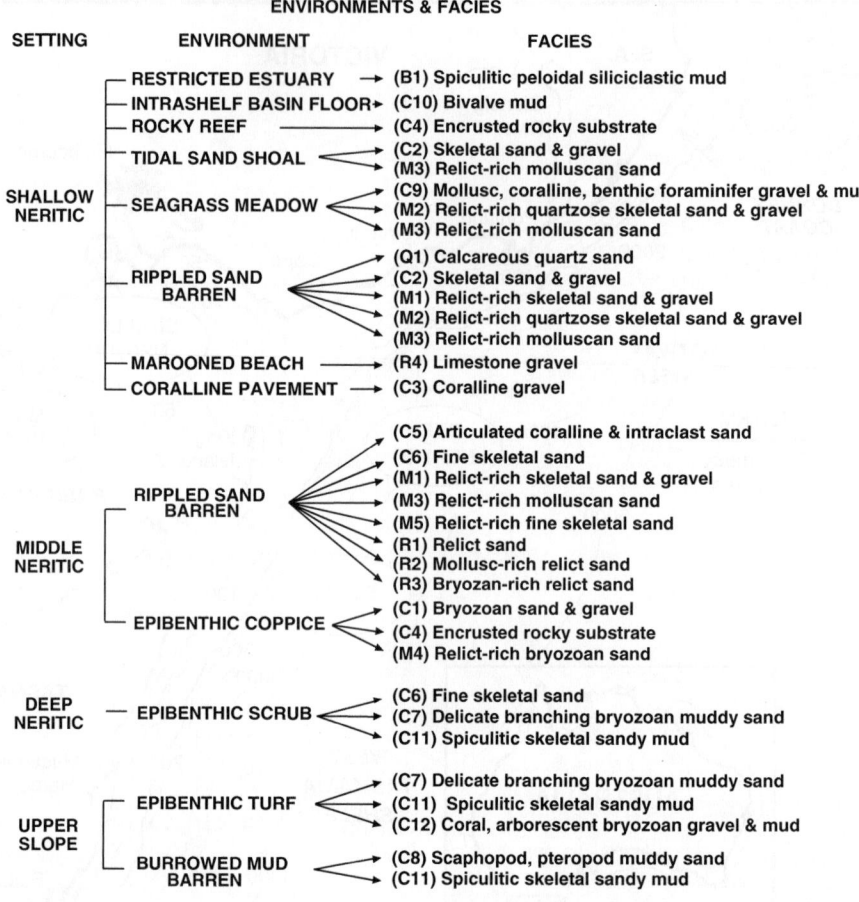

the inboard margin of the Lacepede Shelf (e.g. Sanders Bank, Threshold bank, Carters Knoll—Fig. 7.4).

There are eight major environments in the shallow neritic zone; (1) restricted estuary, (2) intrashelf basin, (3) rocky reef, (4) tidal sand shoal, (5) seagrass meadow, (6) rippled sand barren, (7) marooned beach, and (6) coralline pavement. A wide variety of lithofacies are present in these environments.

1. Restricted Estuary (Fig. 7.7)

This environment is characteristic of the southeastern continental margin (especially the Otway Coast and western Tasmania) where rainfall is higher than in the west and leads to fluvial runoff. The sedimentology is not well documented but sediments are siliciclastic-dominated. One particularly important example is the narrow, restricted estuary of Port Davy-Bathurst Harbour in western Tasmania where low-nutrient, dark tannin-rich brackish waters overlie normal marine seawater. Such tannins block out all light below 2–3 mwd and so the floor of the estu-

ary is completely dark. Sediments here are all facies B1 (spiculitic peloidal siliciclastic mud)

2. Intrashelf Basin Floor (Fig. 7.8)

Spencer Gulf (5520 km²) and Gulf St. Vincent (1530 km²) are located in a zone of ancient crustal weakness between the Paleoproterozoic Gawler Craton massif in the west and the Neoproterozoic -early Paleozoic deformed continental margin-foreland basin complex of the Delamerian Fold Belt in the east (Fig. 2.1). The modern gulfs themselves each lie atop Cenozoic rift basins that initially formed during the separation of Antarctica and Australia. The area is one of tectonic instability with seismicity continuing to the present day (Greenhalgh et al. 1994). Bass Basin is a large subcircular depression located over stretched Mesozoic crust.

Spencer Gulf & Gulf St. Vincent

The two gulfs (Fig. 7.4) are important depocenters on their own. The region also marks the transitional zone between largely warm-temperate, neritic,

Fig. 7.7 The Bathurst Harbour Estuary, southwestern Tasmania. (a) An areal view of the estuary looking inland, (b) an underwater image at a depth of 6 m, taken at mid-day in complete darkness using a flash to highlight the extensive sponge (S), octocoral (O), and bryozoan (B) community; image courtesy of C. M. Reid

Fig. 7.8 Intrashelf basin. (a) Shallow quartzose and biofragmental sand (facies Q1) with scattered seagrass and numerous bivalve shells (5 cm divisions on scale); 6 mwd Adelaide, (b) sparse seagrass, the semi-infaunal bivalve *Pinna* sp. and attached ascidian in facies Q1 sands, 3 mwd, Adelaide; exposed portion of shell is 8 cm high

marine environments to the west and cold-temperate shelf settings to the east and south. The embayments are characterized, because of the regional semi-arid climate, by relatively little terrigenous clastic sediment input, and by strong seasonal anti-estuarine circulation such that waters range from near normal marine at the mouth to hypersaline at the head. This aspect, their overall shallowness, (<60 mwd) and the fact that they are protected from open ocean swells, is reflected in the extensive development of carbonate sediments (Shepherd and Sprigg 1976; Burne and Colwell 1982; Gostin et al. 1984; Cann and Gostin 1985; Belperio et al. 1988; Cann et al. 1988). The gulf floors are largely low energy depocenters (Fig. 7.8). The floors of these basins are covered with facies C10, bivalve muds.

Bass Strait

Bass Strait (Fig. 7.5), located between Tasmania and mainland Australia, has a maximum present depth of 83 m. Two ridges with numerous islands that extend along the eastern and western margins restrict water circulation somewhat in the deeper parts of the basin resulting in a relatively tranquil, albeit shallow, depositional environment. The basin floor is covered with facies C10, bivalve muds, but with a significant pelagic component.

3. Rocky Reef (Fig. 7.9)

Offshore from erosional shorelines, there are generally numerous submerged rocky substrates in the form of elongate ridges, buttresses, or isolated outcrops called 'reefs'. These hard substrates (facies C4) are typically sites of prolific epibenthic growth. The seafloor biota is in most instances dominated by algae, particularly macrophytes, and mobile or epibenthic attached invertebrates.

Shallow subtidal rock surfaces in warm temperate environments west of Cape Jaffa are covered by articulated coralline algae (*Corallina*) to ~3 mwd and encrusting corallines to ~100 mwd. These surfaces are habitats of grazing gastropods (including

Fig. 7.9 Rocky reef environment (facies C4). (**a**) A meter-high rock ridge heavily encrusted by soft red algae, coralline algae, and sponges (S) together with asteroids (sea-stars)–*Tosia australis*, (E) 6 mwd, Edithburgh, Gulf St. Vincent, (**b**) the brown alga *Ecklonia* (E), soft green algae, and coralline algae on a rocky substrate, 4 mwd, Gulf St. Vincent, width of image = 1.2 m, (**c**) oscular sponges growing on a bare rock ridge, Gulf St. Vincent 8 mwd, crab carapace is 12 cm wide, (**d**) rocky substrate heavily encrusted with coralline algae (C) and local bryozoans (B) 4 mwd, Otway Shelf (image is 2 m across), (**e**) grazing echinoid (6 cm wide) on a surface covered with numerous encrustations of coralline algae, 5 mwd, Noarlunga Reef, off Adelaide, (**f**) rock surface encrusted with coralline algae (C) and articulated zooidal bryozoans (B), cm scale, 26 mwd, Lacepede Shelf. Images A, C, E courtesy of J. Bone-George

Haliotis [abalone]), and a few solitary corals, all of which are partially obscured by the prolific canopy of macroalgae (*Ecklonia, Cystopora, Sargassum,* and *Caulerpa*). *Ecklonia radiata*, the most abundant form, grows to 35 mwd in clear waters but shallows to 5 mwd in turbid waters.

Cool-temperate inshore rocky habitats east of Cape Jaffa and around western, southern and eastern Tasmania are, by contrast, more highly vegetated by macroalgae than elsewhere in Australia, probably because of cool temperatures and somewhat higher nutrients. The euphotic zone extends to ~40 mwd

(Harris et al. 1987) with shallow portions dominated by the growth of large kelp not present in the west.

This environment not only forms the seafloor but it also caps a series of banks that rise above the middle neritic siliciclastic sand plain on the Lacepede Shelf (Sanders Bank, Threshold Bank, and Carters Knoll) and ridges in Investigator Strait. Outcrops are also sporadically present on parts of the outer shelf. Images show surfaces veneered by a wide variety of encrusting and rooted (particularly encrusting) bryozoans, multiple layers of crustose coralline and peyssonnelid algae, coppices of rigid branching coralline algae, small clusters of brachiopods, and intergrowths of serpulid worms. Between and rising above this low-level biota is an abundant and extraordinarily diverse fauna of demosponges. In shallow areas of the euphotic zone there are also, rooted between all these organisms and rising into the water column in great profusion, a variety of macroalgae. These algae are the ephemeral substrates for other encrusting calcareous algae, bryozoans, and foraminifera. Particles from this sediment factory are shed onto the surrounding seafloor.

4. **Tidal Sand Shoal**

Shallow seafloor environments swept by strong tidal currents are either bare rock surfaces dusted with a lag of coarse, relict-rich molluscan sand (facies M3) and skeletal sand and gravel (facies C2) or the same sediment swept into impressive subaqueous dunes. They are well developed at the head of Spencer Gulf and along both sides of Bass Strait. The sediment comes from the bare seafloor, seagrass meadows, upstanding seafloor rock ridges (as described above) and growth on the dunes themselves. The sands host numerous vagrant (free-living) bryozoans.

5. **Seagrass Meadow (Fig. 7.10)**

Southern and Western Australia are two of the world's most extensive areas of temperate seagrass growth (Kirkman and Kuo 1990). The most luxuriant of these seagrass meadows occur in the clear, shallow, sheltered gulf waters of Spencer Gulf and Gulf St. Vincent (Lewis et al. 1997; Edyvane 1999) but are also important on inner parts of the shelf.

The carbonate sediment is generated on the grass itself and on the seafloor. The calcareous biota is dominated by coralline algae, infaunal bivalves, gastropods, bryozoans, serpulid worms, small ben-

Fig. 7.10 Seagrass environment (facies C9). **(a)** Extensive *Posidonia* growth at 4 m depth, Port Victoria, Spencer Gulf, grass blades ~1 cm wide, **(b)** a ripped up mat of individual *Amphibolis* plants illustrating the extensive intertwined root system that binds sediment together in the shallow neritic zone, Great Australian Bight, **(c)** seagrass having been eroded from just offshore during winter storms and now piled on the beach, Marion Bay, Investigator Strait

thic foraminifers, and locally by, symbiont-bearing large benthic foraminifers (Burne and Colwell 1982). This results in two types of sediment, carbonate mud from the disintegration of encrusting coralline algae and delicate bryozoans together with lime sand from the breakdown of articulated (geniculate) corallines (especially *Metagoniolithon* and *Amphiroa*), bryozoans, and benthic foraminifers (facies C9). The most abundant calcareous epiphytes are coralline algae with non-geniculate (encrusting) types more abundant than geniculate (articulated) types. The seagrass blades also support a prolific and distinctive benthic foraminifer biota characterized by the encrusting form *Nubecularia*, the large form with symbionts *Peneropolis*, and the small forms *Discorbis* and *Elphidium*. Epiphyte abundance peaks at water depths of ~10 m (James et al. 2009). These, together with articulated coralline algae, scabs of encrusting corallines, serpulid worms, and encrusting bryozoans comprise an active sediment-producing factory. The molluscs living beneath and on the grass blades and stems together with mud trapped by the extensive roots system results in texturally variable sands and muds with floating large bivalves and gastropods. There are local areas of facies M3 (relict-rich molluscan sand) and facies M2 (relict-rich quartzose skeletal sand), reflecting modern grass growth on relict carbonate and quartzose sand respectively.

6. **Rippled Sand Barren (Fig. 7.11)**

These areas of open sand form extensive subaqueous sand plains or occur between rocky substrates and grass beds. They extend into deep water and merge with sands of the mid-neritic zone (see below). Overall the seafloor is devoid of obvious algae or epifaunal invertebrates except for those attached to shells or lithoclasts or semi-mobile brachiopods and large, low diversity bryozoan assemblages attached to the pioneer bryozoans (e.g. *Adeona*) and articulated zooidal species. There is, however, a ubiquitous bivalve infauna. The sediments are typically quartzose or relict-rich and so are either facies Q1 (calcareous quartz sand), facies M1 (relict-rich skeletal sand and gravel), facies M2 (relict-rich quartzose skeletal sand and gravel), or facies M3 (relict-rich molluscan sand). Large areas of the outer neritic environments are floored by facies C2 (skeletal sand and gravel).

7. **Marooned Beach (Fig. 7.12)**

These coarse deposits of stranded and relict limestone pebbles and large bivalves (facies R4) occur in both shallow- and deep-water settings. In shallow water, at discrete depths of 40 and 65 m on the Lacepede and Otway shelves, they form semi-continuous, narrow sublinear, shore-parallel bands. Seafloor images show ridges and sheets of limestone pebbles and bivalve coquinas with bryozoans growing on many of the larger clasts. They are similar in composition to sediments adjacent to the

Fig. 7.11 Underwater images–shallow neritic environment, Great Australian Bight Baxter Sector, compass scale 12 cm diameter, (**a**) rippled sand barren with a sparse sponge and bryozoan biota, 60 mwd, (**b**) rippled sand barren with large ripples composed of coarse bryozoan fragments, 60 mwd

modern shoreline and are thus interpreted as paleostrandlines (James et al. 1992).

The gravels are also present in much deeper water on the Bonney Shelf, between 150 and 400 mwd, as a complex and poorly sorted mixture of relict, shallow water bioclasts, limestone pebbles, quartz sand, and modern deep-water components such as corals, bryozoans, sponges, and brachiopods. They form a dm-thick capping veneer over upper slope muds and are interpreted as a relict facies related to conditions of deposition during the LGM or earlier lowstands (Boreen and James 1993).

8. **Coralline Pavement**

The seafloor in these relatively shallow-water environments can be entirely covered by coralline algae in the form of either rhodoliths or rigid-branching, fruticose, subspherical nodules, both of which can be up to 10 cm in diameter (facies C3—coralline gravel). Rhodoliths are typically an interlayered consortium of corallines (especially *Sporolithon*) and encrusting foraminifers (usually *Gypsina* and related types). These dense or branching corallines can be so densely packed that they form a "billiard

Fig. 7.12 Paleostrandline (facies R4). (**a**) Seafloor image at the foot of a gravel ridge with numerous bivalves (*Plebidonax* sp.), living articulated bryozoan at arrow; 57 mwd, Lacepede Shelf (compass scale 12 cm diameter), (**b**) lithoclast gravel of brown Fe-stained pebbles and cobbles with scattered recent bryozoans and mollusc fragments; 36 mwd, Lacepede Shelf

observed on video under natural light without strobes to 90 mwd. In Spencer Gulf these corallines are best developed in 20–40 mwd along the western margin (Fuller et al. 1994) and around topographic highs on the shelf proper (James et al. 1997). Similar facies surrounding highs on the Lacepede Shelf are composed not only of the branching coralline *Metagoniolithon* sp., but numerous bryozoans, especially large pieces of *Adeona* sp. (James et al. 1992). Epifaunal bivalves are prominent in all areas.

7.3 Middle Neritic Environments

7.3.1 Setting

This vast region that lies between ~60 and ~140 mwd (Figs. 7.3, 7.4, 7.5) is roughly similar everywhere with differences only at the local scale. It is a storm-dominated setting with the seafloor swept by long period swells and winter storm waves. Seafloor gradients range from 0.11° on narrow shelves to 0.03° on wide shelves. There are two major environments: the (1) rippled sand barren and (2) epibenthic coppice.

At the large scale, the shelf is a flat submarine plain that ranges from a few tens of kilometers to several hundred kilometers in width. It extends seaward from ~60 mwd, which is usually the foot of a small, discontinuous but persistent terrace, across the major terrace at ~120 mwd and onto the slope in front. The top of the deeper terrace that marks the position of the LGM eustatic lowstand can be as shallow as 100 mwd, whereas the base can be as deep as 130 mwd. This environment comprises most of the Albany Shelf, much of the Baxter sector and most of the Ceduna sector of the Great Australian Bight where it is locally more than 200 km across. The environment also forms most of the Lacepede Shelf and wide areas of Bass Strait. It is narrow on both the Otway and western Tasmania shelves.

The seafloor is typically a series of variably subdued terraces, 5–30 km in width, each of which has an irregular ridge-like topography with 5–10 m relief at the seaward, distal end. The most conspicuous terraces are at 95 mwd (most prominent terrace), 110 mwd, and 115 mwd.

ball–like" carpet several nodules thick on the seafloor. The broken twigs of fruticose types can lodge between the nodules to form a dense, almost impenetrable pavement. This pavement in turn acts much in the same way as a rocky reef and supports a community of macrophytes, bryozoans (mostly fenestrates), epifaunal bivalves (especially pectens), and gastropods. All such environments are present in semi-protected settings along the western sides of the Great Australian Bight, Spencer Gulf, and the Lacepede Shelf, away from the incessant swell that comes from the southwest.

They are particularly abundant around Cape Pasley and eastward inboard off Israelite Bay, with the densest accumulations at ~30 mwd. Deposits on the western Roe Terrace are both rhodoliths and branching corallines together with a profusion of red, green and brown macroalgae in water depths of 35–50 m. Particularly large rhodoliths were

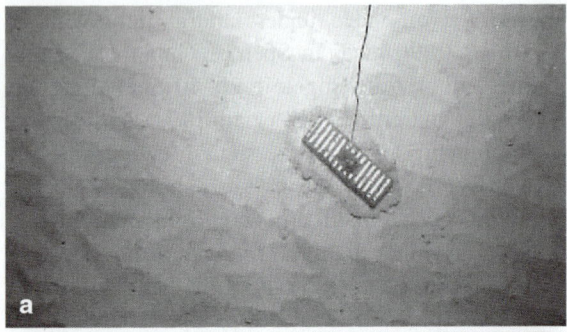

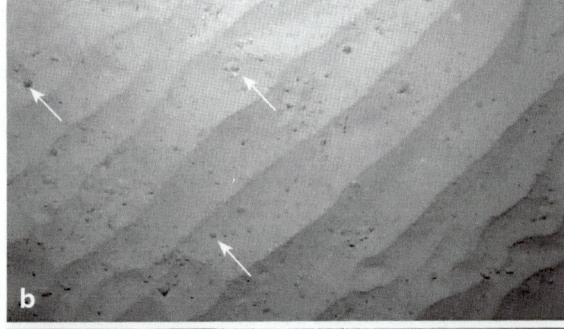

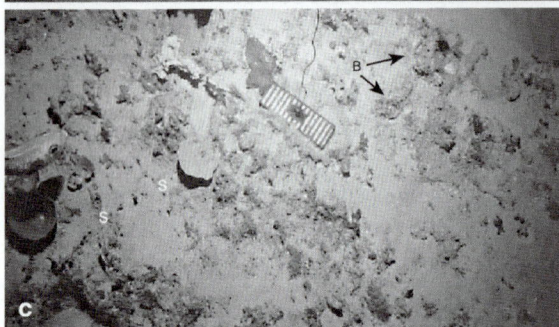

Fig. 7.13 Underwater images–shallow neritic environment, Lincoln Shelf, **(a)** rippled sand barren with no seafloor biota except a few small Free-living vagrant bryozoans (*dark spots*); scale in cm divisions, 96 mwd, **(b)** rippled sand barren of wave-rippled fine sand and numerous vagrant bryozoans (*arrows*); ripple crests are ~20 cm apart; 126 mwd **(c)** rocky reef with abundant sponge (S) and bryozoan (B) growth; scale in cm divisions, 110 mwd

Much of this environment reflects recent tectonic inversion and so is flat to gently seaward-dipping Cenozoic limestone bedrock variably veneered with sediment. Small outcrops of crystalline rock again form islands on this segment of the shelf where there is Precambrian basement onshore. Particularly prominent are the Recherche Archipelago on the Albany Shelf, the islands and outcrops at the entrance to Spencer Gulf and Gulf St. Vincent, and the island complexes marginal to Bass Strait. The islands create small sub-environments of shallow mixed carbonate-siliciclastic sedimentation while the submerged bedrock is an ideal

substrate for numerous calcareous invertebrates and macrophytes.

1. Rippled Sand Barren (Figs. 7.13, 7.14)

Wide areas of the open shelf are covered with rippled sand, in some places wholly carbonate in other parts mostly quartz. A distinguishing feature throughout is the scarcity of epibiota, such that in some regions it has been referred to as a biological desert. This is particularly true on the middle parts of the Lacepede Shelf and the eastern Great Australian Bight where much of the seafloor is rippled quartz

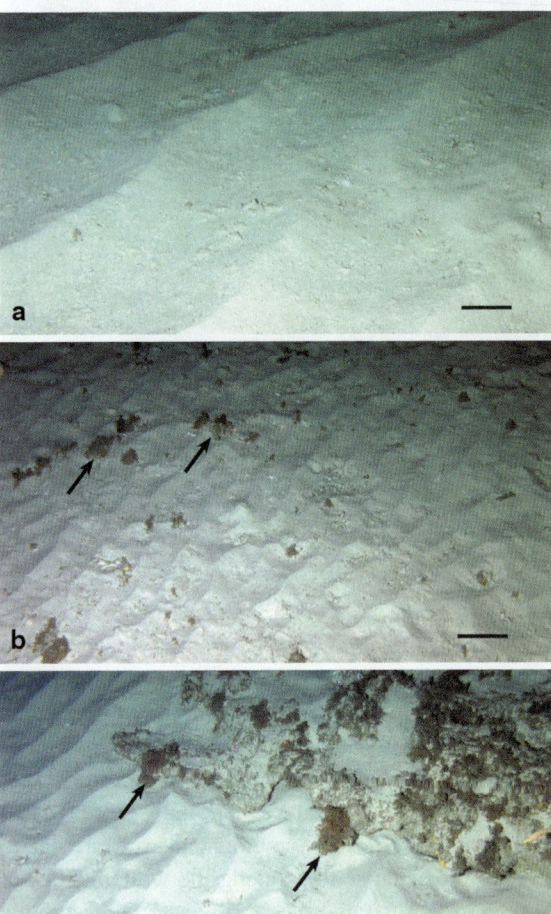

Fig. 7.14 Underwater images–middle neritic environment, NW Tasmania, all scales 20 cm. **(a)** Rippled sand barren with no obvious biota, 138 mwd, **(b)** rippled sand barren–wave rippled sand with scattered bryozoans (*arrows*), mostly articulated zooidal growth forms, 128 mwd, **(c)** rippled sand barren adjacent to a rocky reef with abundant bryozoan growth (*arrows*), 129 mwd. Images courtesy of Commonwealth Scientific and Industrial Research Organization

or biofragmental sand respectively with little obvious biota except free-living (vagrant) bryozoans and brachiopods (mostly *Anakinetica cumingi*) with an infauna dominated by bivalves (especially *Donax* and *Katalysia*). The seafloor is smooth and monotonous. Bottom photographs show clean sediments with small straight-crested wave ripples (10–15 cm apart) and no large organisms other then burrowing anemones and scattered rooted bryozoans (*Adeona* and *Parmularia*), various articulated zooidal types, and local vagrant forms (*Lunulites*).

Small local areas do, however, support a 20–40% organism cover. The most obvious large benthos comprises sponges, generally upright flat planar to digitate, oscular forms growing together, and a wide variety of bryozoans. Such an association suggests that the epibenthic biota is growing on a hard substrate dusted with a thin sediment veneer.

Sediment from three megafacies are represented (Fig. 7.6); megafacies C=facies C6 (fine skeletal sand) and facies C5 (articulated coralline and intraclast sand), megafacies M=facies M1 (relioct rich skeletal sand and gravel, facies M3 (relict-rich molluscan sand) and facies M5 (relict-rich fine skeletal sand), megafacies R=facies R1 (relict sand), facies R2 (mollusc–rich relict sand), and facies R3 (bryozoan-rich relict sand).

2. **Epibenthic Coppice (Figs. 7.15, 7.16, 7.17)**

The seafloor in this environment is a patchwork of rocky reefs and open rippled sands that support a variably prolific epibenthic biota. The environment begins at ~60 mwd and usually extends to ~140 mwd, but can continue all the way through the deep neritic to the shelf edge at ~220 mwd. The term coppice is borrowed from the terrestrial literature where it means a small forest of underwood and small trees grown for periodic cutting (Oxford Dictionary) and is commonly used in the marine paleobiological literature.

The calcareous biota that inhabits the hard substrates (facies C4) produces sediment that is shed onto the surrounding seafloor where it is moved by waves and swells. Such sediment eventually covers the hard seafloor, and if thick enough, the biota is dominated by rooted instead of attached organisms (James et al. 1992, 1997). The calcareous invertebrates also grow as epizooal attachments on the calcareous and noncalcareous epibenthos (cf. Hageman et al. 2000). Although relatively slow growing, these flourishing

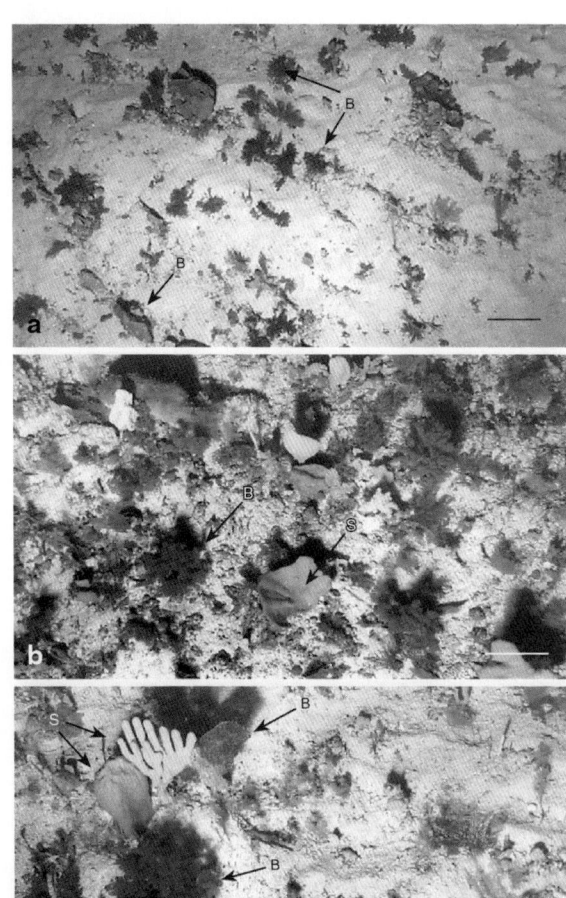

Fig. 7.15 Underwater images–middle neritic environment, Lacepede Shelf. These images from 100 mwd show the range in environments over a short distance; all scales=20 cm. (**a**) Mixed rippled sand and scattered epibenthic growth, mainly bryozoans (B) and sponges, (**b**) prolific epibenthic coppice with bryozoan (B) and sponge (S) epibenthic growth, (**c**) close view of large sponges (S) and bryozoans (B) in the epibenthic coppice

subaqueous coppices and biogenic islands are akin to the reefs of tropical carbonate settings, but they do not accrete in the same way because the invertebrates are primarily sediment producers rather than hard substrate encrusters. The sediment of all grain sizes is generated by the growth of epifaunal molluscs, sponges, bryozoans, hydroids, ascidians, and their epizooal counterparts.

This observation has led to the "shaved shelf" concept (James et al. 1994: see Chap. 12) whereby the rocky substrate produces sediment that is partly swept away on a regular basis by the energetic

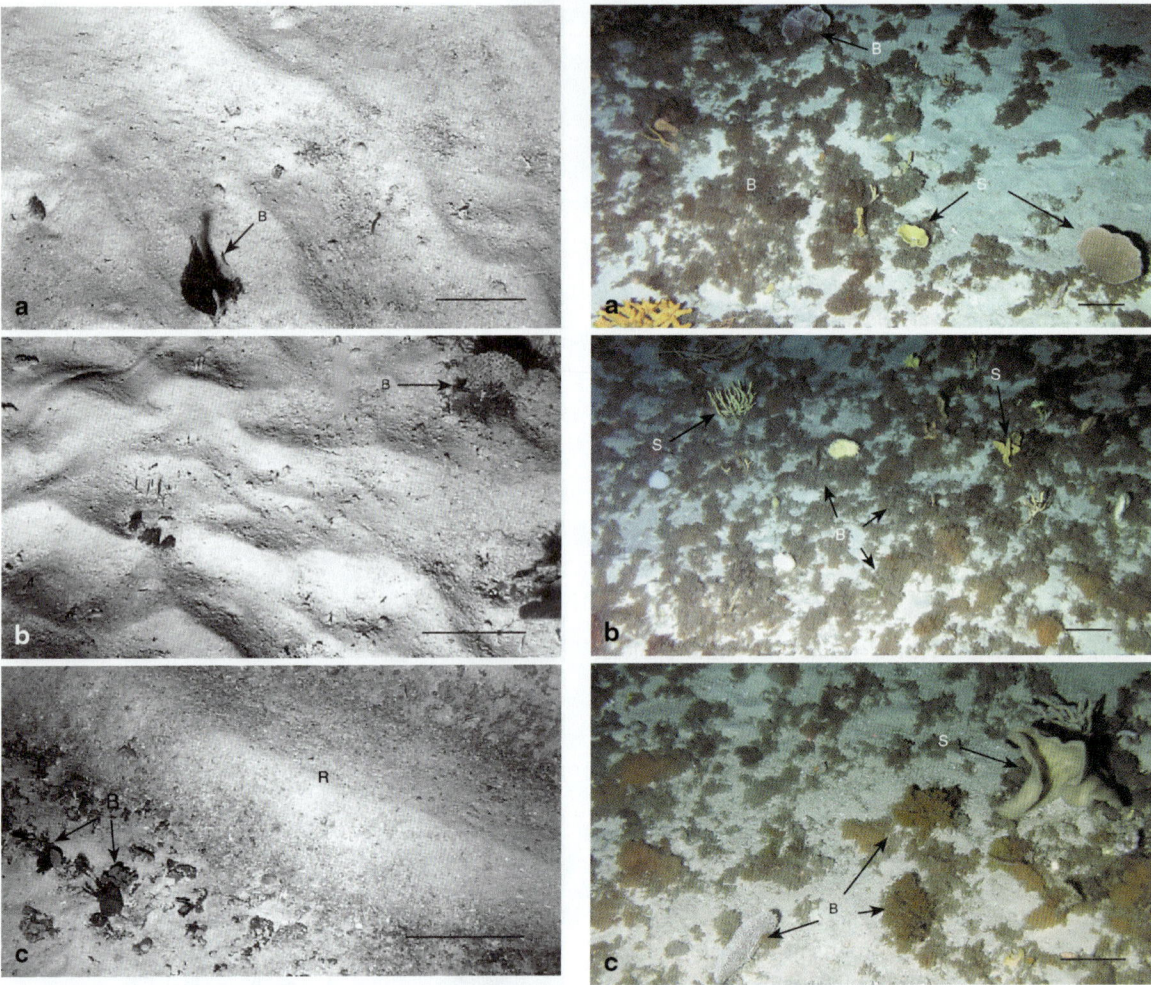

Fig. 7.16 Underwater images–middle neritic environment, Lacepede Shelf; all scales = 20 cm. (**a**) Rippled sand barren with degraded bedforms and a large (10 cm high) fenestrate bryozoan (B), *Adeona* sp., 120 mwd, (**b**) ripped sand barren of somewhat degraded interference ripples and fenestrate bryozoans (B–*upper right*) 120 mwd, (**c**) hard seafloor (rocky reef–*lower right*) with a bryozoan epibiota (B) and a starved wave ripple (R) composed of coarse bryozoan fragments, 100 mwd

Fig. 7.17 Underwater images along a 20 m transect–middle neritic environment, NW Tasmania, scale bar = 20 cm. (**a**) Epibenthic coppice of bryozoans (B) and sponges (S), dominated by articulated zooidal and delicate branching types, together with skeletal sand, 137 mwd, (**b**) extensive epibenthic coppice dominated by bryozoans (B) and subsidiary sponges (S), 134 mwd, (**c**) epibenthic coppice of bryozoans (B) and large sponges (S), 137 mwd. Images courtesy of Commonwealth Scientific and Industrial Research Organization

hydrodynamic environment. In deeper environments, where sediments are intermittently moved, at least during winter months, angularity of Holocene biofragments and lack of fines argues for winnowing and offshelf transport of fine-grained material as the major physical process. The role of renewed hard substrates produced by synsedimentary cementation in this setting (cf. Nelson and James 2000) has yet to be evaluated.

Precision depth profiles on outer parts of the shelf indicate large subaqueous dune fields in a zone of

seaward-dipping clinoforms. Bottom photographs show rock substrates covered by variable thicknesses of coarse sediment with environments ranging from rippled sands to starved ripples to an open hard seafloor dusted with sediment. Open rippled sands are barren except for local areas where the surface is 20–40% covered by organisms. The most obvious benthos is sponges, generally upright flat planar to digitate oscular forms growing together. The sediments themselves are mainly facies C1 (bryozoan sand and gravel) and facies M4 (relict-rich bryozoan sand) reflecting bryozoans as the most important sediment producer and the presence of relict grains in shallow parts of the environment.

7.4 Deep Neritic Environments and Facies

7.4.1 Setting

The outer part of the shelf that extends down from ~140 mwd to the shelf break at ~220 mwd is comparatively narrow (Figs. 7.3, 7.4, 7.5), usually <10 km wide, a more steeply dipping and irregular slope than above, and of comparative complexity. The seafloor is affected only by exceptionally intense storm waves and as a result the sediment is muddy. It is relatively narrow on most shelves except on the Lincoln Shelf, and Eyre and Ceduna sectors of the Great Australian Bight.

The shallower part, between ~140 and 180 mwd, is usually a series of seaward-dipping strata that are imaged on seismic profiles as the updip, truncated edges of prograding clinoforms. Recent sediment veneers these outcrops and is ponded between the hogbacks. Repeated attempts to sample the ridges were mostly unsuccessful but fishers with strong dredges have recovered hardgrounds (see Chap. 11). Discontinuous terraces with irregular topography and small cliffs at their leading, seaward margins locally interrupt these strata at depths of 140–150 mwd.

The lower part of the deep neritic between ~160 and ~220 mwd along most of the margin is a complex region of terraces, mass wasting, or continuing clinoforms. In most areas, the edge is marked by an increase in slope at about 150–160 mwd that becomes locally precipitous at ~220 mwd. There are numerous scarps on the outer part of the edge, which, on seismic, are clearly the exposed hanging walls of normal faults or enormous slumped masses, some of which have moved downslope several hundred meters as documented in detail by von der Borch and Hughes-Clarke (1993) for the Lacepede Shelf. An exception is the shelf edge in the central part of the Great Australian Bight with a prominent terrace at 200 mwd that grades with depth into a prograding upper slope sediment wedge (Feary and James 1998). There is only one environment here, an epibenthic scrub.

7.4.2 Epibenthic Scrub

This environment is one of irregular rocky substrates, intervening areas of sediment, and variable topography all in close proximity. Biogenic carbonate production is also variable. Images of the seafloor depict a burrowed, muddy sediment with scattered bryozoans and sponges to an environment wherein the seafloor is completely covered with these same organisms. Regardless, it is the same community, a low-level, mid-tier, epibenthic assemblage, not as tall as the mid-neritic coppices, but locally as prolific (Figs. 7.18, 7.19, 7.20).

Fig. 7.18 Underwater images–deep neritic environment–Lacepede Shelf. (a) Close view of delicate articulated branching bryozoans and scattered small bryozoans (B) comprising an epibenthic scrub, scale 5 cm, 175 mwd, (b) epibenthic scrub of small bryozoans and a burrowed muddy sediment, scale 10 cm, 175 mwd, (c) burrowed mud barren, 10 cm scale, 140 mwd

Fig. 7.19 Underwater images–deep neritic environment–NW Tasmania; scale on all images = 20 cm. **(a)** Epibenthic scrub and burrowed mud barren of small, low bryozoans (B) and muddy sediment, 175 mwd, **(b)** epibenthic scrub of bryozoans (D) including fenestrates (F), sponges (S), and carbonate mud, 154 mwd with evidence of trawling damage in upper LHS, **(c)** epibenthic scrub of bryozoans (B) and abundant yellow sponges (S), 191 mwd. Images courtesy of Commonwealth Scientific and Industrial Research Organization

Fig. 7.20 Underwater images–deep neritic environment. **(a)** Epibenthic turf of small bryozoans and mud, compass 12 cm diameter, Great Australian Bight, Baxter Sector, 200 mwd, **(b)** bryozoan-dominated (B) epibenthic turf scoured by a fishing trawl (T), and illustrating rapid recolonization by burrowers, NW Tasmania, scale bar 20 cm, 203 mwd, **(c)** dense epibenthic turf of small bryozoans (B) and sponges (S), NW Tasmania, scale bar 20 cm; 207 mwd. Images B and C courtesy of Commonwealth Scientific and Industrial Research Organization

Availability of different local environments all below fair-weather wave base that are only occasionally perturbed by storms results in a diverse benthic biota that generates a spectrum of muddy carbonate sediment that largely accumulates in place. These sediments are nevertheless augmented by material swept seaward from the mid-neritic environment. Thus, although much of the sediment is autochthonous, some is also allochthonous.

The biota is a cornucopia of small sponges, bryozoans, molluscs, hydroids, and ascidians, with locally numerous ophiroids but dominated by sponges and bryozoans.. There is a complete spectrum of bryozoan growth forms but delicate branching forms, robust forms such as *Adeona* sp., and articulated branching (particularly articulated zooidal) types are most conspicuous. Sponges and bryozoans act as substrates for numerous other epibionts resulting in a strongly tiered community. The bryozoan *Celleporaria* sp., for example, typically encrusts and grows over oscular sponges, resulting in a distinctive hollow, tubular morphology

(Hageman et al. 1998). The resulting muddy sediment contains a diverse association of calcareous biofragments, large and small, that are augmented by a minor pelagic component. Resulting sediments form facies C6 (fine skeletal sand), facies C7 (delicate branching bryozoan muddy sand), and facies C11 (spiculitic skeletal sandy mud).

7.5 Upper Slope Environments and Facies

7.5.1 Setting

The shelf edge as such is either the transition into a near-vertical upper slope, a complex zone of irregular topography that passes downward into the upper slope, or an imperceptible change in seafloor gradient. Regardless, it is not marked by a specific environment, but instead exhibits a change in the seafloor biota.

The shallowest parts of the slope to 500 mwd, the deepest sites investigated, comprise two segments, (1) an upper part (<400 mwd) with a calcareous benthic biota, and (2) a lower part (>400 mwd) that is only sediment. The upper slope deeper than ~200 mwd is in places the extension of a collapsed margin that has been incised by submarine canyons. In the Eyre sector, however, the upper slope is a large prograding wedge of Plio-Pleistocene carbonate (James and von der Borch 1991; Feary and James 1998; Feary et al. 2000b), with buried bryozoan mounds (James et al. 2000) that downlaps on to and form part of the Eyre Terrace. These mounds are also likely present, on the basis of mounded seafloor reflectors, off the Lacepede Shelf. A Pliocene–Holocene system canyon system (von der Borch 1968; Leach and Wallace 2001; Exon et al. 2005; Hill et al. 2005) is incised into the upper slope in many sectors.

Surficial sediments beyond the shelf edge are a mosaic of deeper water recent muddy deposits that locally overly or are mixed with older relict deposits that formed during the LGM lowstand and during the early stages of the subsequent sea level rise. The most widespread of these facies is composed of numerous erect, rigid robust bryozoans, especially *Celleporaria*, that form a floatstone to rudstone texture with an otherwise diverse coral and bryozoan biota (facies C12).

The other stranded facies is a series of inner shelf calcareous quartz sands (facies Q1). The deeper-water, fine-grained Holocene facies range from delicate branching bryozoan muddy sands (facies C7) to spiculitic skeletal sandy muds (facies C11). The coarse-grained stranded facies are locally sites of abundant azooxanthellate coral growth.

7.5.2 Epibenthic Turf

Below storm and swell wave base on the upper slope to ~400 mwd the seafloor is monotonously similar, an undulating, muddy substrate with a short, diminutive, low diversity, but abundant bryozoan and sponge biota (facies C7 and C11). This community decreases with increasing water depth (Figs. 7.21, 7.22).

7.5.3 Burrowed Mud Barren

The seafloor below 400 mwd is muddy sediment partly pelagic, partly autochthonous, and partly derived from above that is burrowed by a variety of infaunal and neritic organisms (facies C8 and C11). There are few obvious epibenthic invertebrates (Fig. 7.22).

7.6 Synopsis

The seafloor in Spencer Gulf, Gulf St. Vincent, Bass Strait, and on the open shelf is herein divided, on the basis of modern hydrodynamics, into four major environments (1) shallow neritic, 0–60 mwd and above storm wave base, (2) mid neritic, 60–140 mwd, and above swell wave base, (3) deep neritic, 140–220 mwd and largely below any hydrodynamic disturbance, and (4) upper slope, >220 mwd. Shallow neritic seafloor environments are the most diverse. The mud-line lies at ~140 mwd on the open shelf. The various facies present in each large environment are outlined in Fig. 7.6.

1. The shallow neritic segment comprises eight environments. Carbonate muds with a conspicuous bivalve biota blanket the seafloor in the two gulfs and Bass Strait. Although there is not much information about estuaries, the restricted estuary at

Fig. 7.21 Underwater images–upper slope environment–NW Tasmania; all scales=20 cm. **(a)** Epibenthic turf of numerous small bryozoans (B) and sponges (S), 263 mwd, **(b)** epibenthic turf and irregularly burrowed carbonate mud, 355 mwd, showing little difference from that 100 m above, **(c)** epibenthic turf scoured by trawling (T) but exposing the underlying mud substrate, 261 mwd. Images courtesy of Commonwealth Scientific and Industrial Research Organization

Fig. 7.22 Underwater images–upper slope environment–NW Tasmania–all scale bars = 20 cm. **(a)** Epibenthic turf of scattered diminutive bryozoans (B) and small sponges (S), 359 mwd, **(b)** burrowed mud barren–sparse epibenthic turf and flathead fish, 425 mwd, **(c)** burrowed mud barren, 500 mwd. Images courtesy Commonwealth Scientific and Industrial Research Organization

Bathurst Harbour in western Tasmania is floored with spiculitic siliciclastic muds, largely the result of carbonate dissolution. Seagrass meadows, rooted in muddy sediments rich with molluscs, coralline algae fragments, and benthic foraminifers occur in the two large gulfs and just offshore in many open shelf settings. Coralline algal pavements are present in the two gulfs and in nearshore environments, often in the same locations as seagrass beds. Tidal currents locally sweep the carbonate sediments into subaqueous dunes, especially in areas of focused

tidal flow, such as the upper reaches of Spencer Gulf and Investigator Strait.

Rock outcrops are common in these shallow locations and are populated by a profuse epibenthic growth of macroalgae and invertebrates, especially bryozoans and sponges that commonly act as substrates for another higher, more elevated tier of calcareous organisms. The biota on these rocky reefs shed carbonate particles onto the surrounding seafloor. Paleostrandlines composed of beach cobbles, boulders, and molluscs form slightly elevated, hard habitats for epibenthic growth on an otherwise open sandy seafloor and likewise act as islands of carbonate sediment production. These locations are surrounded by vast areas of barren rippled carbonate and siliciclastic sand, especially on the open shelf wherein there is little epibenthic growth, but a locally abundant infaunal bivalve fauna.

2. The middle neritic seafloor, in spite of its vast extent, comprises only two environments, an open sand barren similar to the one present in the inner neritic environment and an epibenthic coppice. The coppice is similar to the shallower rocky reef environment but more irregular with hard substrates and sand deposits in close proximity over large areas. It has been called a 'shaved shelf system' wherein epibenthic growth is razored off by the incessant vigorous swell regimen to produce calcareous sediment.

3. The deep neritic environment extends across and over the shelf edge. The shelf edge ranges from a gradual deepening of the seafloor, to eroded rocky limestone clinoforms, to numerous erosional scarps.

The environment here is called an epibenthic scrub because it is a low-level, mid-tier biota that is mostly surrounded by muddy burrowed sediment. Again, sponges and bryozoans dominate the living biota with bryozoans comprising a conspicuously mixed robust and delicate biota. Sediments have a noticeable pelagic component.

4. The upper slope environment is one of muddy sediment with a progressively decreasing epibenthic biota downslope. An epibenthic turf of short diminutive scattered sponges and bryozoans interspersed with burrowed muddy carbonate forms a monotonous seafloor across wide areas. This substrate passes downward on the slope into muddy burrowed sediment that is partly allochthonous from above and partly pelagic.

This muddy seafloor is locally interrupted by relict sediment in the form of (1) bryozoan gravels (stranded deep-water reef mounds) with modern deep water corals growing on them and, (2) spreads of relict quartzose–carbonate, shallow neritic sediments and beach sands.

Chapter 8
The Southwestern Shelf

8.1 Introduction

The western part of the southern Australian continental margin is a huge shelf that extends some 2000 km from Cape Leeuwin (115° E) in the west to Coffin Bay (135° E) in the east (Figs. 8.1, 8.5). The shelf stretches from 33° to 35° S. It comprises two quite different sectors; the 800 km long relatively narrow Albany Shelf and the 1200 km long wide Great Australian Bight. The region is dominated by two massifs, the Yilgarn Craton in the west and the Gawler Craton in the east, with the Eucla Basin in between.

Waters that cover the shelf range from warm temperate to sub-tropical with most currents flowing from west to east. Neritic waters are strongly influenced by seasonal variations in flow of the Leeuwin Current, summer heating and evaporation and local seasonal upwelling. There is virtually no freshwater flow onto the shelf except for small ephemeral streams along the Albany coast; there is no fluvial input in the Great Australian Bight.

Surface sediments are highly variable and almost wholly heterozoan carbonate except near the coast where they are locally quartzose. Photozoan elements are present in areas on the Albany Shelf. The Great Australian Bight is distinguished by large areas of relict and stranded sediment.

8.2 The Albany Shelf

8.2.1 General Attributes

The Albany Shelf stretches from Cape Leeuwin to Cape Pasley. The relatively narrow shelf, 40–75 km across, lies at a latitude of 34°–35° S. The shelf is a transitional environment between the long meridonal continental shelf facing the Indian Ocean off Western Australia to the west and the vast latitudinal Great Australian Bight to the east. The seafloor in most places descends quickly to 60 mwd and is relatively deep throughout except in the east where the shelf is dotted by more than 100 islands and 1500 shoals of the Recherche Archipelago (Fig. 8.1).

The shelf west of Albany is narrow and terraced. The narrow shoreface is fronted by an escarpment that descends steeply to a prominent terrace at ~50 mwd. This relatively flat seafloor is, in turn, terminated by another escarpment that drops from 50 to 120 mwd and another terrace that lies between 120 and 170 mwd at the shelf edge only ~50 km from shore. The shelf is much wider (75 km) east of Albany across Esperance Bay and in the Recherche Archipelago.

The shelf has not been sampled in the same detail as other parts of the continental margin. Nevertheless, when combined with recent surveys in the Recherche Archipelago (Ryan et al. 2007, 2008) and local shallow water studies (Cann and Clarke 1993) a general picture of the sedimentology emerges.

8.2.2 Oceanography

The narrow continental shelf is exposed to the most extreme wave energy of the entire continental Australian coastline (Hemer 2006). The relatively warm Leeuwin Current dominates Albany shelf waters during winter when it flows strongest and there is a distinctive west-to-east temperature gradient. Inshore seawater temperatures range from ~22°C in summer to a spring minimum of 16° (Ryan et al. 2008). Winter shelf-edge currents may exceed 50 cm s^{-1} (1.0 knot). The absence

N. P. James, Y. Bone, *Neritic Carbonate Sediments in a Temperate Realm*,
DOI 10.1007/978-90-481-9289-2_8, © Springer Science+Business Media B.V. 2011

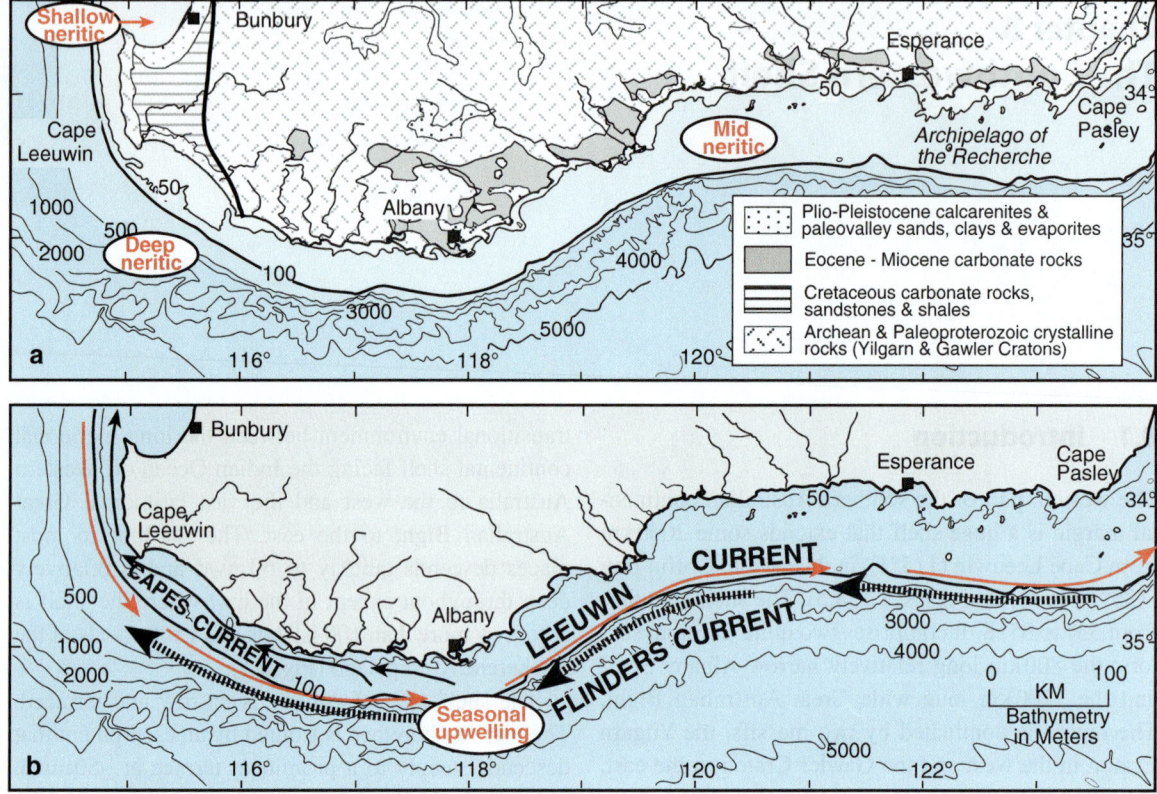

Fig. 8.1 Charts of the Albany Shelf: **(a)** Geology of the mainland, localities mentioned in the text, and bathymetry. **(b)** Major ocean currents in the region

of upwelling despite a favourable summer wind system has been attributed to blocking by the Leeuwin Current. There is, nevertheless, cool water on the inner shelf during summer and it flows westward from near Albany, around Cape Leeuwin and northward up the coast of western Australia: the Capes Current (Pearce and Pattiaratchi 1999). These waters are ~1°C cooler than the Leeuwin Current overall, come from the base of the Leeuwin Current itself and are upwelled onto the inner shelf (Gersbach et al. 1999), 5 to 9 times each summer. Such waters are low in both nitrogen and phosphorous. Chlorophyll-a values range from 0.1 mg m^{-3} outboard to 0.3 mg m^{-3} inboard in summer and 0.4 to 0.5 mg m^{-3} in winter (low mesotrophic) (Fig. 8.2).

8.2.3 Marginal Marine

The low-lying coast of mostly Archean and Proterozoic crystalline rocks at the southern end of the Yilgarn Craton is locally veneered with Paleogene sediments and sedimentary rocks (Gammon et al. 2000; Gammon and James 2003). The thin, irregular cap comprises Bremer Basin Eocene siliciclastic, calcareous, carbonaceous, and spiculitic sediments and Eucla Basin limestones. Eocene paleovalleys inland are now large saline lakes. Gentle Neogene uplift has elevated the Paleogene strata such that they now form a gently seaward dipping plane.

The coast is a series of somewhat sheltered quartz sand beaches (Fig. 8.3) between crystalline basement promontories (Sanderson et al. 2000). Prolonged Phanerozoic erosion of basement rocks has created a series of inselberg-like hills of crystalline rock that are onlapped by Cenozoic strata onshore and surrounded by Holocene sediments offshore. Small rivers flow to the coast but they have low discharge rates because of the Mediterranean like semi-arid climate wherein annual rainfall (~619 mm year^{-1}) is exceeded by evaporation (1716 mm year^{-1}). Estuaries are small and inactive during the dry summers.

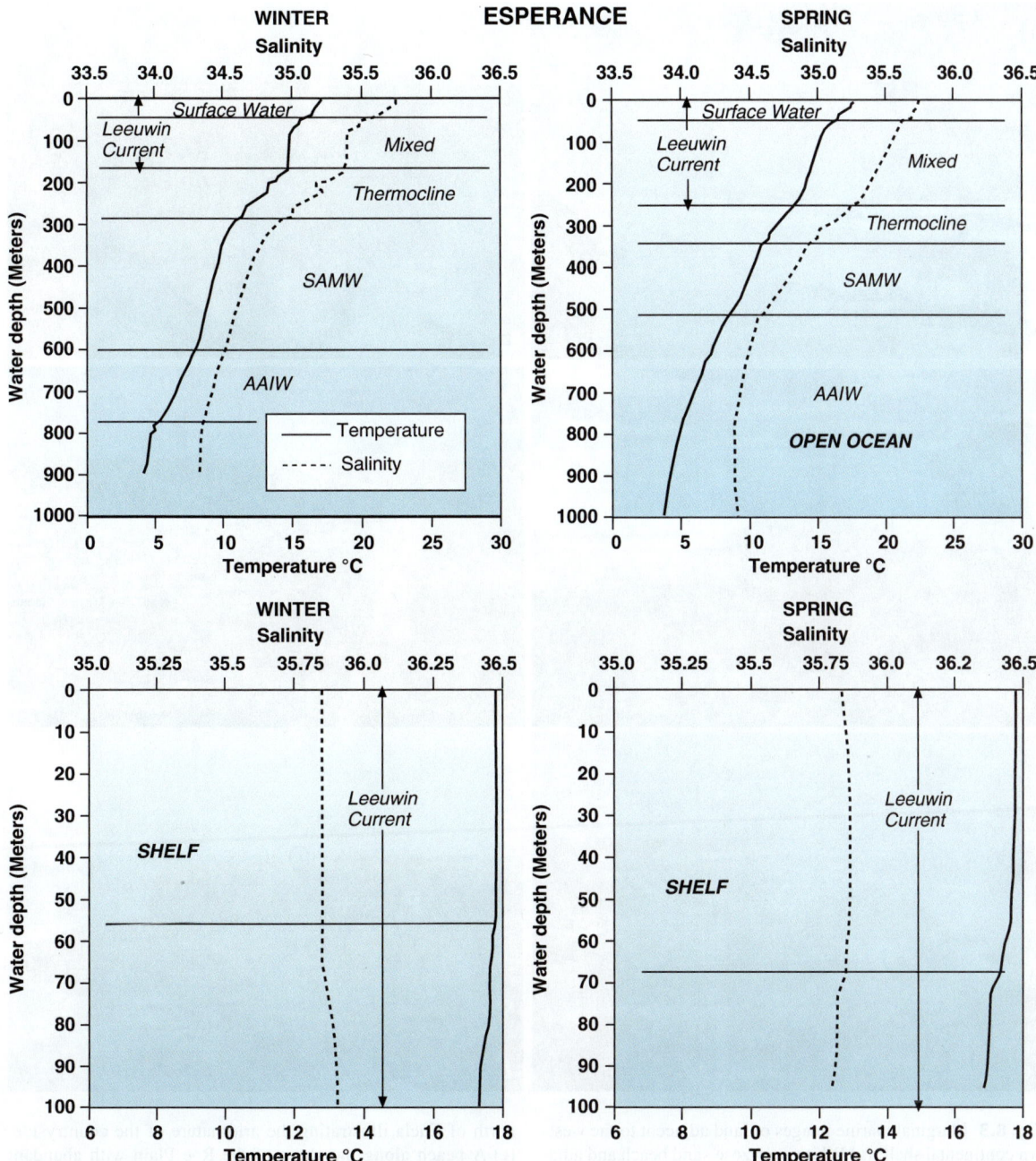

Fig. 8.2 Temperature and salinity profiles for the upper 1000 m of the open ocean water column (*top*) and the upper 100 m of the shelf water column (*bottom*) during summer and winter seaward of Esperance. *AAIW* Antarctic Intermediate Water, *SAMW* Subantarctic Mode Water. The mixed and surface layers are composed of sub-tropical surface water modified by input from the Leeuwin Current, summer heating, and winter cooling. (Compiled from CSIRO (2001) Data Trawler, Marine and Atmospheric Research (CMAR) Data Center, Commonwealth Scientific and Industrial Research Organization)

Fig. 8.3 Marginal marine images of land adjacent to the western continental shelf. (**a**) White quartzose sand beach and adjacent low dunes along Israelite Bay. (**b**) The steep 90 m high Nullarbor Cliffs eroded into relatively soft Eocene to Middle Miocene limestone. (**c**) Degraded 60 m-high paleoseacliffs (Madura Cliffs) some 30 km inland from the present shoreline ~30 km west of Eucla. (**d**) Surface of the Nullarbor Plain north of Eucla illustrating the arid nature of the countryside. (**e**) A beach along the shore of the Roe Plain with abundant saltbush and other halophytes (shrubs ~2 m high). (**f**) Seacliff ~25 m high composed of gently dipping Jurassic sandstone (*J*) overlain by Pleistocene Bridgewater Formation aeolianite (*P*), person for scale (*circled*); near Streaky Bay on the coast of Eyre Peninsula

8.2.4 Shelf Sediments

Overview: The megafacies distribution is relatively simple. Beaches are mainly composed of white ortho-quartzitic sand but such sediments do not extend far offshore because the shallow neritic zone is so narrow. Much of the middle neritic seafloor is veneered by relict-rich sediment with local areas of wholly relict

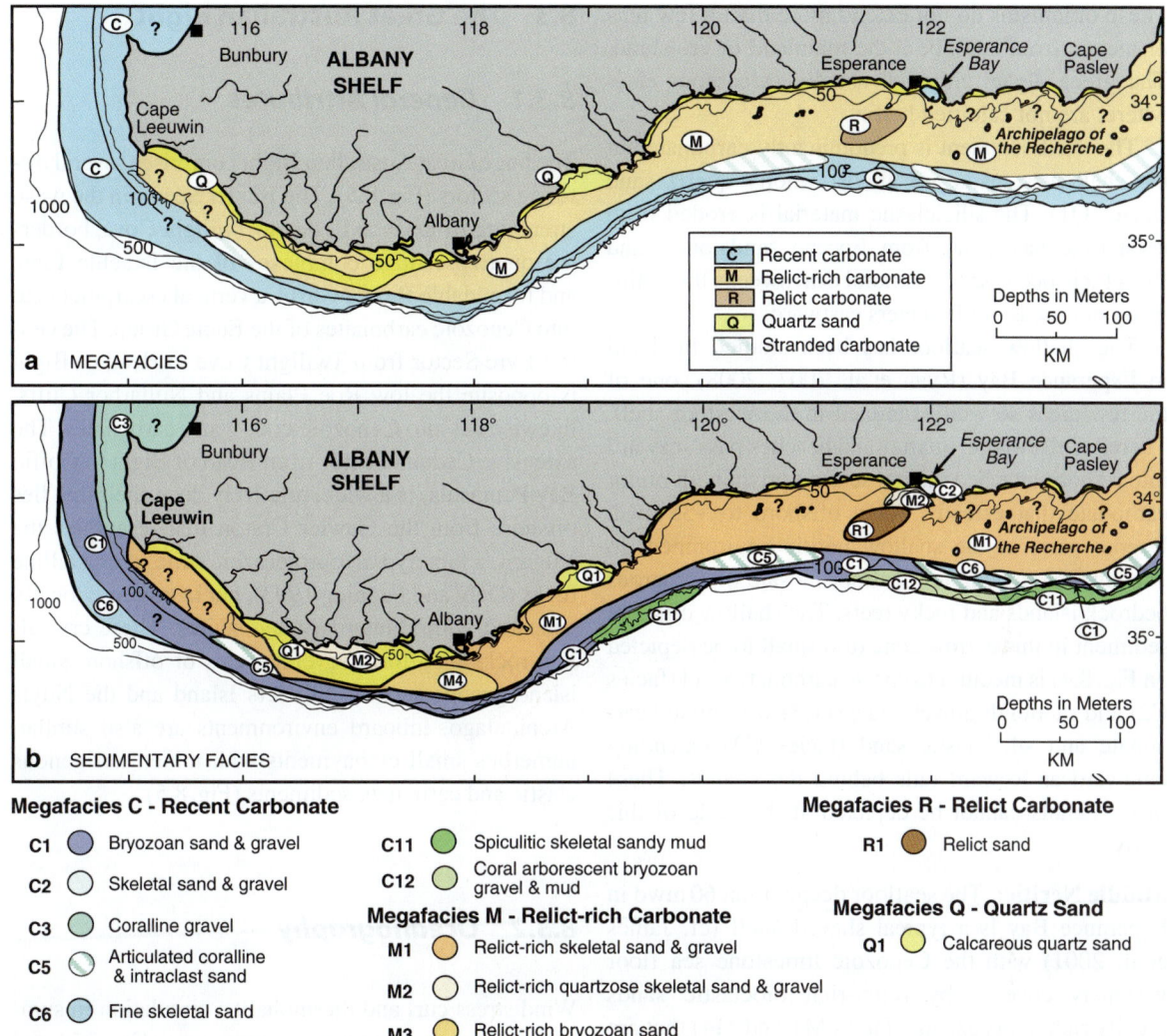

Fig. 8.4 Charts of neritic sedimentary facies on the Albany Shelf. **(a)** Megafacies illustrating the quartz-rich sediments along and adjacent to the coastline, the inner neritic relict-rich deposits and the mid-to outer neritic zone covered with recent carbonates, some of which are locally rich in stranded particles. **(b)** Composition of surface sediment facies

material. Most of the outer neritic shelf is covered by recent carbonate with areas of stranded sediment (Fig. 8.4).

Shallow Neritic: The seafloor ranges from sublinear and irregular rocky reefs to sediment covered sand plains to islands surrounded by mixed siliciclastic and carbonate sediment, particularly in the Archipelago of the Recherche. Extensive rocky reef communities are widespread and seagrass banks lie in sheltered inshore lagoons leeward of offshore islands and rocky reefs. Barnacles are particularly conspicu-

ous on rocky intertidal headlands. Macroalgae thrive on rocky substrates from intertidal environments to ~50 mwd as extensive populations of *Ecklonia radiata* and *Sargassum* sp. with mixed assemblages of foliose, green, red, and brown algae. Seagrass banks are present in sheltered environments from 0 to 45 mwd. Many areas shallower than 20 mwd, especially in Esperance Bay, support a sub-tropical biota of platy zooxanthellate corals (*Plesiastrea versipora*, *Coscinaraea* ssp.), large benthic foraminifers (e.g. *Marginopora vertebralis*), green algae (*Halimeda cuncata*, *Rhipiliopsis peltata*) and large gastropods.

These organisms do not extend more than a few tens of meters from land, be it the mainland or an island. The green algae, although usually calcareous elsewhere, are not calcified here.

The shelf sediment is predominantly carbonate but the shallowest parts are almost wholly quartz sand (facies Q1). The siliciclastic material is eroded from crystalline basement, from Eocene sandstones, and from Pleistocene sands. Littoral drift sweeps the grains many hundreds of kilometers eastward.

The shallow seafloor has been studied in detail in Esperance Bay (Ryan et al. 2007, 2008), one of the few areas so well examined in the western shelf. There, as elsewhere, quartz sand beaches pass seaward into a shore-attached quartzose sand prism with minor carbonate that extends ~4 km offshore to ~35 mwd. Deeper parts of the shallow neritic environment to ~60 mwd are a complex seafloor mosaic of numerous bedrock islands and rocky reefs. The shallow offshore sediment in this narrow zone (too small to be depicted on Fig. 8.4) is medium to coarse carbonate sand (facies C2) and rhodolith gravels (facies C3) with mixed carbonate and siliciclastic sand (facies C2) extending landward as leeward tails behind the islands. These facies details cannot be depicted at the scale of this study.

Middle Neritic: The seafloor deeper than 60 mwd in Esperance Bay is a typical shaved shelf (cf. James et al. 2001) with the Cenozoic limestone sea floor variously covered by relict-rich bioclastic sands locally rich in bryozoans (facies M1 and M4) that are swept into large subaqueous dunes (Ryan et al. 2008, this study).

Deep Neritic: Clean bryozoan sands (facies C1) characterize the outer neritic seafloor in most regions. This sector contains one of the most prolific sponge-bryozoan communities of any along the southern margin. Areas of stranded coralline algal sands (facies C5) with variable proportions of bryozoan fragments and pecten shells occur east, south, and west of Esperance. Outermost shelf and upper slope facies are mostly fine bioclastic sands (facies C6), spiculitic skeletal sandy muds (facies C11) and coral, arborescent bryozoan gravel and mud (facies C12). Planktic sub-tropical and temperate foraminifers overlap in this zone, highlighting the complex effects of the warm Leeuwin Current and the cold Flinders Current.

8.3   The Great Australian Bight

8.3.1   General Attributes

The huge Great Australian Bight comprises several different sectors (Fig. 8.5). The Baxter Sector in the west, stretching from Cape Pasley to Twilight Cove, borders the quartzose dune complexes of the Israelite Plain and formidable Baxter Cliffs, a vertical escarpment cut into Cenozoic carbonates of the Eucla Group. The central Eyre Sector from Twilight Cove to Head of Bight is opposite the low Roe Plains and Nullarbor Cliffs, likewise cut into Cenozoic cool-water carbonates. The extensive Ceduna Sector from Head of Bight to Coffin Bay Peninsula, is a wide, relatively deep shelf that lies offshore from the Gawler Craton that resembles the Yilgarn, a largely Paleoproterozoic suite of crystalline rocks (Daly and Fanning 1993), but in this case locally veneered with Quaternary aeolianites. These crystalline rocks also form several series of offshore small island groups such as Flinders Island and the Nuyts Archipelago. Inboard environments are also similar, numerous small embayments with mixed terrigenous clastic and carbonate sediments (Fig. 8.5).

8.3.2   Oceanography

Wind stress curl and thermohaline circulation in summer can result in an anticyclonic gyre (Herzfeld and Tomczak 1997; Middleton and Platov 2003) with coastal currents flowing northward along the Eyre coast and west across the top of the Great Australian Bight (Marshallsay and Radok 1972). These coastal winds also act to lower sea-level, drive the westward currents, and upwell water toward the coast. The thermocline water in summer has temperatures above 15°C and is upwelled near the coast but downwelled above the shelf break (Figs. 8.5, 8.6, 8.7).

By contrast, the prevailing and stronger winter westerlies act to raise sealevel near the coast, drive a coastal current (15–30 cm s^{-1}) from west to east, generate an east-flowing current at the shelf edge (Godfrey et al. 1986), and give rise to downwelling. Passing fronts and low pressure systems every 3–12 days lead to mixing and cooling of the waters resulting in a deep 150–250 mwd

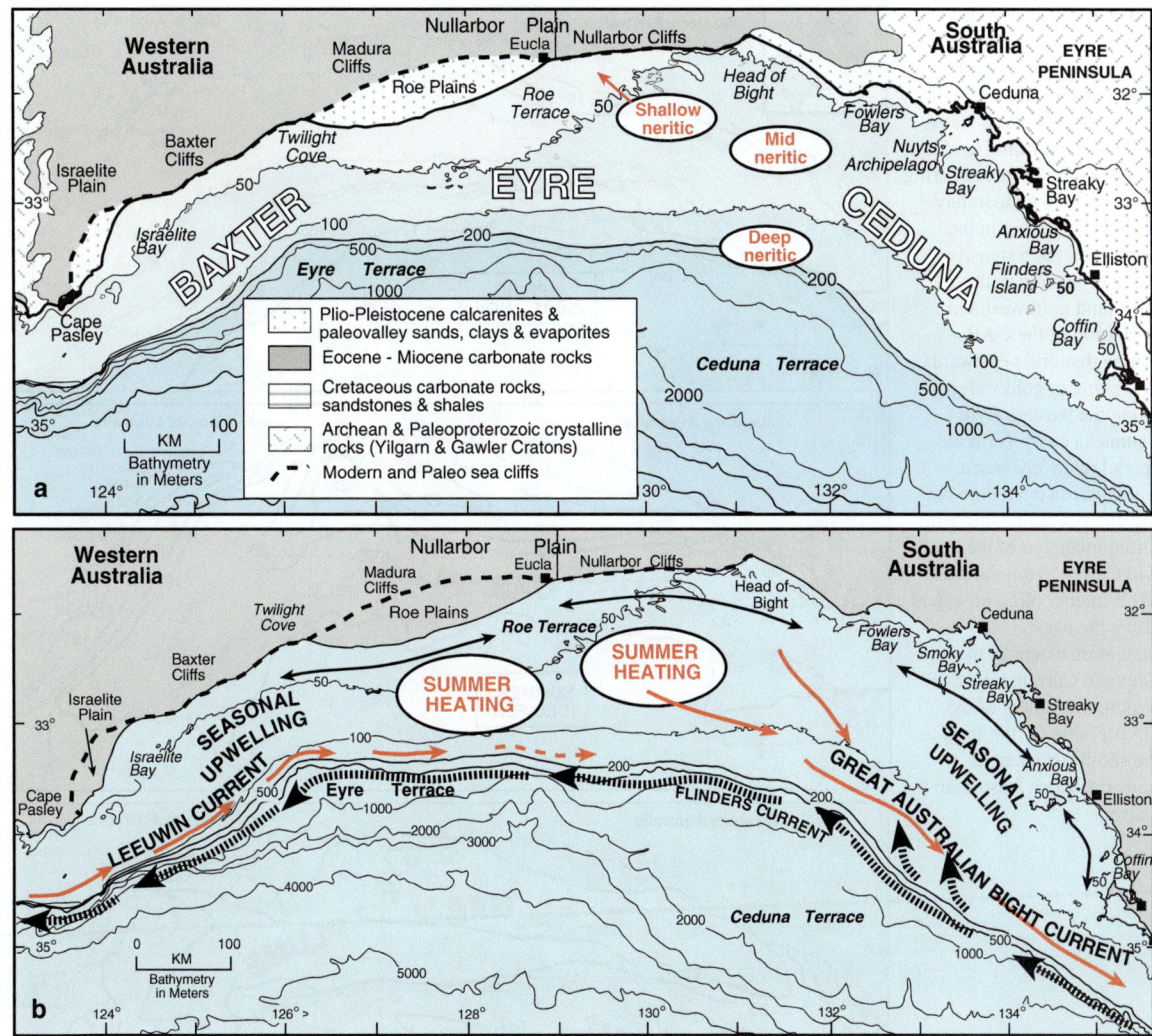

Fig. 8.5 Charts of the Great Australian Bight. (**a**) Geology of the mainland, bathymetry, and named sectors. (**b**) Main ocean currents, water circulation patterns, areas of summer heating, and seasonal upwelling

surface mixed layer. Waters near the shelf break are downwelled during winter to about 200 mwd and a surface mixed layer extends to ~75 m; water temperature over the shelf break is ~15°C. There is a seasonal thermocline at 50–70 mwd during the transition to summer. Shelf water seasonality is detailed by Herzfeld (1997).

Spring: Spring waters on the shelf are well mixed overall, with sea surface temperature (SST) ~17°C. There is strong heating on the shelf, which is especially manifest in the NW and in embayments along the Ceduna Sector, where waters are warm (18°C–19°C) to depths of ~30 m.

Summer: Continued heating during summer generates a zone of warm and saline water, much warmer than the rest of the shelf, all along the northern and western inboard margins. Heating is augmented by hot air bursts from the north, with air temperatures > 40°C, as much as 20°C above SST (Godfrey et al. 1986). Resultant salinities there commonly exceed 36‰. This water becomes permanent during midsummer as the GAB Plume and expands eastward with some parts becoming detached. These waters are presumably replaced by flow up onto the shelf in the Baxter Sector from the open ocean to the south. Meanwhile, cold water develops in the Ceduna Sector

Fig. 8.6 Charts of the
Great Australian Bight
illustrating the seasonal
change in surface and
near-surface water patterns
(summarized from Herzfeld
1997). (**a**) Widespread spring
uniformity in temperature
with local heating in the
northwest. (**b**) Extensive
summer heating in the
north and northwest and
formation of the GAB
Plume that drifts eastward;
upwelling of cold waters
along the western Eyre
Peninsula coast. (**c**) Widespread relatively warm
water across the region during autumn coinciding with
initial incursion of the Leeuwin Current from the west.
(**d**) Winter cooling of waters
along the coast,
maximum extent of the
Leeuwin Current into the
region, movement of the
GAB Plume off the shelf to
the southeast where it
joins the South Australian
Current

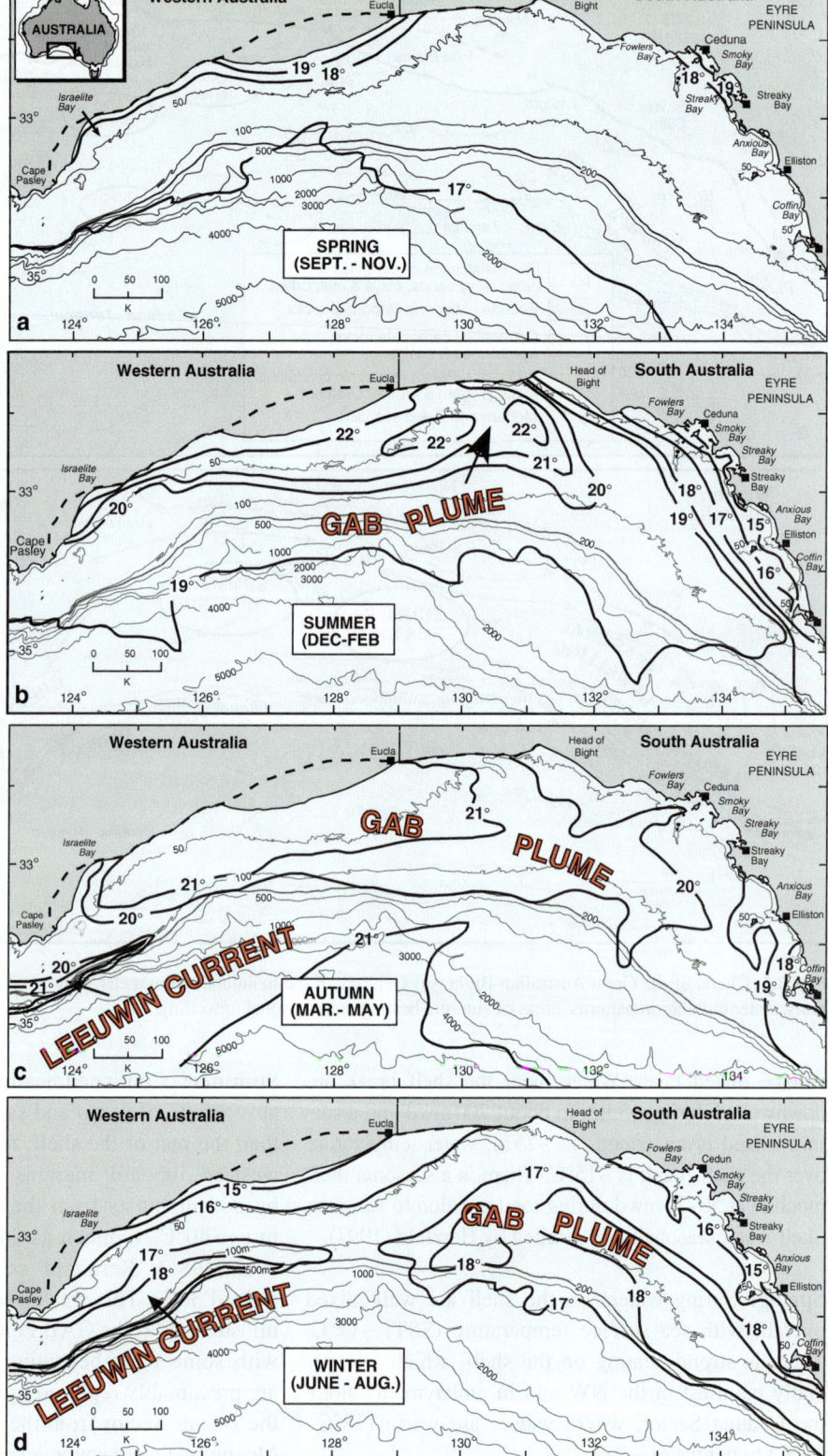

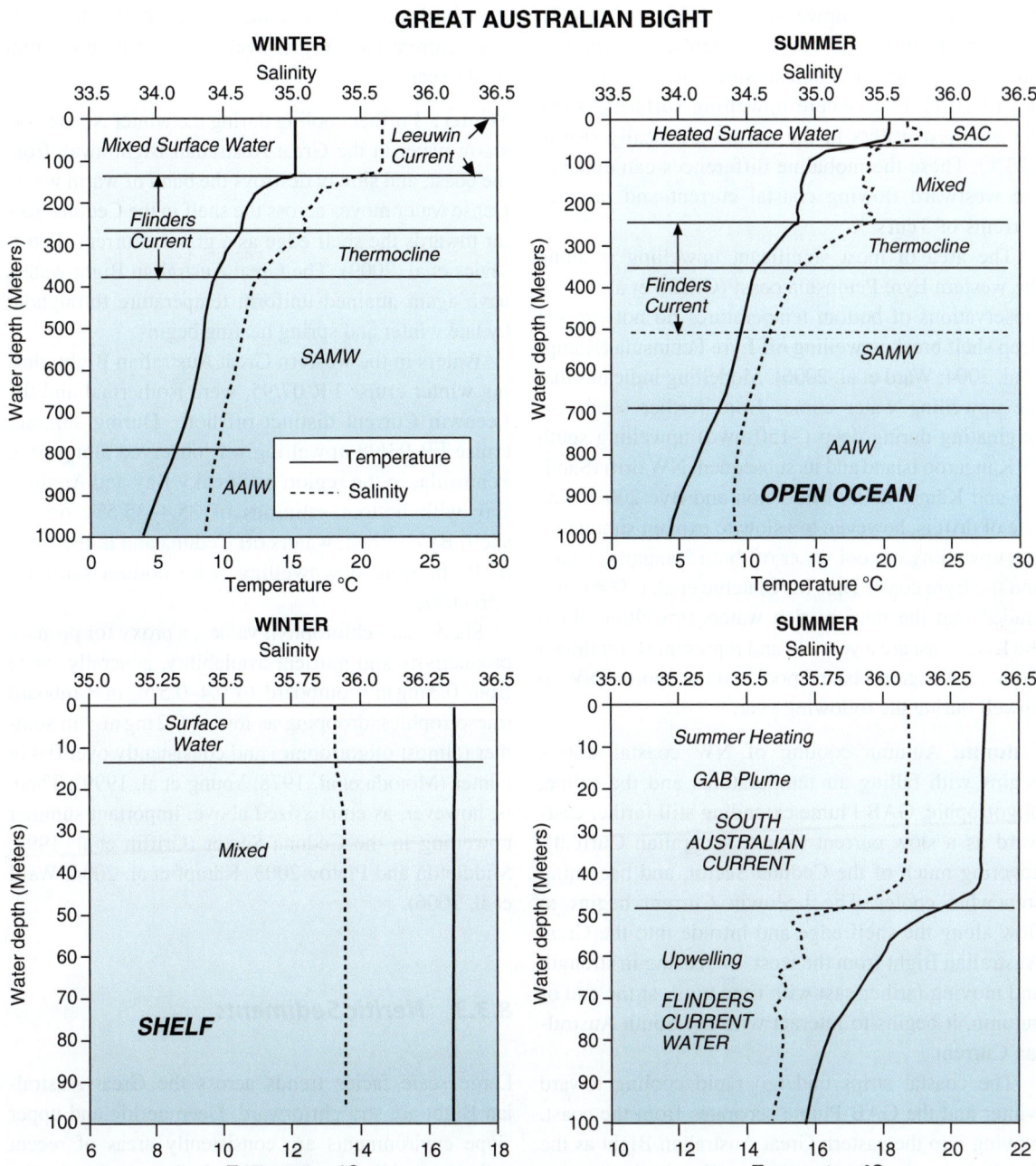

Fig. 8.7 Temperature and salinity profiles for the upper 1000 m of the open ocean water column (*top*) and the upper 100 m of the shelf water column (*bottom*) during summer and winter seaward of Eucla. *AAIW* Antarctic Intermediate Water, *SAMW* Subantarctic Mode Water, *SAC* South Australian Current. The mixed and surface layers are composed of sub-tropical surface water modified by input from the Leeuwin Current, sum-

mer heating, and winter cooling. The Flinders Current flows strongest in summer and forms an undercurrent to the Leeuwin Current in winter. (Compiled from CSIRO (2001) Data Trawler, Marine and Atmospheric Research (CMAR) Data Center, Commonwealth Scientific and Industrial Research Organization) because of coastal upwelling (Griffin et al. 1997). Both the heating and cooling intensify as summer progresses, with the coldest waters in the east falling to 10°C–14°C where upwelling surfaces, while the warmest waters in the northwest locally rise to ~23°C. These thermohaline differences can enhance the westward flowing coastal current and produce currents of $3\,cm\,s^{-1}$.

The area of most significant upwelling is along the western Eyre Peninsula coast (Griffin et al. 1997). Observations of bottom temperatures do not support deep shelf break upwelling off Eyre Peninsula (Kämpf et al. 2004; Ward et al. 2006). Modelling indicates that the upwelling water comes from further southeast, originating during deep (~150 mwd) upwelling south of Kangaroo Island and its subsequent NW drift (Sandery and Kämpf 2005; Middleton and Bye 2007). The rate of drift is, however, too slow to explain simultaneous upwelling of cool water off both Kangaroo Island and the Eyre coast. Thus McClatchie et al. (2006) concluded that the nutrient-rich waters upwelling along the Eyre coast are a year old and represent water drawn from the Kangaroo Island pool and transported NW to upwell during the following year.

Autumn: Autumn cooling of NW coastal waters begins with falling air temperatures and the saline, oligotrophic, GAB Plume extending still farther eastward as a slow current (South Australian Current), covering much of the Ceduna Sector, and becoming somewhat cooler. The Leeuwin Current begins to flow along the shelf edge and intrude into the Great Australian Bight from the west, increasing in strength and moving farther east with time until, at the end of autumn, it begins to interact with the South Australian Current.

The coastal strips undergo rapid cooling toward winter and the GAB Plume separates from the coast, moving into the eastern Great Australian Bight as the South Australian Current and cooling as it does so. These dense waters appear to sink down the slope off Coffin Bay. The Leeuwin Current connects with the South Australian Current and by mid winter there is a warm band of water formed by the Leeuwin Current and South Australian Current from Cape Leeuwin along the entire outer margin of the shelf. Outer shelf waters are now considerably warmer than those at the coast. The Leeuwin Current waters likely act as a wall, preventing any intrusion of Flinders Current waters onto the shelf, much as they do off western Australia

(Gersbach et al. 1999; James et al. 1999). The shelf-edge current has progressively higher salinities from west to east.

Winter: Further cooling during the winter isolates the warm water in the Great Australian Bight away from the coast, and slowly destroys the band of warm water. Dense water moves across the shelf in the Ceduna Sector towards the shelf edge as a gravity current (Petrusevics et al. 2009). The Great Australian Bight waters have again attained uniform temperature throughout by late winter and spring heating begins.

Waters in the western Great Australian Bight, during winter cruise FR 07/95, were isothermal and the Leeuwin Current distinct offshore. During summer cruise FR 03/98 upwelling was observed along Eyre Peninsula, in the region of Streaky Bay and Anxious Bay with bottom salinities of 35.4–35.5‰ on the shelf. By contrast, waters off Ceduna and in the Head of Bight were downwelling, with bottom salinities >36.00‰.

Shelf water chlorophyll values, a proxy for primary productivity and nutrient availability, generally range from $0.3\,mg\,m^{-3}$ outboard to 0.4–$0.5\,mg\,m^{-3}$ inboard (mesotrophic), dropping as low as $0.2\,mg\,m^{-3}$ in summer (almost oligotrophic) and consistently over 0.4 in winter (Motoda et al. 1978; Young et al. 1999). There is, however, as emphasized above, important summer upwelling in the Ceduna Sector (Griffin et al. 1997; Middleton and Platov 2003; Kämpf et al. 2004; Ward et al. 2006).

8.3.3 Neritic Sediments

Large scale facies trends across the Great Australian Bight are straightforward. Deep neritic and upper slope environments are consistently areas of recent carbonate sedimentation. The shallow neritic in the eastern Great Australian Bight (Ceduna Sector) is very narrow and the site of mostly recent carbonate or quartzose sand facies, locally enriched in relict sediment. The wide and laterally extensive shallow neritic zone in the central and western parts of the shelf is generally covered with recent carbonate except in the far west where the sediments are relict rich. The extensive middle neritic zone is covered with mainly relict or stranded sediment containing relatively little recent material (Figs. 8.8, 8.9).

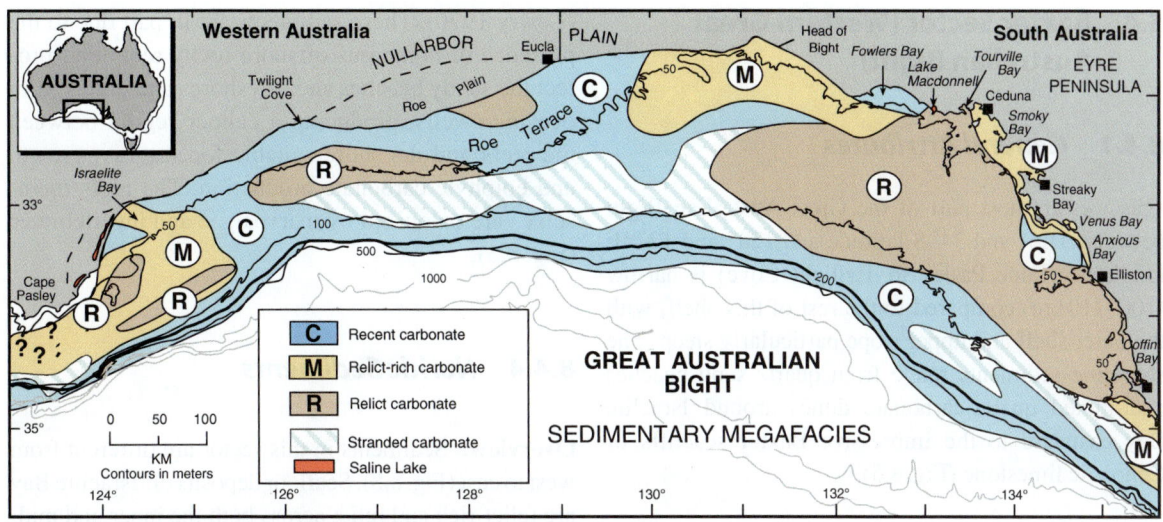

Fig. 8.8 Chart of surface sediment megafacies in the Great Australian Bight illustrating the discrete areas of relict sediment in the middle neritic zone, inner neritic carbonate and relict rich deposits and the deep neritic environment dominated by recent carbonate material; the zone of stranded, coralline-rich carbonate sediment is highlighted

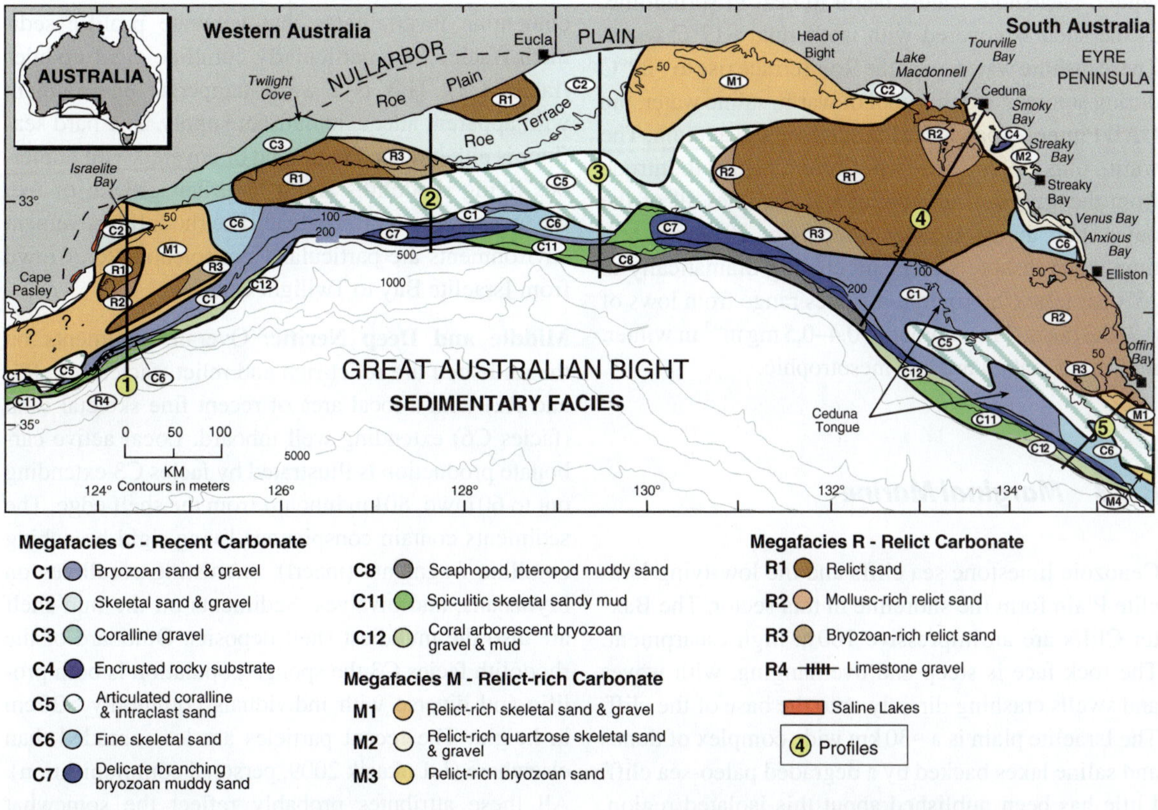

Fig. 8.9 Composition of surface sediment facies in the Great Australian Bight. Lines of section are given in Figs. 8.10 and 8.11, with the number designated here

8.4 Baxter Sector (Western Great Australian Bight)

8.4.1 General Attributes

This westernmost part of the Great Australian Bight between 31.5° and 34°S latitude and 124° and 127°E longitude (Cape Pasley to Twilight Cove) is narrow (100–110 km) compared to the rest of this shelf, with the outer shelf and upper slope particularly steep. The shoreline sediments range from quartz sand beaches and mixed quartz-carbonate dunes around Israelite Bay eastward to the impressive Baxter seacliffs of Cenozoic limestone (Fig. 8.5).

8.4.2 Local Oceanography

The shelf is somewhat protected from the predominant SW swells during most of the year by Cape Pasley, but is still exposed to strong storm waves and surge during winter. Nearshore waters begin to heat in springtime, as the shelf is covered with isothermal ~17°C water. These inshore waters over the Roe Terrace rise to ~22°C during summer and this pool of warm, saline water, the GAB Plume, expands during summer and autumn. The warm, oligotrophic Leeuwin Current begins to intrude from the west during autumn and occupies all of the outer shelf during winter to depths of 200 mwd, at a time when inshore waters are cooling dramatically to 15°C or less. Chlorophyll-a values range from lows of 0.2–0.3 mg m^{-3} in summer to 0.4–0.5 mg m^{-3} in winter, indicating that all are low mesotrophic.

8.4.3 Marginal Marine

Cenozoic limestone sea cliffs and the low-lying Israelite Plain form the shoreline in this sector. The Baxter Cliffs are an impressive 100 m high escarpment. The rock face is steep and overhanging, with waves and swells crashing directly onto the base of the cliff. The Israelite plain is a ~30 km wide complex of dunes and saline lakes backed by a degraded paleo-sea cliff. Little has been published about this isolated region. The dunes are quartzose and the lakes evaporitic

(Lowry 1970). These sediments are in part due to the presence of numerous offshore rocky reefs that protect the sandy beaches and foreshore sand dunes from oceanic swells, producing a calmer region between the reefs and the shore suitable for seagrass growth and epiphyte carbonate production. The reefs themselves are colonized by macroalgae and invertebrates (Fig. 8.3).

8.4.4 Neritic Sediments

Overview: Sediments in this sector are different from west to east (Fig. 8.8). Seafloor deposits off Israelite Bay are relict-rich and relict across both the inner and middle neritic parts of the shelf, with local areas of wholly relict sediments (Fig. 8.10–line 1). This is in stark contrast to the seafloor off the Baxter Cliffs that are almost wholly recent carbonate, suggesting more active recent carbonate sediment generation (Figs. 8.8, 8.9).

Shallow Neritic: The Roe Terrace is a nearshore zone of flourishing macrophytes, grasses, and sessile epibenthic invertebrates that generate prolific sediment (facies C2), particularly coralline algal nodules (facies C3). This is a warm-temperate environment with apparent adequate nutrient supply. The hard seafloor is either dusted with sand or covered with numerous living rhodoliths and a luxuriant growth of red, green, and brown macroalgae. The rhodolith pavement environments are particularly conspicuous to 30 mwd from Israelite Bay to Twilight Cove.

Middle and Deep Neritic: Overall, sediments on the mid-shelf are relict-rich and relict (facies M1, R1, and R2) with a local area of recent fine skeletal sand (facies C6) extending well inboard. Local active carbonate production is illustrated by facies C3 extending out to 60 mwd, 80 km inboard from the shelf edge. The sediments contain conspicuous living rigid-branching coralline fragments (maerl), encrusting corallines on bryozoans, and bivalves. Sediments on the mid-shelf are also the muddiest shelf deposits. Outboard of the rhodolith facies C3 the sponge population is both prolific and diverse with individuals commonly >20 cm in height. The recent particles are less abraded than elsewhere (J. Lukasik 2009, personal communication). All these attributes probably reflect the somewhat protected position of this sector. Bryozoan sands and

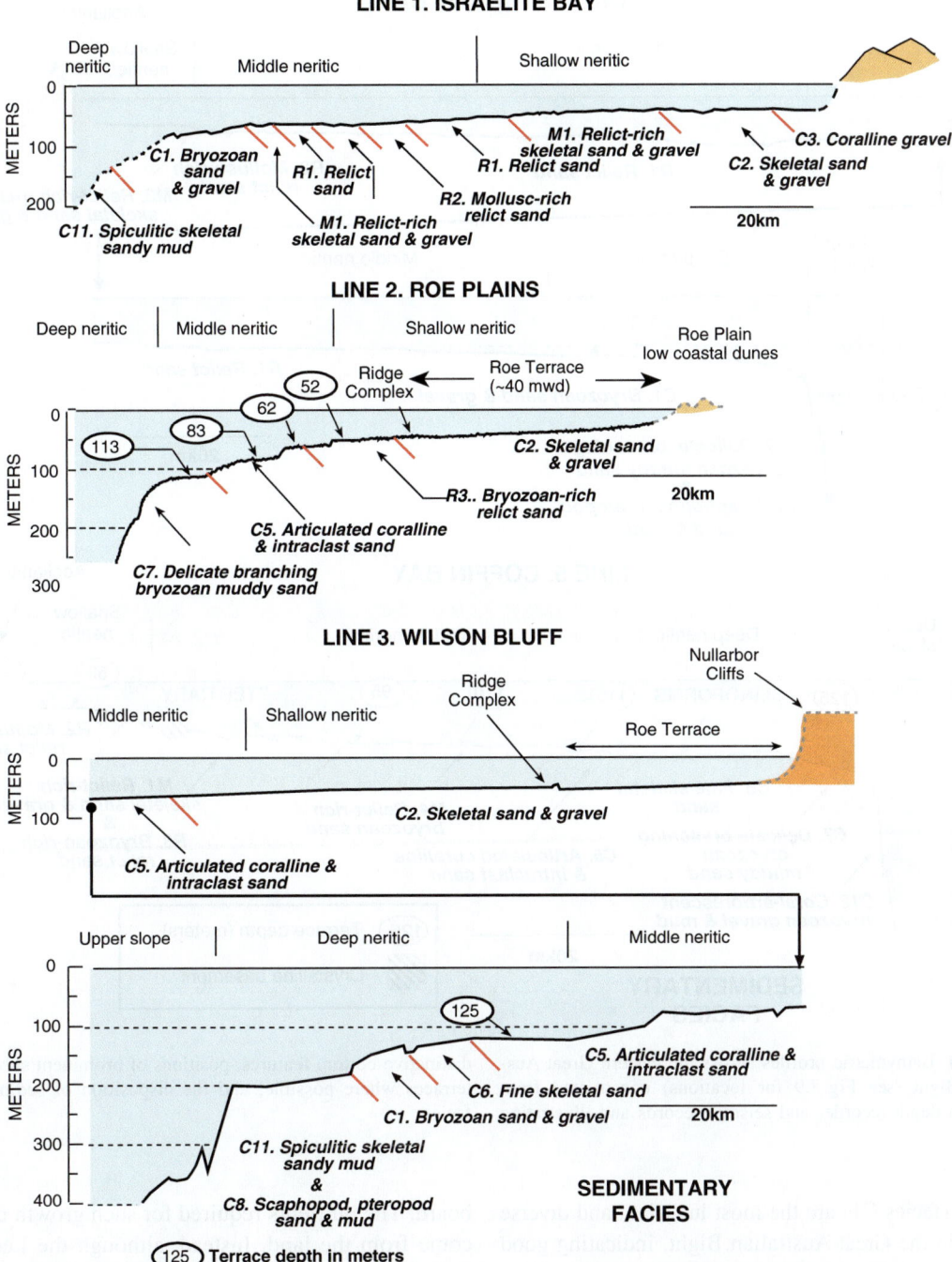

Fig. 8.10 Bathymetric profiles across the western Great Australian Bight (see Fig. 8.9 for locations) constructed from precision depth recorder and seismic records and illustrating distinctive bottom features, positions of prominent submarine terraces where possible, and the disposition of sedimentary facies (see Fig. 8.9 for facies compositions)

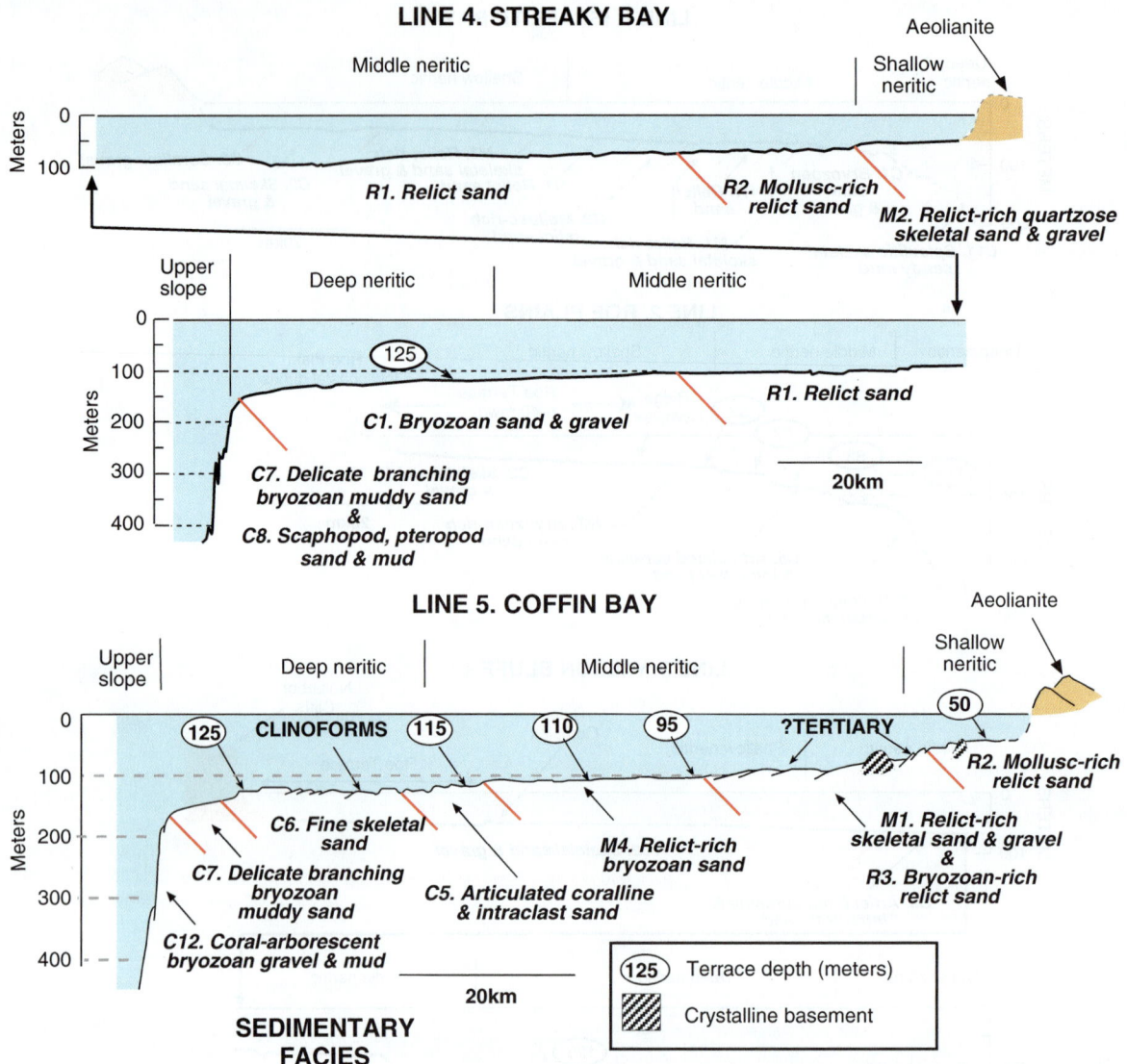

Fig. 8.11 Bathymetric profiles across the eastern Great Australian Bight (see Fig. 8.9 for locations) constructed from precision depth recorder and seismic records and illustrating distinctive bottom features, positions of prominent submarine terraces where possible, and the disposition of sedimentary facies

gravels (facies C1) are the most luxuriant and diverse of any in the Great Australian Bight, indicating good trophic resources in the epibenthic coppice environment (cf. McKinney and Jackson 1989; Bone and James 1993).

All of these traits are reminiscent of the Rottnest Shelf off Western Australia (Collins 1988; James et al. 1999; Pearce and Pattiaratchi 1999). The waters inboard support abundant macrophyte and coralline algal growth whereas bryozoans are prolific out-

board. The nutrients required for such growth cannot come from the land. Instead, although the Leeuwin Current blocks any upwelling, especially during the winter months, it weakens during the summer and thus allows short-term onshelf incursion of nutrient-rich slope waters. Summer shoreline upwelling is not reported from the Baxter Sector but incursions of the nutrient-rich Flinders Current onto the shelf are likely (J. Bye and P. Baxter 2000, personal communication).

8.5 Eyre Sector (Central Great Australian Bight)

8.5.1 General Attributes

This sector forms most of the central Great Australian Bight between 31.5° and 34.5° S latitude, and 127° and 131° E longitude, Twilight Cove and Head of Bight have attributes that are quite unlike the rest of the continental margin. The shelf is much wider overall than in the west or east and the shallow Roe Terrace extends out to almost the geographic middle of the shelf. The shoreline ranges from the extensive quartzose-carbonate sand-covered Roe Plain in the west (James et al. 2006; James and Bone 2007) to the steep Nullarbor Cenozoic limestone cliffs in the east. The continental slope downlaps onto and forms part of the Eyre Terrace in ~700–1000 mwd. The shelf margin is a large prograding wedge of fine-grained Quaternary carbonate sediment (Feary and James 1998; Feary et al. 2000b) (Fig. 8.6).

8.5.2 Oceanography

The sector is exposed to winter storms and feels the full force of year-round SW swells. While the shelf is covered with ~17°C water in spring, solar heating warms shallow waters over the Roe Terrace. This heating extends into summer when water temperatures rise to more than 23°C inshore, the warmest in the Great Australian Bight, and become somewhat saline via evaporation. Summer water column stratification is prevalent and downwelling is prominent. Winter sees dramatic cooling of inshore waters, falling to as low as 15°C along the shoreline and incursion of the warm Leeuwin Current from the west to join with the GAB plume on the outer shelf-upper slope. Chlorophyll-a values vary from 0.2–0.3 mg m^{-3} in summer to >0.5 mg m^{-3} in winter (low mesotrophic) (Figs. 8.6, 8.7).

8.5.3 Marginal Marine

This region is dominated by sea-cliffs. The degraded paleoseacliffs behind the Roe Plains pass laterally into impressive modern seacliffs east of Eucla to the Head of Bight as the Nullarbor Cliffs. The Nullarbor Cliffs are steep, overhanging, and largely inaccessible.

The Roe Plains are a Pliocene erosional terrace cut into Cenozoic limestones in much the same way as similar limestones are being eroded today in front of the Baxter and Nullarbor cliffs (Lowry 1970; James et al. 2006). Eroded Oligocene limestones on the Roe Plains are veneered by the m-thick, Pliocene, mollusc-rich Roe Calcarenite (James et al. 2006; James and Bone 2007) that is in turn overlain in the west by moribund quartz aeolian dunes. Beaches throughout the western and central Great Australian Bight are quartz-rich due to western longshore transport. Dunes at the eastern end of the Roe Plains near Eucla (the Dessinor Sandhills) are, however, more carbonate-rich and have locally blown up and over the tops of the cliffs. The seafloor directly offshore the Roe Plain is a series of active seagrass meadows whose blades and calcareous epiphytes are swept onshore to form thick piles of dead seagrass on the beach (Fig. 8.3).

8.5.4 Shelf Sediments

Overview: The Eyre Sector seafloor is partitioned into shallow recent neritic carbonates, middle neritic relict and stranded deposits, and recent deep neritic carbonates (Fig. 8.10, lines 2 and 3). The shallow neritic seafloor is mostly on the Roe Terrace, an area of extensive seagrass meadows and active recent carbonate productivity. Unlike most other locales along the southern margin shelf, there is little distinctive clean bryozoan sand in deep neritic environments but instead the seafloor between 100 and 200 mwd is covered with spiculitic mud containing scattered floating delicate branching bryozoans, far shallower than anywhere else (Figs. 8.8, 8.9).

Inner Neritic: Inner shelf environments are locally rich in macrophytes and grasses, with abundant modern carbonate production. The inshore non-calcareous biota is less prolific than in the Baxter Sector and decreases eastward, suggesting comparatively lower trophic resources overall. Nevertheless, the sediments are largely skeletal sands (facies C2) throughout.

Middle-Neritic: The seafloor outboard of the Roe Terrace is a monotonously rippled sand barren, with few living epibenthic organisms and the sediment is largely composed of stranded coralline algal rods (facies C5).

Ripple troughs are littered with pebble-size lithic intra-clasts, bivalves, and bryozoan fragments. Free-living bryozoans and brachiopods (particularly *A. cumingi*) are locally abundant. Scattered bright orange and yellow sponges and articulated zooidal bryozoan shrubs up to 5 cm high cover ~5% of the surface. The few hard, bioeroded substrates are subaqueous islands or epibenthic coppices of profuse epibenthos (Facies C4, too local to depict on Fig. 8.9), particularly sponges (up to 1 m high), ascidians, hydroids, bryozoans (*Parmularia*, *Adeona*, articulated zooidal forms), brachiopods, pectenid bivalves, and rare cowries.

Deep Neritic: These outermost shelf deposits comprise only a narrow zone of bryozoan sand and gravel (facies C1). By contrast, large areas are covered by rigid delicate branching bryozoan muddy sands (facies C7) and fine skeletal sand (facies C6) forming an epibenthic turf that extend well onto the shelf and up into the middle neritic zone. The water is clear. The seafloor is a muddy, locally microbially veneered, uniform substrate supporting the 1–2 cm high turf of epibenthic organisms. The burrowed fine sand and mud in shallower water, ~125 mwd, is textured with degraded ripples and conspicuous ray feeding traces. Overall epibenthic growth in these areas is less than in facies C1 and covers only ~30% of the surface. The sediment is intensively burrowed with numerous pits and depressions. There are local small clusters of demosponges ~3 cm high and isolated anemones. About 1/5 of the sediment surface is littered with free-living and rigid delicate branching bryozoan skeletons.

The shelf edge and upper slope is mostly covered with fine spiculitic skeletal muddy sand (facies C11) and scaphopod, pteropod muddy sand (facies C8). This facies disposition is interpreted to reflect year-round downwelling and scarce trophic resources (James et al. 2001). The high proportion of infaunal benthic foraminifers (Li et al. 1999) supports this interpretation.

8.6 Ceduna Sector (Eastern Great Australian Bight)

8.6.1 General Attributes

This extremely wide shelf between 31.5° and 35° S latitude and 131° and 136° E longitude (Head of Bight to Coffin Bay Peninsula) (Fig. 8.6a) has a very narrow inner neritic environment mostly in the form of irregular bays, Precambrian crystalline rock outcrops, and high erosional cliffs of Pleistocene aeolianite. The middle neritic zone is up to 150 km wide, the most extensive in the Great Australian Bight.

8.6.2 Oceanography

The whole region feels the full force of SW swells as reflected by the thick and extensive onshore Pleistocene aeolianites composed of shallow neritic carbonate particles that are swept onshore. The middle neritic shelf plain, most of which is within the zone of wave abrasion, also experiences fluctuating water composition. Water movement is in the form of an anticyclonic gyre (Richardson et al. 2009) that shows progressive mixing, heating, and evaporation of cool upwelled waters as it flows in a northwesterly direction along the coast. These waters mix with GAB Plume waters at the Head of the Bight and then flow southeastwards toward the shelf edge (Figs. 8.5, 8.6, 8.7).

A total of five water masses have been identified in the Ceduna Sector (Richardson et al. 2009); end members are the Flinders Current and GAB plume with the others being mixtures of these waters. Spring waters over the shelf are, as everywhere else, ~17°C and isothermal. Surface waters warm significantly in summer as the GAB plume expands and moves eastward, while at the same time sub-thermocline waters become cold through upwelling. Shelf water stratification weakens and disappears during autumn and winter, when saline, nutrient-depleted GAB plume waters cover the entire Ceduna sector. This water cools and moves eastward during the winter, cascading off the shelf as an outflow south of Coffin Bay. The influence of this low-nutrient, saline water body is profound: sediment in the mid-neritic environment is almost all relict with virtually no modern carbonate production and accumulation (James et al. 2001) (Fig. 8.8).

Summer upwelling that brings cold waters, across the shelf and to the shore, locally counterbalances this effect. Integrated physical and chemical oceanography suggest that the upwelled water is a combination of the Flinders Current and the Great Australian Bight outflow; i.e. the GAB plume water flows off the shelf, mixes with the Flinders Current and is incorporated into upwelling water along Eyre Peninsula. This upwelling water comes from ~150 mwd at the shelf break but seems to incorporate some additional nutrient-rich deep upwelled water from Kangaroo Island that has been advected westward along the shelf and into the Ceduna Sector.

Winter chlorophyll-a values in the area are low ($<1\,mg\,m^{-3}$) and comparable to those in the western GAB (Ward et al. 2006). Summer values outside upwelling areas are similar (~$0.2\,mg\,m^{-3}$). Values within upwelling regions, however, can be as high as $4.5\,mg\,m^{-3}$ (mesotrophic).

8.6.3   Marginal Marine

This is the most laterally variable succession of marginal environments in the Great Australian Bight because of the changing geology afforded by the Gawler Craton, and the spectacular aeolianites formed by winds and waves from the southwest (Belperio 1995). The numerous, variably protected embayments are due to erosional breaching of Pleistocene dune systems when sea level rose to near modern levels at ~7 ka. The bays themselves are located in former back-beach corridors. Sedimentation has been so active since that time that many bays have been partially to completely filled by either Holocene marine sediments or seaward prograding dunes.

The coast is a series of shoreline calcreted Pleistocene aeolianites overlying Precambrian basement in the form of unprotected eroding cliffs, islands and shoals, headland-attached barriers, and large prograding dunefields. Dune sedimentation appears to have been most active during early stages of the Holocene highstand and many dune systems are now eroding; in extreme examples clifftop dunes have had their sand ramp completely removed. The bays have a variety of facies that are similar to those in the large gulfs to the east. They typically contain intertidal sand flats, mangroves, cyanobacterial-halophyte marshes, and *Posidonia australis* seagrass banks. The most important sediment producers are benthic foraminifers, bryozoans, and molluscs.

Although they were once open, broached embayments such as Tourville Bay and Smoky Bay are now largely filled in with peritidal mudflats and mangrove marshes that have prograded seaward over *Posidonia*

grassbanks (facies C9) yielding excellent shallowing-upward successions (Belperio et al. 1988; Belperio 1995). Those bays with somewhat restricted entrances (Streaky Bay, Venus Bay, Coffin Bay) are partially filled by flood-tidal deltas and associated sand shoals that protect nearshore shallow muddy depressions. The sediments are swept into the bay from the inner shelf, eroding aeolianites, and from in situ grass bank production (facies C9). Finally, those embayments that are now completely cut off from the ocean and where seawater seeps into coastal depressions (Sleaford Mere, Lake MacDonnell) are sites of gypsum and carbonate (aragonite) precipitation (Warren 1982a).

8.6.4   Shelf Sediments

Overview: Most of the Ceduna Sector, in contrast to the nearby Eyre Sector, is relatively deep and lies in the middle neritic zone. It is distinguished by relict sediments, the largest such area along the southern continental shelf (Fig. 8.8). Living encrusting corallines are present on bioclasts and lithoclasts to depths of 115 m, defining the effective base of the photic zone. The narrow inboard shallow neritic is locally rich in carbonate sediments and quartzose facies. Superimposed on this bathymetric facies trend is a an area of active epibenthic growth, the 'Ceduna Tongue' of James et al. (2001) that appears to coincide with an area of seasonal upwelling. Deep neritic sediments are wholly carbonate with local stranded facies in the southeast (Figs. 8.8, 8.9).

Shallow Neritic: Inner-shelf sediment ranges from very coarse sand and gravel (facies M2) near islands or headlands to medium and coarse sand to almost all recent fines in gentle depressions off Anxious Bay (facies C6). Deposits locally contain up to 15% terrigenous material, generally quartz, feldspar, and crystalline rock fragments. Near islands it may rise up to 50% coarse igneous rock fragments, or up to 10% large cemented clasts (facies M2).

Middle-Neritic: The sediment mosaic is complex and characterized by a high proportion of relict grains. Inboard parts of the zone are covered with relict-rich sediment ranging from almost wholly relict sand (facies R1) in the west to relict-rich skeletal sands and gravels (facies M1) in the northwest to bryozoan and mollusc relict sands in the southeast (facies R2).

Facies R1 is a sand barren environment that images show to be a sea floor of extensively rippled sands. Wave ripples (l=~50 cm; a=~10–20 cm) are sinuous-crested and somewhat degraded with bioeroded and abraded shells in the troughs. Epibenthic growth is scattered and not clustered; it covers only ~10% of the seafloor as 20–30 cm high individuals. Most organisms are long thin demosponges, bryozoans (especially *Adeona, Parmularia*, and flexible articulated zooidal forms), and abundant hydroids.

Images of the seafloor to the south off Elliston show sand barrens of rusty-coloured relict biofragmental sands and gravels rich in molluscs (facies R2) and little epibenthos. Wave ripples are straight-crested to locally sinuous (l=~5–60 cm; a=~5–10 cm) whose troughs are filled with bivalve shells (with occasional secondary smaller ladder ripples). About 10% of the surface is covered with numerous white (dead) or brown (live) free-living bryozoans. The only upright epibenthos are a few small hydroids.

The deeper outboard part of the middle neritic zone is clean bryozoan sand, but it still contains a substantial relict component (facies C1, M4 and R2). Sediments are, however, distinguished by 20–40% stranded articulated coralline algal particles (facies C5). These vast spreads of uniform sand have minor coarse material, generally bryozoans, ubiquitous slit shells (Siliquarids, implying abundant sponges), scattered calcarenite clasts encrusted with the red foraminifer *M. miniacea* and serpulid worms, epifaunal echinoid spines, and a few rhodoliths in deeper waters.

The Ceduna Tongue south-west of Anxious Bay (Fig. 8.9) is, in contrast to the relict rich molluscan sands and relict sands described above, a conspicuously different region of epibenthic growth. The stranded facies C5 (articulated corallines and intraclast sand) are overprinted by facies C1 (bryozoan sand and gravel).

Seafloor images in this area from deepest water show prolific epibenthic growth in sands and on rocky reefs. At midday in the core of the clean bryozoan sand facies (facies C1), although dim, there is enough light at 120 mwd to see the seafloor and operate the video camera without lights. Roughly 20% of the seafloor is hard substrate (epibenthic coppice), 10% is open, rippled and barren sand, and 70% is rippled sand with scattered upright filter feeders (sand barren). The rock is partly mantled with a centimeter-thick veneer of medium to very coarse sand and gravel or locally completely buried by the same sediment. The epibenthic coppice is a site of sublinear rocky outcrops, a meter or two wide and 5–10 m lateral extent that dip gently seaward. These hard substrates are epibenthic islands covered by a dense growth of numerous demosponges, especially vase, globular and invaginated forms, that are typically covered with epizooidal bryozoans, innumerable hydroids, conspicuous shrubs of articulated zooidal bryozoans, ascidians and brachiopods. Sponges also host the gastropod *Tenagadus* whereas ascidians and sponges locally bind the sandy substrates, thus providing a suitable substrate for further epibenthic colonization. Bryozoans are diverse, the most conspicuous forms being *Adeona*, the smaller fenestrate *Iodictyum*, and articulated zooidal shrubs. Organisms are either dispersed or together in small clusters in the sandy substrate but densely packed on rocky highs.

The sands between epibenthic coppices are not totally barren. The sharp wave ripples are sinuous to straight-crested (l=~30–80 cm; a=~5–30 cm) locally starved, and confused in places. The clean sand is moderately sorted and rich in recent angular, medium sand to gravel-size bryozoan fragments. The sand supports scattered isolated colonies of *Adeona* and hydroids. Densities of large *Adeona*, up to 20 cm high, can reach 5 individuals per square meter in the sandy substrates.

In shallower water, well onto the shelf, up to 100 km inboard from the shelf edge, in ~80 mwd (facies R3) sponges are still prolific but the bryozoans, although numerous, are less diverse. Sediments and living bryozoans are dominated by fenestrate, foliose, vagrant and locally articulated zooidal growth forms. Stranded coralline algal rods and dendritic rhodoliths are present in sediments to 90 mwd (facies C5).

Deep Neritic and Upper Slope: Sediment patterns are intricate. Clean bryozoan sands (facies C1) extend to ~180 mwd where they grade downward into deeper water rigid delicate bryozoan muds and sands (facies C7). Rhodoliths (facies C3) form roughly 30% of the sediment, between 120 and 250 mwd; maximum development is between 160 and 220 mwd. These coralline algal nodules are clearly stranded and out of equilibrium with their current growth environment. Coralline algal rods (facies C5), that are also stranded, have a similar distribution they are highest between 85 and 200 mwd where they comprise between 20% and 40% of the sand-size grains.

Slope sediments below 200 mwd are principally delicate branching bryozoan muds (facies C7) or spiculitic skeletal sandy muds (facies C11). Images of the upper slope show a muddy seafloor covered by a uniform growth of tiny filter feeders, none more than 2 cm high, that overall resembles a sparse lawn (an epibenthic turf). They are mostly hydroids, sponges and rigid delicate branching and free-living brozoans, with variable numbers of flexible articulate zooidal bryozoans. Erect rigid fenestrate growth forms are rare. There are widely spaced small (4 cm high) tubular sponges, some of which are encrusted with *Celleporaria*. The sediment surface, partially covered with a microbial scum and littered with a few bryozoan fragments, was stirred up every time the camera landed. There are occasional bare patches and burrowing is locally intense. Scours across the seafloor attest to intensive commercial trawling. This pattern is interrupted by a distinctive zone of Facies M, robust bryozoan-coral gravel between ~250 and 350 mwd. Sediments below 350 mwd are a burrowed mud barren, (mainly facies C8–scaphopod, pteropod muddy sand).

8.7 Synopsis

Against a background of generally high-energy and cool-water oceanography, the region has several distinctive characteristics, in particular the flow of the seasonal Leeuwin Current, local upwelling, and profound solar heating. Most of the shelf is a warm-temperate (cf. Betzler et al. 1997; Fornos and Ahr 1997), marginally-tropical depositional setting inboard with cold-temperate deep neritic and upper slope environments outboard. Robustness of the carbonate factory is largely dependent upon local oceanography and adequate trophic resources because the main carbonate-producing organisms are infaunal and epifaunal filter feeders.

1. Siliciclastic sands dominate marginal marine sediments along the coast east of Cape Leeuwin. This is because of the strong wave and swell regimen that keeps the material close to shore and drives particles eastward via longshore drift. The sands are orthoquartzitic along the Albany and Baxter sectors but become progressively diluted by carbonate material eastward such that in the Eyre sector they are roughly equal parts of carbonate and quartz.
2. The shelf adjacent to a coast of Precambrian crystalline rock has offshore island complexes that compli-

cate the simple 'graded shelf' facies pattern. These islands act as sources of local terrigenous clastic sediment and generate a series of protected environments in otherwise shallow- and mid-neritic settings.
3. Warm-temperate shallow to mid-neritic environments in the Albany and western Great Australian Bight are characterized by abundant corallines, especially rhodoliths and dendritic forms. This situation is in contrast to cool-temperate environments at similar depths to the east where the most abundant shallow neritic sediment producers are molluscs.
4. The near oligotrophic system with few planktic microorganisms and minor suspended material results in transparent ocean waters wherein light penetrates much deeper than in those environments with higher nutrient levels. Thus, seagrasses grow to 40 mwd, macroalgae to 40 mwd, and coralline algae to 102 mwd. They are important sediment producers to ~40 mwd. Measurements indicate summer, midday light levels at 80–90 mwd are ~5% of those at the surface. When taking videos at this time there was enough ambient light to see clearly at 95 mwd and dimly at 120 mwd. Using living encrusting corallines as a proxy for light, they are present across the Great Australian Bight to 65 mwd with the deepest growing at 66 mwd (Baxter Sector), 95 mwd (Eyre Sector), and 102 mwd (Ceduna Sector).
5. The sponge-bryozoan coppice assemblage is widespread and prolific in the Baxter sector especially from mid-neritic depths of ~80 m across the deep neritic zone to the shelf edge. This situation implies somewhat elevated trophic resources, at least seasonally.
6. This is in stark contrast to the Eyre sector in the centre of the GAB where upper slope muddy facies extend well onto the shelf and in the mid-neritic zone, suggesting low nutrient levels brought about by year-round downwelling.
7. The marginal marine zone along western Eyre Peninsula in the Ceduna Sector is complex, comprising bedrock and aeolianite headlands with intervening embayments or beach complexes. These bays are mostly breached Pleistocene aeolianite ridges with the lagoons themselves located in former interdune corridors. They range in style from normal marine to restricted and evaporitic. Many bays are now being filled or have been totally filled with recent sediment from both the adjacent open ocean and associated seagrass beds.

8. The middle-neritic zone across much of the central and eastern Great Australian Bight is almost of devoid of recent carbonate. The sediments are instead dominated by relict and stranded particles. This is interpreted to be the result of the long residence times of the GAB Plume composed of somewhat saline and nutrient–depleted waters in this region and so relatively low carbonate productivity.

9. By contrast, the deep neritic zone seaward in the Ceduna sector is an area of exceptional sponge-bryozoan growth, much like the Baxter Sector described above. This abundant carbonate production is correlated with seasonal summer upwelling that locally reaches the coast.

10. Recent benthic foraminifers that constitute a steady 15–20% of the Holocene sand-size carbonate fraction show a distinctive west-to-east, warm-to-cold gradation in species across the Great Australian Bight (Li et al. 1999). The large foraminifer *Marginopora vertebralis*, a signature of warm-temperate, euphotic environments (Hallock 1984) is particularly instructive. Although generally present in low numbers, dead tests occur in ~50–90 mwd and extend eastward across the Great Australian Bight to about Elliston, and not much farther (James et al. 1997). *Heterostegina*, indicative of similar conditions, is present in a few of the same localities. No living examples of either of these large benthic foraminifers were found in any of our bottom samples across the Great Australian Bight. At this point it is unclear if these tests are stranded, but regardless, they do not represent modern oceanographic conditions.

11. Several species of zooxanthellate corals and calcareous green algae (together with the above large benthic foraminifers) live in the shallowest Esperance Bay environments where waters remain warm during summer. The green algae are not calcified, much like the situation along the western coast of Australia (James et al. 1999).

Chapter 9
The South Australian Sea

9.1 Introduction

The seascape of elongate gulfs that penetrate into the continental interior, the corridor of Investigator Strait, the sweeping arc of Lacepede (Encounter) Bay, and the bedrock of Kangaroo Island together generate a wide variety of neritic depositional environments (Fig. 9.1). The cliffed coast around Kangaroo Island and Yorke Peninsula is mostly bedrock, as it is along the southern tip of Fleurieu Peninsula with bays cut into softer Pleistocene Bridgewater calcarenites. The curving sweep of the Lacepede Bay shoreline is a prograding aeolianite coast and high-energy beaches that have undergone Quaternary uplift. It is now a series of stranded Pleistocene beach-dune complexes that in some places may stretch back to the late Pliocene. The entrances to Spencer Gulf and Gulf St. Vincent are bounded either by bedrock or by aeolianites whereas the interiors have local low energy tidal flats. There are no rivers delivering any significant siliciclastic sediment in this relatively arid region (Figs. 9.1, 9.2, 9.3).

The oceanography is in many ways similar to the Great Australian Bight, being dominated by the Leeuwin Current System, local evaporative heating, and seasonal upwelling. The southeast-flowing Leeuwin Current System is here dominated by outflow of the Great Australian Bight Plume that forms the South Australian Current that is driven eastward by prevailing winds. Both gulfs are inverse estuaries.

Investigator Strait connects Spencer Gulf and Gulf St. Vincent to the open ocean and to one another and the Lincoln Shelf. Although waters flow from Gulf St. Vincent eastward via Backstairs Passage onto the Lacepede Shelf during winter, their affect on oceanography is minimal. The River Murray today does not reach the ocean because of anthropogenic activity

but may have delivered fresh water to the sea in the past. Thus, the Lacepede Shelf is normal open marine in character. The South Australian Current generally blocks upwelling from the base of the Flinders Current system, although favourable summer winds lead to significant upwelling on the southeast part of the Lacepede Shelf and off southern Kangaroo Island.

There is a clear overall macrofacies trend from shallow neritic carbonates to relict sediment in the middle neritic zone. These relict sediments are gradually mixed with recent carbonate outboard such that the deep neritic environment is almost wholly recent carbonate sediment. This pattern is perturbed somewhat by active carbonate sedimentation adjacent to Kangaroo Island and the vast siliciclastic blanket on the Lacepede Shelf. These latter quartzose megafacies were mostly deposited during and just following the LGM lowstand (James et al. 1992; Hill et al. 2009). The gulfs are floored by recent sediment.

9.2 Spencer Gulf

9.2.1 General Attributes

Spencer Gulf lies in a broad, 30 km-wide meridonal depression, is up to 200 km wide at the mouth, and extends northward for almost 400 km. It is flanked by Eyre Peninsula to the west and bounded by the Flinders Ranges and Yorke Peninsula to the east. Basement along the western edge comprises Paleo- to Middle Proterozoic crystalline rocks intruded by Neoproterozoic mafic dikes (Daly and Fanning 1993). The northern part of the eastern shore is mostly formed by the Flinders Ranges, a modest range of hills that rise ~800 m above a Pleistocene alluvial fan apron. The southern

N. P. James, Y. Bone, *Neritic Carbonate Sediments in a Temperate Realm*, DOI 10.1007/978-90-481-9289-2_9, © Springer Science+Business Media B.V. 2011

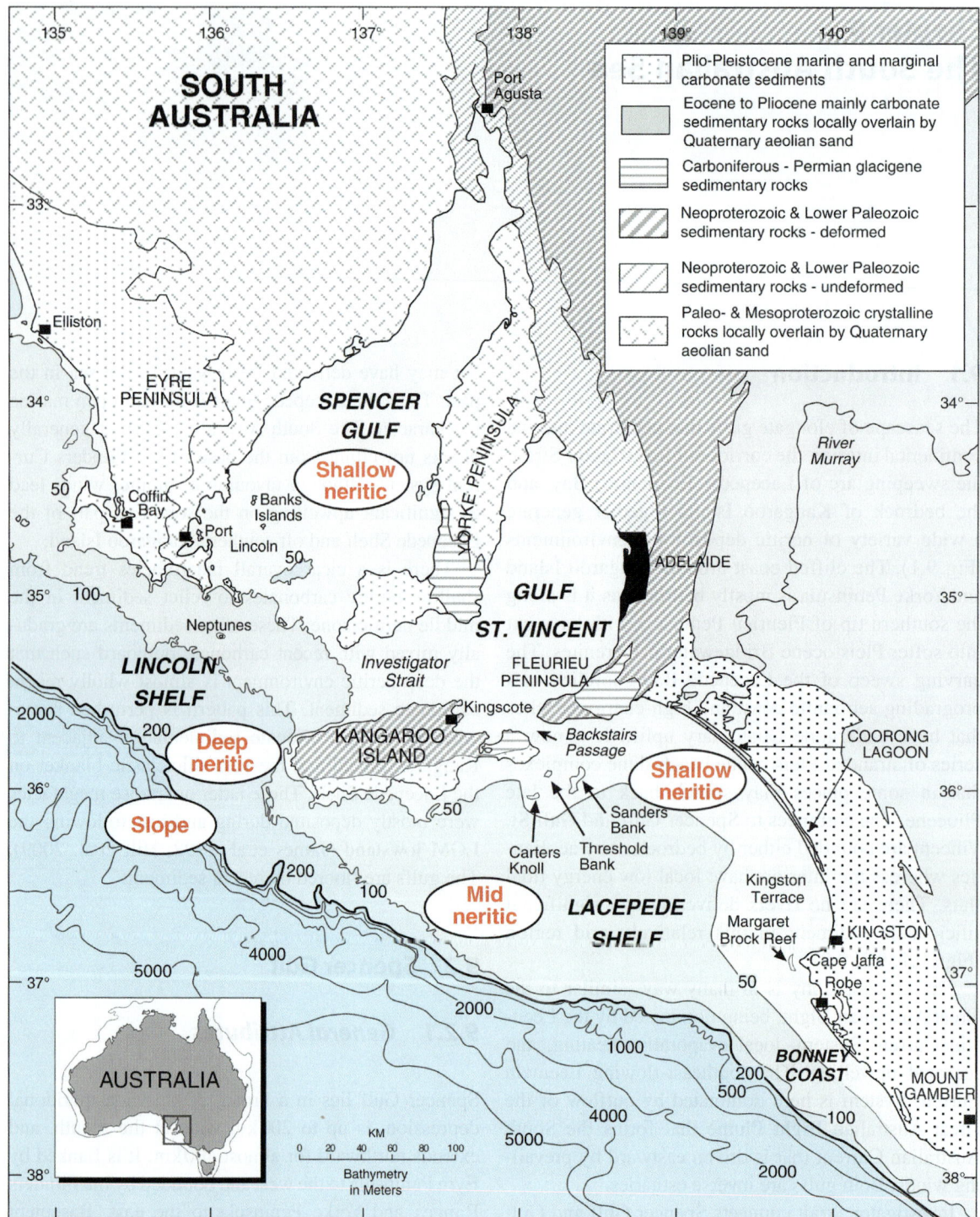

Fig. 9.1 Chart and map of the South Australian Sea illustrating important localities, geology of the adjacent mainland, bathymetry, and different neritic zones

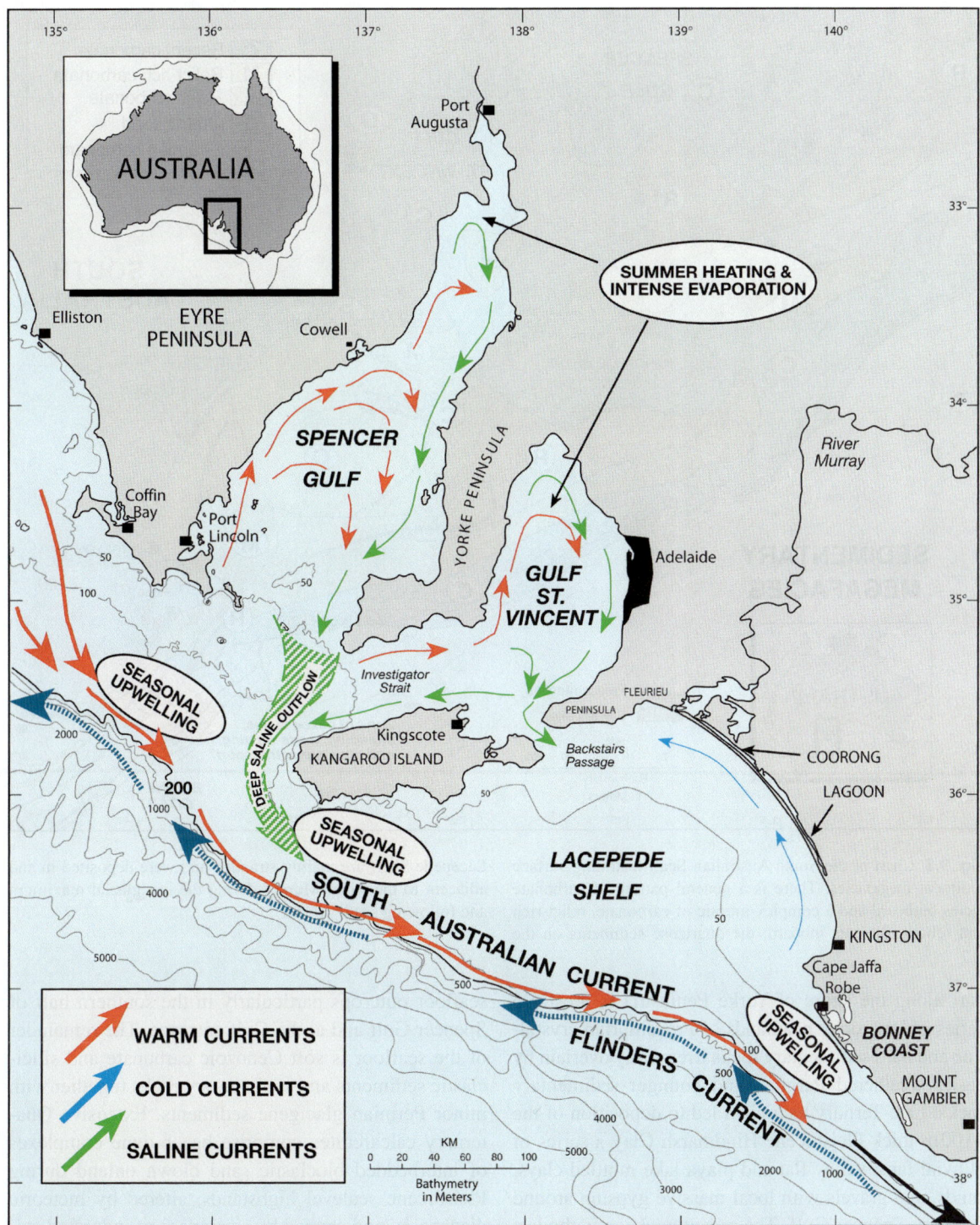

Fig. 9.2 Chart of the South Australian Sea illustrating major ocean currents, local seasonal currents, and areas of seasonal upwelling

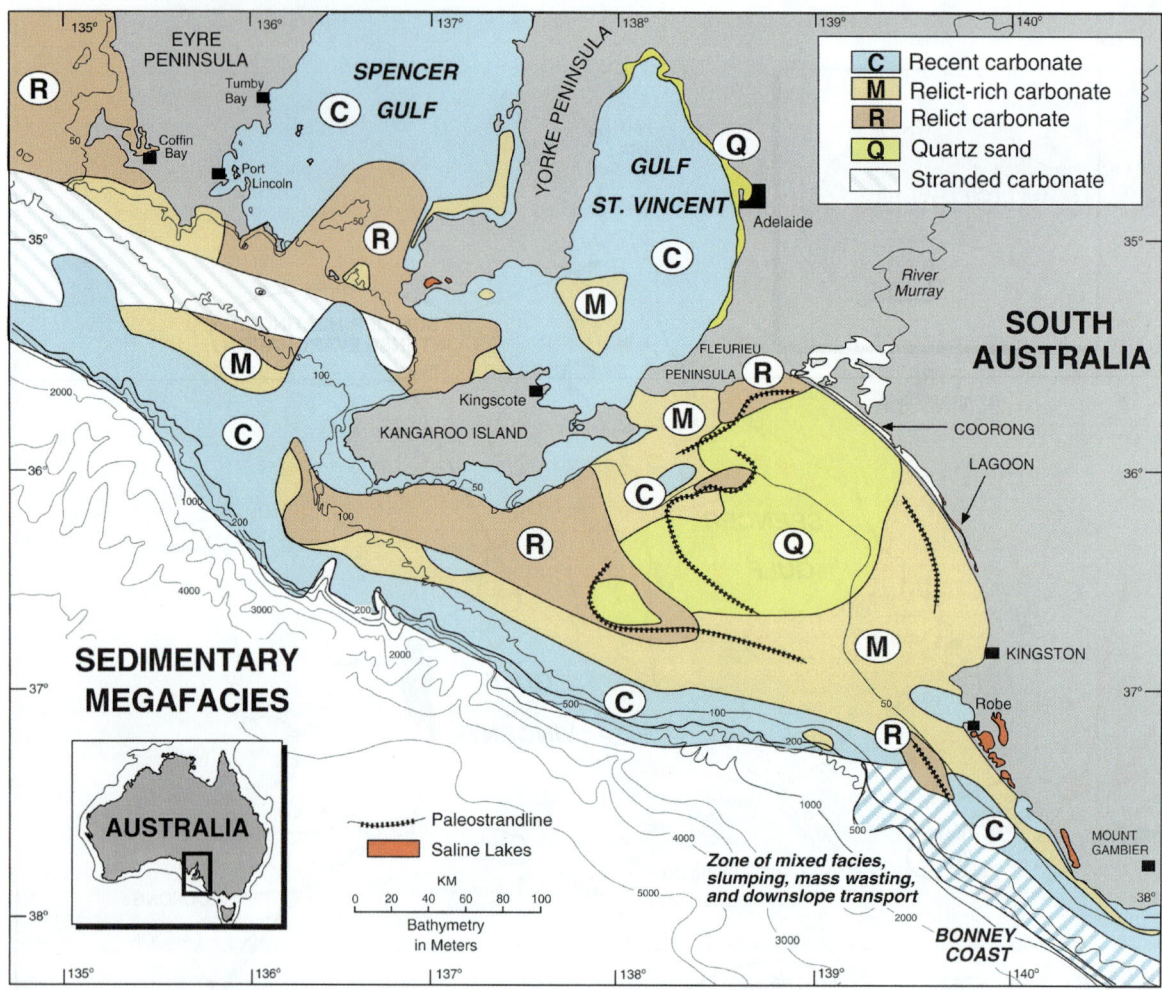

Fig. 9.3 Chart of the South Australian Sea illustrating surface sediment megafacies. There is a general pattern of carbonate facies outboard and a complex mosaic of carbonate, relict-rich and relict sediments inboard; the quartzose sediments on the Lacepede Shelf are mainly stranded and were deposited in and adjacent to the River Murray during the last glacial maximum and following sea level rise

part along the shore of Yorke Peninsula is mostly a series of low rocky headlands of Precambrian crystalline and hard sedimentary rocks irregularly overlain by easily weathered Permian and younger sedimentary rocks. Late Tertiary tectonism led to deposition of the ~100 m-thick Pleistocene Hindmarsh Clay a series of alluvial fans, valley flat and playa lake mottled clays, sands, and gravels with local massive gypsum around northern Spencer Gulf. The recent package is the last of four discrete Pleistocene sea-level highstand sedimentation events, most of which did not extend to the current shoreline (Belperio 1995) (Figs. 9.1, 9.4).

Crystalline Proterozoic rocks form parts of the shallow sea floor and also poke up as scattered islands and seafloor outcrops particularly in the southern half of Spencer Gulf and at the Gulf entrance. The remainder of the seafloor is soft Cenozoic carbonate and siliciclastic sediments and sedimentary rocks together with minor Permian glacigene sediments. Extensive Quaternary calcarenites comprise beach-dune complexes of interbedded bioclastic sand blown onland during Pleistocene sealevel highstands, altered by meteoric diagenesis, and capped by a carapace of indurated calcrete. Topographic depressions within aeolianite complexes that are underlain by the Glanville Formation and St. Kilda Formation or Permian glacigene rocks form modern playa lakes and are sites of restricted carbonate and evaporite precipitation (Belperio 1995).

Fig. 9.4 A chart of Spencer
Gulf with an enlargement of
the northern sector at right,
illustrating surface sediment
facies and environments.
Compiled from our studies,
Gostin et al. (1984) and Fuller
et al. (1994)

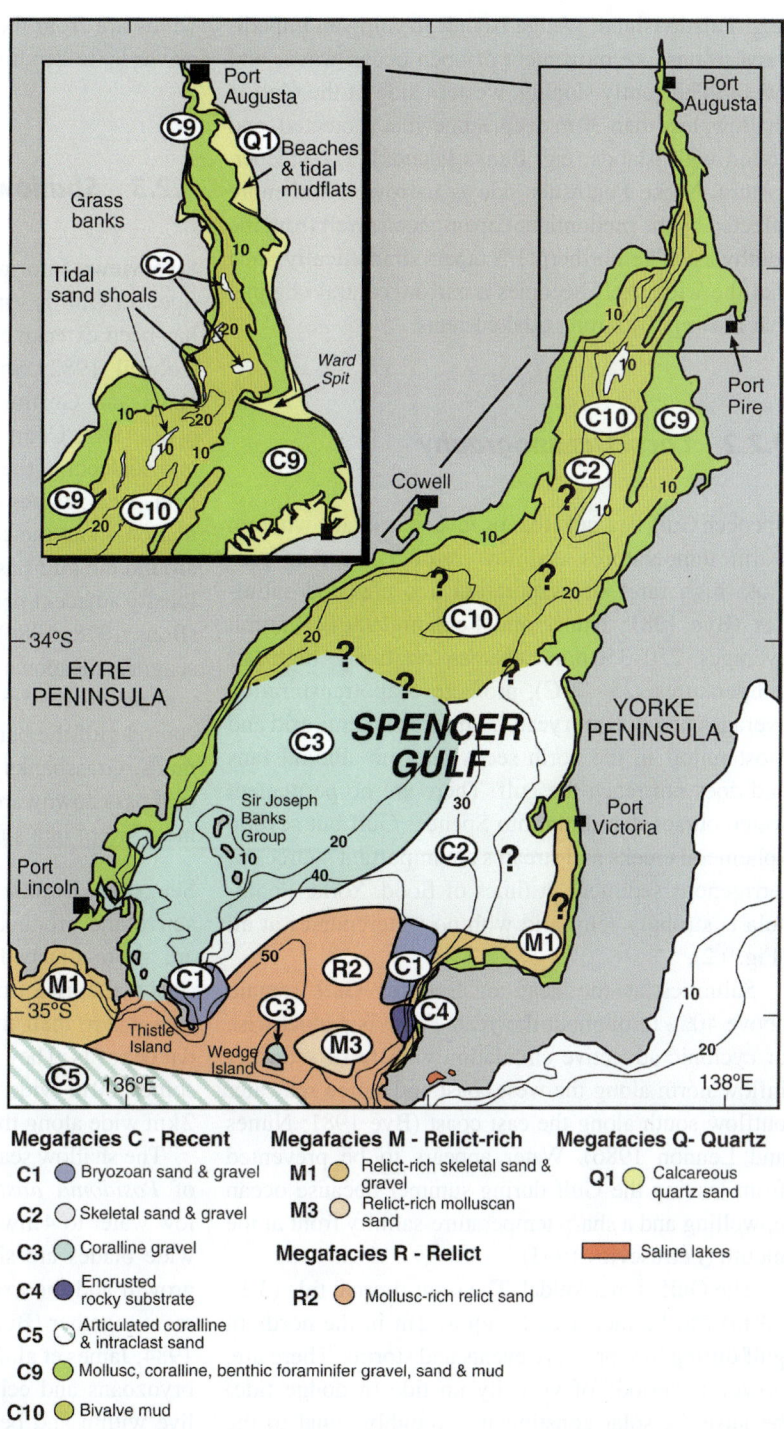

Megafacies C - Recent

C1 ⬤ Bryozoan sand & gravel

C2 ◯ Skeletal sand & gravel

C3 ◯ Coralline gravel

C4 ⬤ Encrusted
       rocky substrate

C5 ◯ Articulated coralline
       & intraclast sand

C9 🟢 Mollusc, coralline, benthic foraminifer gravel, sand & mud

C10 🟡 Bivalve mud

Megafacies M - Relict-rich

M1 ◯ Relict-rich skeletal sand &
       gravel

M3 ◯ Relict-rich molluscan
       sand

Megafacies R - Relict

R2 ◯ Mollusc-rich relict sand

Megafacies Q- Quartz

Q1 🟡 Calcareous
       quartz sand

🟧 Saline lakes

Overall Holocene deposition can be summarized as a
series of shoreline-attached sediment wedges passing
seaward into a thin deep-water sediment blanket.

The seafloor comprises a shallow, <10 mwd, 5–
10 km-wide shelf around the perimeter that drops gently

to a relatively flat seafloor at 30–60 mwd. The central
seafloor deepens southward such that the deepest part
lies just inside of the gulf mouth. The entrance from
the Southern Ocean is marked by a ridge at 50 mwd,
from which numerous islands rise to the surface

(e.g. Thistle Island, Wedge Island) forming an impediment to onshore movement of open ocean waves and swells. The gently sloping, western side of the Gulf is shallow, less than 30 m deep, somewhat protected, and dotted with islands (e.g. Banks Islands). The steeper, eastern, Yorke Peninsula side is narrower, and more affected by the predominant open-ocean swell from the southwest. The northern 1/3 tapers dramatically such that the wide shelf becomes a narrow central channel that is swept by strong tidal currents.

9.2.2 Local Oceanography

Spencer Gulf is a negative or inverse estuary wherein warm temperatures and low rainfall together promote high rates of evaporation and elevated salinities (Bye 1981; Nunes and Lennon 1986). Rainfall averages 250–350 mm whereas relatively high air temperatures (25–32°C), promote evapotranspiration averages of 2,400 mm year^{-1}. The area is semi-arid and most runoff in the north seeps into the alluvial fans and does not reach the gulf. There are no permanent watercourses that drain into Spencer Gulf but several ephemeral creeks and streams are important sources of terrigenous sediment in times of flood. Yorke Peninsula is strongly semi-arid with no watercourses at all (Fig. 9.2).

Salinities at the head of Spencer Gulf remain above 40‰ throughout the year. There is a clockwise or cyclonic advective circulation with fresh seawater inflow north along the west coast and saline seawater outflow south along the east coast (Bye 1981; Nunes and Lennon 1986). Water appears to be prevented from leaving the Gulf during summer because ocean upwelling and a sharp temperature-salinity front at the mouth (Petrusevics 1993).

The Gulf is macrotidal. The semi-diurnal tide (3.1–3.9 m) can be increased by up to 2 m in the northern gulf during low-pressure events and storms. There are, however, periods of virtually no tide (a dodge tide) because the solar constituent is roughly equal to the lunar constituent. Tidal currents reach peak velocity during spring tides of ~1 m s^{-1} (2 knots). At the head of the gulf, where the largest tides in the region occur, they are largely damped out by peritidal samphire and mangrove. Water movement overall results from tidal currents and wind-driven wave action; prevailing winds are from the north and south and can generate waves up to 2 m high in channels.

9.2.3 Shallow Neritic Facies

Overview: Information about recent sediments in Spencer Gulf is variable. The northern part of the gulf has been extensively studied (Burne 1982; Burne and Colwell 1982; Shepherd 1983; Gostin et al. 1984; Hails and Gostin 1984; Gostin et al. 1988; Barnett et al. 1997; Cann et al. 2000, 2002), the seafloor at and adjacent to the entrance has been surveyed (Fuller et al. 1994; James et al. 1997), but there is little information about the centre of the gulf proper. Shorelines around the gulf pass seaward into seagrass banks with locally adjacent peritidal beaches or peritidal mudflats (Bone 1978, 1984). In the southwestern corner these seagrass meadows grade outboard into coralline algal gravels whereas in most of the southern and possibly central gulf the banks descend downward into bivalve muds. Grassbanks in the narrow northern part of the gulf pass downward into a tidal channel of coarse sediment swept into subaqueous dunes (Fig. 9.4).

Seagrass Banks & Peritidal Facies: The most active and productive carbonate systems in the gulfs are shore-attached seagrass banks. These structures (facies C9) rise from water depths of 10 m to the surface where their inner parts are veneered with peritidal deposits. They are asymmetric in their distribution some 200 m wide along the western shore but up to 2 km wide along the eastern coast.

The shallow seafloor is covered with dense growths of *Posidonia australis* and *P. sinuosa* from spring low water to 4 mwd and *P. sinuosa* to 8–10 mwd. The wide blades are sites of prolific calcareous epiphyte growth and the extensive root systems bind the sediment together (Burne and Colwell 1982; Gostin et al. 1984; James et al. 2009). Molluscs, together with some bryozoans and echinoderms, ascidians, and hydroids live within and beneath the grass canopy and on the muddy sediment. These are locally mixed with Pleistocene Glanville Formation components, especially the warm-water bivalve *Anadara* and the large benthic foraminifer *Marginopora*. Sediments are grey, poorly sorted terrigenous and calcareous sandy muds with fragmentary gastropods and bivalves. An abundant epifaunal contains numerous benthic foraminifers, especially encrusting and symbiont-bearing types. The seagrass banks are cut by steep-sided sinuous channels up to 8 m deep that pass landward into intertidal creeks. Coarse sands and abundant shell lags floor channels and creeks. Sedimentation rates vary from 0.2 to 2.7 mm year^{-1} with accumulations of up to 6 m in the last 6000 years.

Peritidal environments are generally wide and form a veneer over the inner parts of the grassbanks because of progradation over the last 7000 years, driven largely by hydro-isostasy (Lambeck and Nakada 1990). Intertidal and supratidal muds are locally covered by cyanobacterial mats, particularly in mangrove and samphire environments. They are best developed in shallow depressions that stay moist but split and curl during long periods of desiccation.

The intertidal zone is a complex lateral mosaic of sand flats and beaches along exposed coasts with mangrove forests and samphire flats in more protected areas. The *low intertidal zone* comprises barren or *Zostera* and *Heterozostera* seagrass-covered sandflats. Sediment is grey to light grey poorly sorted terrigenous and calcareous shelly muddy sand with numerous bivalves and gastropods. In high-energy areas these sand flats terminate in shelly sand beaches (facies C2). The *middle intertidal zone* is dominated by mangrove woodland, a low forest that extends from mean sea level to spring high-tide and is crossed by a network of tidal channels. The mangroves are present southward to Franklin Harbour at Cowell on the west coast and to 20 km south of Port Pirie on the east coast. The distribution of mangroves is controlled by pneumatophores on the root system, which require air to flush salts from the tree. The sediment here is a mixture of terrigenous and carbonate mud that is covered to varying degrees by cyanobacterial mats. The siliciclastic clay fraction is both aeolian and eroded Hindmarsh Clay. The brown or bluish grey, muds are riddled with fragments of roots sheaths and fibres. The sediment is extensively burrowed by crabs and inhabited by gastropods, bivalves, and innumerable polychaetes, foraminifers, and diatoms. The *high intertidal zone*, landward of the mangroves, is a broad undulating plane with samphire-microbial communities. The sediment is pale brown to light grey, commonly mottled calcareous and terrigenous clay rich muds with scattered small gastropods, bivalves, foraminifers, and gypsum.

The *supratidal zone* comprises bare evaporative mud flats with local gypsiferous dunes and clay lunettes in protected settings with beach ridges and dunes in higher energy locations. Beach ridges are shelly quartz sands to gravels (facies Q1). Coastal dunes are quartz-rich shelly sands (commonly decalcified) with abundant gypsum silt. The poorly drained, saline, gypsiferous flats with cyanobacterial mats in ponded areas are weakly layered, fenestral, calcitic mud and mottled sandy clays with discoidal and layers of gypsum. Dolomite occurs as white aggregates along the seaward side in the uppermost 10 cm. A regional continental groundwater system flows seaward from the Flinders Ranges (Ferguson and Burne 1981) with aragonite and gypsum precipitation, megapolygons, and teepees present where the waters emerge as springs. Overall, however, there is relatively little authigenic carbonate in the supratidal sediments (Burne and Colwell 1982). The total thickness of the upper intertidal and supra-tidal sediments is less than a meter.

Rhodolith Pavements: Grass banks (facies C9) at the Gulf mouth grade seaward into relict-rich skeletal sands and gravels (facies M1) or skeletal sands and gravels (facies C2). Coralline algal gravels (facies C3) are developed in 20–40 mwd along the southwestern part of the gulf. Images show a seafloor covered with rhodoliths with associated *Chlamys* or *Pinna* shells and fenestrate bryozoan fragments. Most rhodoliths are nodular (not frucitose–short branching) with ellipsoidal to elliptical shapes and occur as a pavement wherein all rhodoliths are touching. The algal nodules are frequently turned over by currents and fish feeding.

Channel Sand Dunes: The narrow channel or trough between grass beds on both sides of northern Spencer Gulf is 10–25 mwd and floored by either subaqueous dunes (Facies C2) or by a thin veneer of sediment (facies C10) overlying stiff Hindmarsh Clay alluvial clays (Hails et al. 1984b). The linear belts and broad areas of tidal sand dunes are elongated parallel to the channel margin. Dunes are characterized by long strait crests, wavelengths = 2–20 m and heights = 0.5–1.3 m (Shepherd and Hails 1984). Individual dunes migrate 2–8 m year^{-1}. They are populated by a foliose rooted bryozoan (*Parmularia*) and an ascidian-dominated benthic community. The scoured channel floor between dunes supports a varied soft red algal, ascidian-gorgonian community on a shelly or sandy bottom. Intervening rocky substrates or hammer oyster (*Malleus*) clusters are typified by a sponge–anthozoan community that includes sponges, worms, echinoids, bryozoans (*Celleporaria*, *Anmathia*), as well as ascidians. Overall the biota is sparse and carbonate production is relatively low. It would appear that much of the coarse sediment in the dunes is allochthonous and from adjacent grass banks.

Gulf Floor: The central part of Spencer Gulf is poorly documented but appears to be mostly bivalve muds and sands (facies C10). Where studied in the deepest part inboard of the entrance (Fuller et al. 1994), the 30–50 m depression is floored by medium to fine grained sands of highly variable composition but mostly facies R2, mollusc-rich relict sand and facies C2, skeletal sand and gravel.

9.3   Gulf St. Vincent

9.3.1   General Attributes

This somewhat smaller ~200 km long embayment lies in a fault-bounded trough on top of the Cenozoic St. Vincent Basin (Belperio 1995). The juncture between the two major underlying Precambrian-Paleozoic crustal elements runs roughly down the axis of the gulf and just of the north coast of Kangaroo Island (Foden et al. 2006). The Tertiary rift basin was a large open gulf facing the Southern Ocean during Paleogene and early Neogene time (Lindsay and Alley 1995). It is now bounded on the east by the Mt. Lofty Ranges, part of the Adelaide fold and thrust belt. This eastern margin is a series of half-graben embayments formed by blocks tilting to the SE against arcuate faults. Along the southern margin it is framed by hard rock cliffs, with most of the intervening shoreline a series of low cliffs formed by soft sandy, calcareous and clayey Permian and Cenozoic sediments and sedimentary rocks (Figs. 9.1, 9.5, 9.6).

Gulf bathymetry is similar to that of Spencer Gulf. A shallow inboard shelf, <10 m deep surrounds the margin north of Adelaide and Edithburg, forming a 10–25 km-wide shallow platform, called the Orontes Shelf in the west (Shepherd and Sprigg 1976). Shallow portions of the gulf reflect adjacent geology. Linear

ridges of bedrock are common adjacent to hardrock outcrops and form subaqueous rocky reefs. Most of the modern gulf is a relatively flat seafloor with a maximum depth of ~40 m.

9.3.2   Local Oceanography

Gulf St. Vincent is characterized by an easterly low-salinity open marine flow up the west coast (Bye 1976). In summer this water rapidly increases in salinity northward due to evaporation to ~38–39‰ at the head (up to 47‰). Strong northward longshore drift retards the southward movement of this high salinity water mass until winter when it flows southward along the east coast. The annual range in surface water temperature is 10–22°C (50 year average). The gulf is mesotidal with the tidal range at Outer Harbour, Adelaide being ~2 m. The diurnal component of the tides vanishes at the equinox so the water level is constant for a whole day (Dodge Tide) (Fig. 9.2).

9.3.3   Shallow Neritic Sedimentary Facies

Overview: Gulf St. Vincent has a facies disposition similar to that in the larger Spencer Gulf, strandline carbonate-siliciclastic sand beaches and local muddy tidal flats that pass seaward into grassbanks (Fig. 9.6). Bivalve muds that are rich in terrigenous clays cover the shallow gulf floor.

Seagrass Banks and Peritidal Flats: The seagrass banks (facies C9), beaches (facies Q1) and peritidal mud flats that occur along the shores of the northern half of the gulf are similar to those in Spencer Gulf (Belperio 1995). The ongoing progradation seems to have, however, been more extensive and rapid (Cann and Gostin 1985; Cann et al. 2009).

Rhodolith Pavements: Coralline algal gravels (facies C3) are only present on MacIntosh Bank (east of Edithburgh), a ridge that rises to 15 mwd from a depth of 26 mwd, above a seafloor that appears as a virtual desert of facies M3 and M5 except for scattered molluscs and seapens (Shepherd and Sprigg 1976). The

Fig. 9.5 Marginal marine environments adjacent to the South Australian Sea. (**a**) Proterozoic crystalline rocks overlain in the distance by modern aeolianites, southern Yorke Peninsula; foreground width of image ~8 m. (**b**) The western cliffed shore of Fleurieu Peninsula in southern Gulf St. Vincent (cliffs are ~80 m high) composed of deformed Cambrian metasedimentary rocks, (photograph courtesy N. George). (**c**) Cliffs of Eocene Tortachilla Limestone and Blanche Point Formation overlain by Pliocene and Pleistocene sands and clays (cliffs ~25 m high), at Maslins Bay south of Adelaide on Gulf St. Vincent. (**d**) Bridgewater Formation aeolianites, southern shore of Kangaroo Island; cliff in background ~12 m high. (**e**) Extensive beach and adjacent low dunes composed of carbonate sand from adjacent grass beds, Marion Bay, southern Yorke Peninsula along the northern shore of Investigator Strait. (**f**) Saline lake ~200 m wide in farmland ~80 km southeast of Adelaide (Fig. 9.1)

coralline algal pavement environment, the top of which is swept by currents that can exceed 1.6 cm s⁻¹ (3 knots) and is capped by coarse calcareous sand (facies C2) and rhodoliths.

Gulf Floor: Sediments on the gulf floor deeper than ~10 mwd are somewhat better known than those in Spencer Gulf (Shepherd and Sprigg 1976; Cann et al. 1993; Bone et al. 2006; Fox et al. 2007). They are

Fig. 9.6 Chart of Gulf St.
Vincent illustrating the different surface sedimentary
facies; compiled from our
information and Shepherd
and Sprigg (1976)

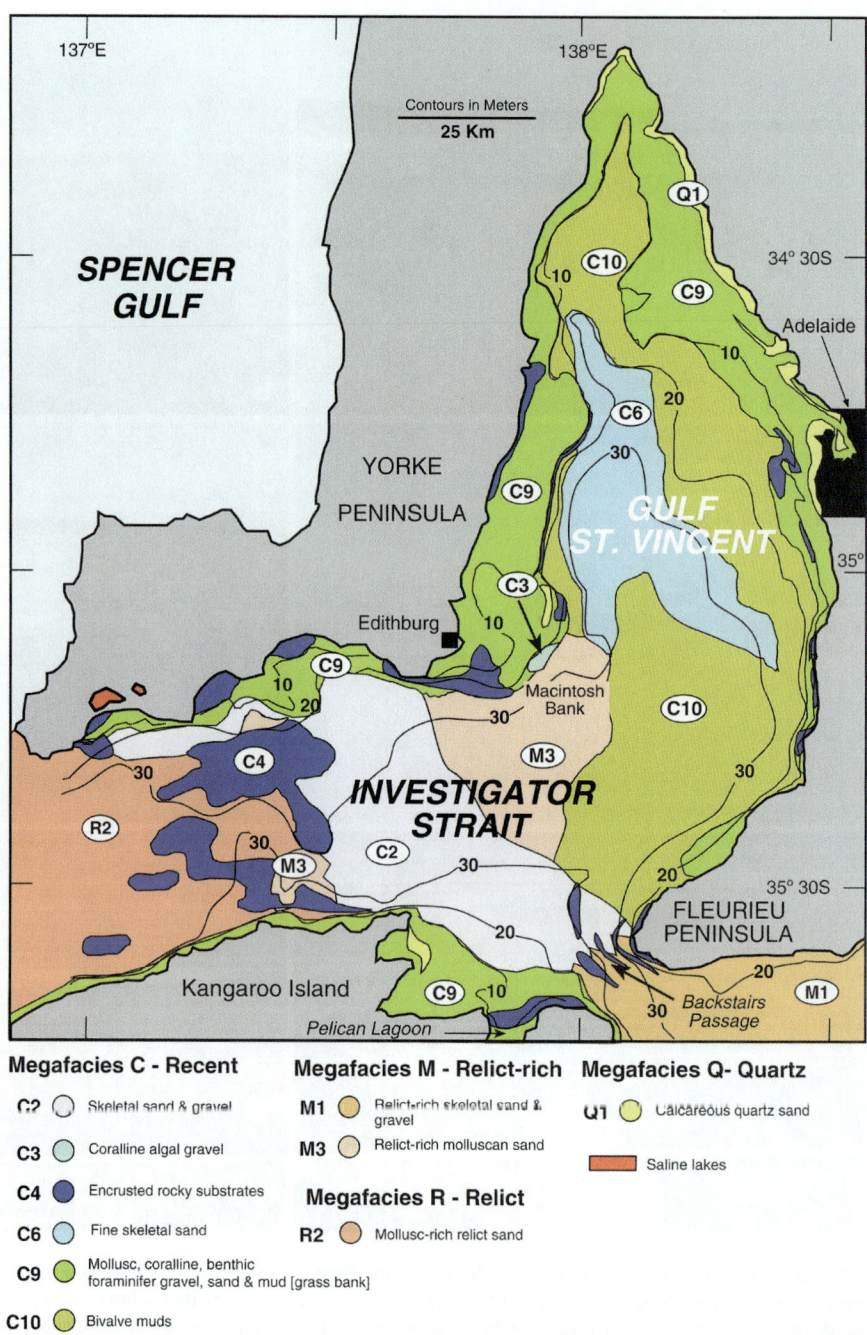

predominantly fine grained (facies C10); it is a sediment sink. The whole basin seafloor, however, lies in
the photic zone although there is increased turbidity
towards the head of the gulf as reflected by the depth
of *Posidonia australis* growth; 30 mwd in Investigator
Strait, 20 mwd in the lower gulf, and 10 mwd in the
upper gulf. Both *Heterozostera tasmanica* and *Halophila ovalis* grow to 35 mwd throughout.

Deposits are a mixture of siliciclastic and bioclastic
carbonate mud. The fine carbonate is a combination
of autochthonous and allochthonous, disarticulated
and bioeroded skeletons, whole benthic foraminifers,
and pelagic skeletons. Siliciclastic sands and muds
result from episodic runoff from the Mount Lofty
Ranges and erosion of Cenozoic sediments at the
shoreline. The most obvious macrobiota throughout

are bivalves together with bryozoans, sponges, ascidians, ophiroids, crinoids and innumerable burrowing echinoderms. Carbonate in the muds is dominated by Mg-calcite.

Silty, very fine grained biofragmental, calcareous sand, with <20% siliciclastic sand and clay particles (facies C6) forms the seafloor in the northern and central parts of the gulf. Coppices of living and piles of dead razor clams support a rich epizoic community of small sponges, ascidians, and bryozoans. In deeper waters between 25 and 35 mwd, the fine sediment still contains a few *Pinna* but they are exceeded by numerous epizoic bryozoans, ophiroids, and crinoids.

By contrast, siliciclastic-rich, fine, biofragmental sands, silts and muds, with up to 50% quartz and clay (facies C10) lie along the eastern side of the Gulf. In shallow water, 15–30 mwd, the fine sands have a conspicuous ascidian and scallop epifauna whereas in the deepest part of the Gulf, 32–40 mwd, the fine sediments have a prominent bivalve (*Malleus* and *Pinna*) biota. These deeper environments are not well understood but as well as a sparse infauna there is a scattered epifauna of scallops, small sponges, and some bryozoans, particularly *Parmularia* sp. All sediment is conspicuously burrowed.

9.4   Investigator Strait and Backstairs Passage

9.4.1   General Attributes

Gulf St. Vincent is connected to the open ocean via two straits, the broad Investigator Strait between Kangaroo Island and Yorke Peninsula (27–35 m deep, 40 km wide) and the narrow Backstairs Passage (35–70 m deep, 15 km wide) that separates Kangaroo Island from Fleurieu Peninsula (Fig. 7.5). Investigator Strait is deep and is floored by seafloor reliefs of Pleistocene aeolianite dunes. Backstairs Passage is a trough whose bottom is a planed off Permian muddy sandstone surface on which large ripples of mixed Permian siliciclastic and recent carbonate sands up to 0.5 m high are formed by strong tidal currents. Coarse sediments in Investigator Strait are winnowed by both storm waves from the SW and strong tidal flows, with the fine fraction swept eastward to settle

out in the quieter, deeper waters of the Gulf proper (Fig. 9.1).

9.4.2   Local Oceanography

Low salinity open ocean waters flow eastward along the northern margin of Investigator Strait (Bye 1976). In winter, warm saline waters flow down the eastern side of the Gulf and exit as a well mixed stream via Backstairs Passage onto the Lacepede Shelf. There are, however, variations with periods during the summer when the outflow moves as an underflow along the north coast of Kangaroo Island. The floor of Investigator Strait is swept by strong tidal currents of 50–200 cm s^{-1} (1–4 knots) that are weakest near the coast and strongest in the centre of the strait. Similar currents across Troubridge Shoals at the eastern end of Investigator Strait are ~1 m s^{-1} (2 knots). Tidal currents in Backstairs Passage are ~1–2 m s^{-1} (2–4 knots) (Fig. 9.2).

9.4.3   Shallow Neritic Sedimentary Facies

Investigator Strait: The northern coast, at the southern end of Yorke Peninsula, is a series of prominent Middle Proterozoic crystalline rock headlands capped by spectacular Bridgewater Formation aeolianites and intervening bays with aeolianites and sand beaches that pass oceanward into seagrass banks. The southern erosional coast is hard, Neoproterozoic and Cambrian rocks and local outcrops of Miocene and Permian sediments. Open embayments are filled with seagrass meadows (facies C9) that pass landward into beaches and low dunes (Fig. 9.6).

The Strait shallows to less than 30 mwd as a N-S bar at the mid-point and deepens again to 40 mwd at the southern end of Gulf St. Vincent. Lithified and calcretized shell beds form much of the seafloor as rocky reefs with up to 5 m or more relief. More specifically, SCUBA observations (Shepherd and Sprigg 1976) indicate rocky reefs veneered with coarse, shelly sand. Submerged Pleistocene aeolianite dunes are prominent, forming rough bottom topography, particularly along the north shore of Kangaroo Island, and supporting a ota of algae, massive erect orange sponges, crinoids, and epifaunal molluscs (facies C4). Oscillatory action of the waves forms small subaqueous dunes (wavelength = 1–2 m) and straight to interference wave ripples (wavelength = 10 cm) in facies T sediments at the western end of the Strait. *Posodonia* seagrass meadows extend to depths of 30 m, some 20 km out from shore along the southern shore of Yorke Peninsula and eastern Kangaroo Island with deeper areas to 35 mwd supporting scattered *Heterozostera* grass communities. Sediments here (facies C2 and M3) are grey in colour with many grey-black stained particles due to the reducing conditions amongst the seagrass roots.

The sand biota throughout is classified by Shepherd and Sprigg (1976) as belonging to a *Heterozostra* (seagrass) and *Lunulites* (vagrant bryozoan) and brachiopod (*Anakinetica* sp.) community with few other animals. The 5–60% coarse fraction is a cornucopia of bivalves (some of which are grass dwellers), large *Chlamys*, oysters, laterite, blackened calcrete fragments, white Permian sandstone clasts, and oblate, poorly lithified, calcarenite lumps, many of which were beach cobbles (facies C2 and C9). Pelican Lagoon off northern Kangaroo Island contains a rich sponge community (Millikan 1994; Wirtz 1994).

Backstairs Passage: The bedrock surface is covered by a thin layer of coarse sand, small flat pebbles, and biogenic carbonate (facies C2, C4 and M1). There are many big sponges, large bryozoans, and numerous brachiopods (*Anakinetica cumingi*) at depths greater than 40 m.

9.5   Lincoln Shelf

9.5.1   General Attributes

This region of the continental shelf between 35° and 36° S lies south of Spencer Gulf at the entrance to Investigator Strait and west of Kangaroo Island. Kangaroo Island which is part of the Mount Lofty Ranges (Flottmann and James 1997) extends well onto the shelf and nearly to the shelf edge. As such the island, together with the seasonal outflow of saline waters from the gulfs, act as a major oceanographic and biological barrier (Fig. 9.1).

9.5.2   Oceanography

This shelf sector is affected by shelf bathymetry and the nearby gulfs and like the Great Australian Bight is best summarized in terms of season. In *early summer* surface waters come from the Great Australian Bight, are relatively warm (18–19°C), and of moderate salinity (36.0–36.5‰). There is a pronounced thermocline at 30–50 mwd below which offshore cold (<15°C) oceanic waters intrude onto the shelf and such that upwelling near the mouth of Spencer Gulf effectively isolates gulf waters during the summer months. The thermocline deepens to 100 m by mid-summer. Waters inside the gulfs are generally well mixed. By *mid-late autumn* oceanic waters retreat off the shelf, waters in the gulfs undergo convective overturning (Nunes and Lennon 1986) and the thermocline on the shelf is destroyed. During the *winter* shelf waters are well mixed, warm, and saline and flow southeastward along the shelf (Figs. 9.2, 9.7).

Cooling of saline waters at the head of Spencer Gulf in winter leads to a dense water mass (Bowers and Lennon 1987; Nunes Vaz et al. 1990) which descends, moves out of Spencer Gulf, and flows across the shelf. This warm (>17°C) 20 m-thick, 100 km-long, bottom-hugging, saline (36–37‰) current is many hundred meters wide and has velocities of 0.1 cm s^{-1} (0.2 knots). It is denser (>27 kg m^{-3}) than the shelf water (26.8 kg m^{-3}) and the resultant gravity current or plume that flows out of the eastern mouth of Spencer Gulf and then south and east along Kangaroo Island descends to reach neutral buoyancy at 250–300 mwd (Godfrey et al. 1986; Lennon et al. 1987). The water, which is rich in particulate organic matter, but depleted in dissolved P, N, and C (Smith and Veeh 1989), may influence biological productivity at the shelf edge as it flows southeastward. Chlorophyll-a levels are consistently high >0.5 mg m^{-3}, (mesotrophic) likely due to a combination of seasonal upwelling and outflow from the gulfs.

In summary, warm waters of the South Australian Current flow eastward onto the Lincoln Shelf, parts flowing into the gulfs, along their western margins, and parts deflected outboard by Kangaroo Island, to flow southeastward at, or along, the shelf edge. Waters that emerge from Spencer Gulf flow onto the Lincoln Shelf as cool and saline bottom currents during winter. Thus, the Lincoln Shelf is an area of complex, mixed waters today and has likely been so since waters flooded the gulfs in the early Holocene.

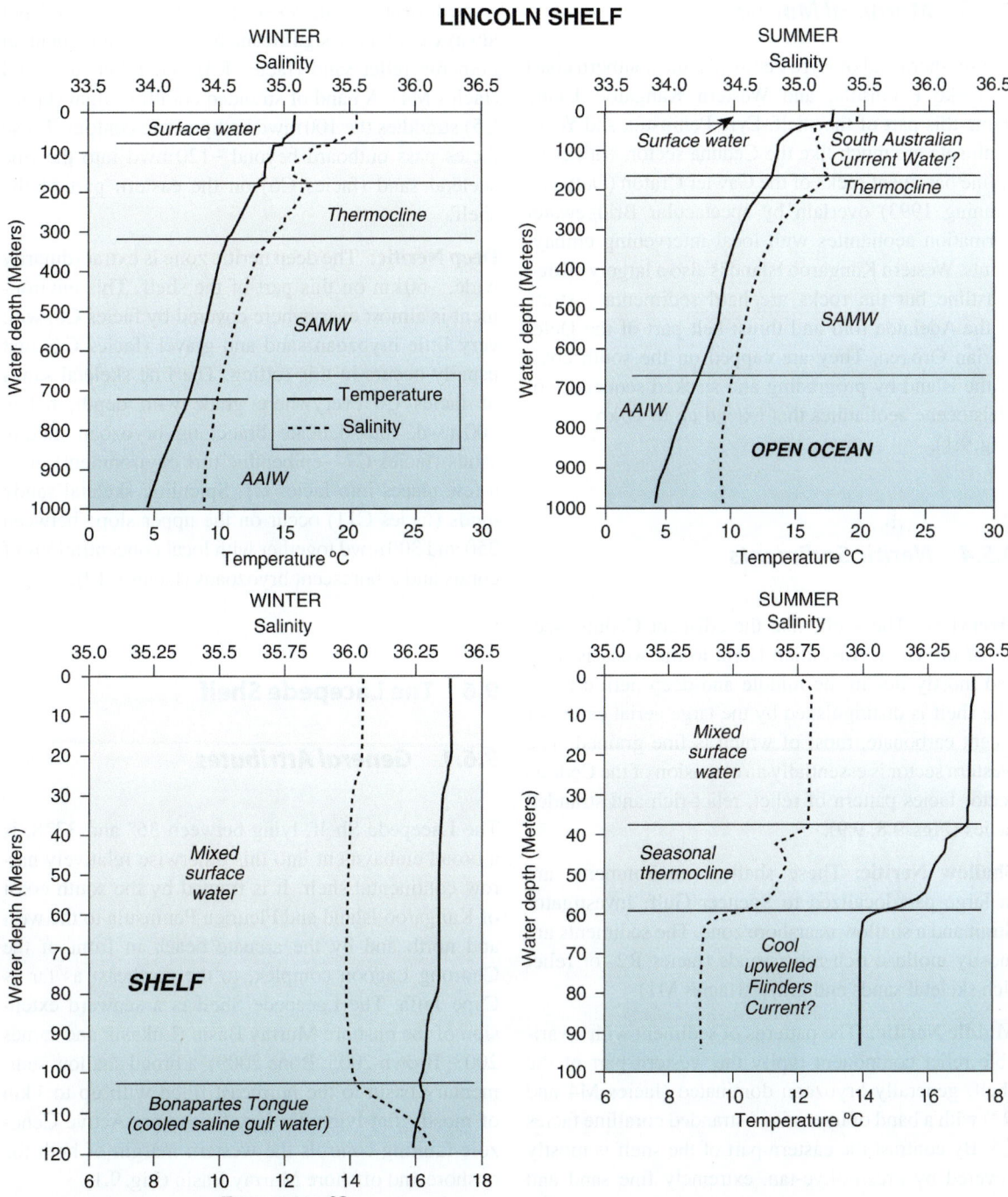

Fig. 9.7 Temperature and salinity profiles for the upper 1,000 m of the open ocean water column (*top*) and the upper 100 m of the shelf water column (*bottom*) during summer and winter on the Lincoln Shelf. *AAIW* Antarctic Intermediate water, *SAMW* Subantarctic Mode Water. The thermocline and surface layers are composed of Sub-tropical Surface Water modified by summer heating and winter cooling. Saline bottom waters flow out from Spencer Gulf and across the shelf during winter; summer upwelling is thought to be derived from waters moved by the Flinders Current. (Compiled from CSIRO (2001) Data Trawler, Marine and Atmospheric Research (CMAR) Data Center, Commonwealth Scientific and Industrial Research Organization)

9.5.3   Marginal Marine

The southern end of Eyre Peninsula, the southern coast of Yorke Peninsula, and Western Kangaroo Island frame this part of the shelf. Eyre Peninsula and Yorke Peninsula are much like the Ceduna sector, with crystalline basement rocks of the Gawler Craton (Daly and Fanning 1993) overlain by spectacular Bridgewater Formation aeolianites with local intervening embayments. Western Kangaroo Island is also a largely cliffed coastline but the rocks are hard sedimentary strata of the Adelaide fold and thrust belt part of the Delemarian Orogen. They are capped on the south coast of the island by prograding and stacked sequences of Pleistocene aeolianites that extend up to 10 km inland (Fig. 9.1).

9.5.4   Neritic Sediments

Overview: The shelf, like the adjacent Ceduna sector of the Great Australian Bight to the west, is deep and mostly lies in the middle and deep neritic zone. The shelf is distinguished by the large aerial extent of recent carbonate, most of which is fine grained. The western sector is essentially an extension of the Ceduna sector facies pattern of relict, relict-rich and stranded facies (Figs. 9.8, 9.9).

Shallow Neritic: These shallow environments are in large part localized to Spencer Gulf, Investigator Strait and a shallow nearshore zone. The sediments are mostly mollusc rich relict sands (facies R2) or relict rich skeletal sands and gravels(facies M1).

Middle Neritic: The patterns of sediment with a variable relict component typify the western part of the shelf, generally bryozoan dominated (facies M4 and R3) with a band of intermixed stranded coralline facies C5. By contrast the eastern part of the shelf is mostly covered by green-olive-tan, extremely fine sand and silt, but with no mud and a minuscule coarse fraction (facies C6). The same facies occupies a wide region just west of Kangaroo Island. Bottom images of the outer shelf show interference wave ripples (wavelength = 10 cm) partly degraded by bioturbation, scattered elongate, branching sponges. Environments range from rippled sand barrens (albethey fine grained) to local epibenthic scrub.

Sediments in the west are more complicated but always contain a significant bryozoan component in both the relict sand (facies R3) and relict-rich sand (facies M4). A band of stranded coralline algal (facies C5) straddles the 100 mwd bathymetric contour. These facies pass outboard beyond ~120 mwd into the fine skeletal sand (facies C6) on the eastern part of the shelf.

Deep Neritic: The deep neritic zone is extraordinarily wide, ~60 km on this part of the shelf. This environment is almost everywhere covered by facies C6, with very little bryozoan sand and gravel (facies C1) that usually occurs in this setting. The fine skeletal sands of facies C6 everywhere grade with depth, below 200 mwd, into delicate branching bryozoan muddy sands (facies C7—epibenthic turf environment) or in a few places into facies C1. Spiculitic skeletal sandy muds (facies C11) occur on the upper slope between 250 and 300 mwd together with local concentrations of corals and arborescent bryozoans (facies C12).

9.6   The Lacepede Shelf

9.6.1   General Attributes

The Lacepede Shelf, lying between 36° and 37° S, is a broad embayment into this otherwise relatively narrow continental shelf. It is framed by the south coast of Kangaroo Island and Fleurieu Peninsula to the west and north and by the arcuate beach in front of the Coorong Lagoon complex, to the northeast as far as Cape Jaffa. The Lacepede Shelf is a seaward extension of the onshore Murray Basin (Lukasik and James 2003; Brown 2005; Bone 2009), a broad shallow sedimentary basin to the northeast filled with up to 1 km of mostly flat-lying Tertiary limestone. Active Cenozoic faulting controls the western margin of both the onshore and offshore Murray Basin (Fig. 9.1).

Kangaroo Island and the Mt. Lofty Ranges of Fleurieu Peninsula are horst-like features of deformed Proterozoic or Paleozoic basement with Delamerian elements of the Encounter Bay granitoid suite in the west and the Padthaway suite in the east. The south coast of Kangaroo Island consists of metamorphosed Neoproterozoic sedimentary rocks facing the open ocean exposed positions, with Bridgewater Formation

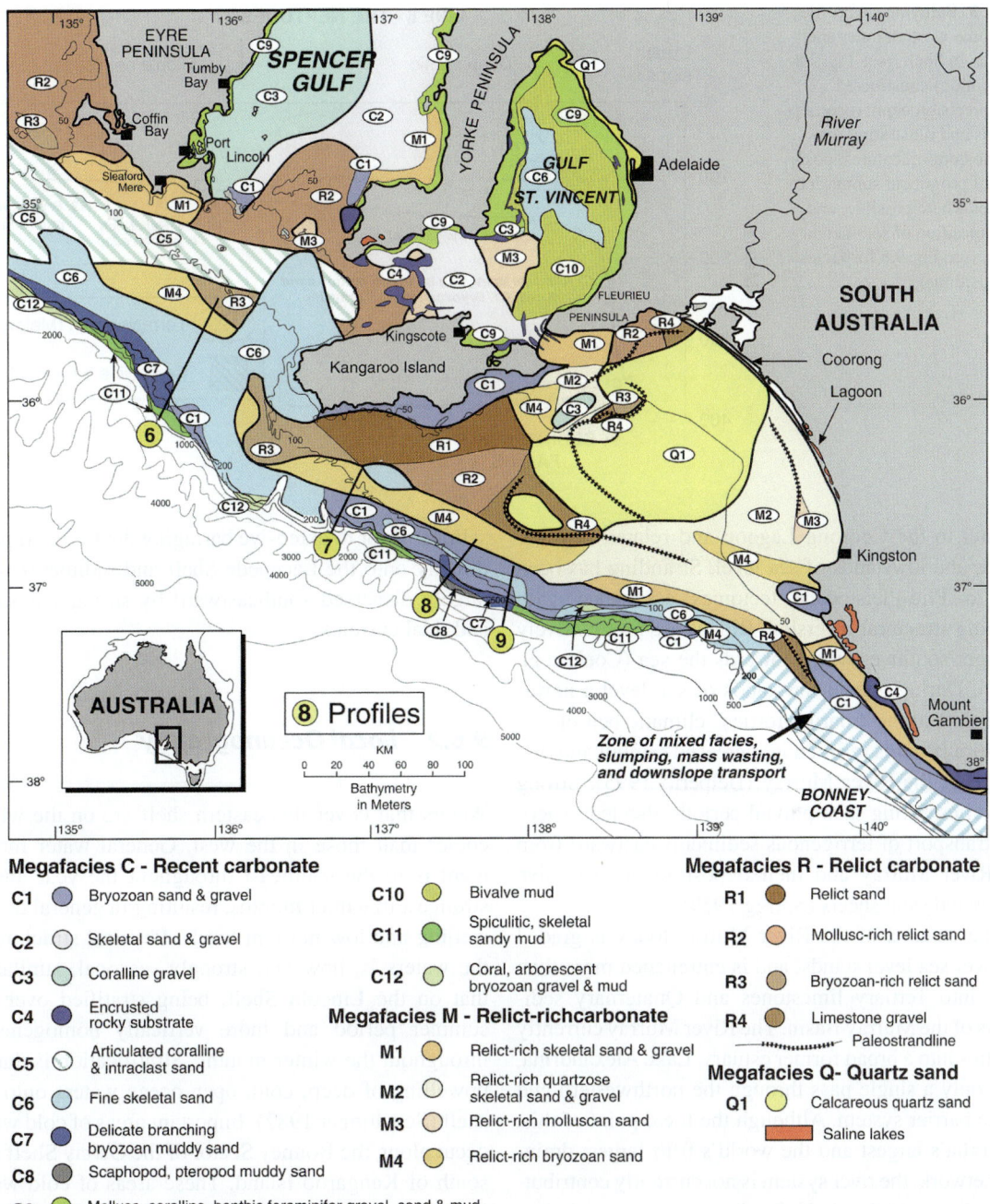

Fig. 9.8 Composition of surface sediment facies in the South Australian Sea. Lines of section in Figs. 9.9 and 9.11 are designated by number

aeolianites in somewhat sheltered embayments. The south coast of Fleurieu Peninsula is composed of Cambrian sedimentary rocks. Rocks immediately underlying the Lacepede Shelf surface are Eocene-Miocene bryozoan-rich carbonates (Sprigg 1952; James and Bone 1989; Hill et al. 2009).

The relatively flat-lying Tertiary carbonates immediately onshore are veneered by a thin succession of stranded Plio-Pleistocene beach-dune ridges that extend inland for 200 km (Sprigg 1952, 1979; Cook et al. 1977; Schwebel 1983). Sprigg (1979) identified 30 linear paleoshoreline systems, each one roughly

Fig. 9.9 Bathymetric profile across the western outer shelf of Lincoln Shelf, (see Fig. 9.8 for location) constructed from precision depth recorder records and illustrating distinctive bottom features, positions of prominent submarine terraces where possible, and the disposition of sedimentary facies. (see Fig. 9.8 for facies compositions)

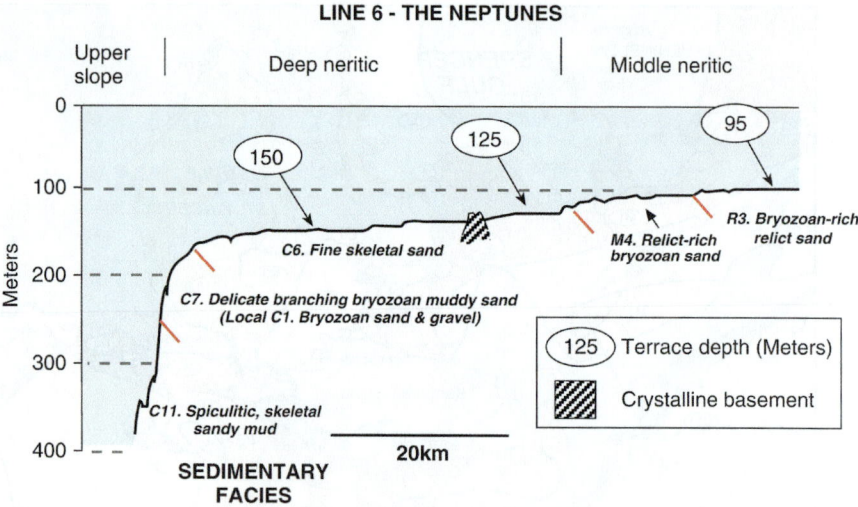

parallel to the Coorong Lagoon and related to high-stands and lowstands of sea level. Stranding has been due to Plio-Pleistocene tectonics associated with ongoing structural inversion. Ridges are progressively younger southwestwards towards the sea (Cook et al. 1977). During glacial lowstands of sea level a north-ward shift of the "roaring forties" climatic belt of low pressure systems resulted in increased rainfall and dis-charge of the River Murray (Belperio 1995). Strong westerlies during such pluvial periods also led to aeo-lian transport of terrigenous sediments eastward from the River Murray and their deposition as extensive downwind sand sheets (Sprigg 1979).

The channel of the River Murray today is graded to lower sea level stands, and is entrenched more than 50 m into Tertiary limestones and Quaternary sedi-ments of the Murray Basin. The River Murray currently empties into a broad former estuary, Lake Alexandrina, with only a single pass through the northwestern end of the barrier system. Although the focal point of this, Australia's largest and the world's fifth largest drain-age network, the river system is not currently contribut-ing sediment to the shelf. Gradients are extremely low along the distal 600 km and flow is relatively sluggish due to over-extraction of water throughout the entire drainage basin, with the result that most of the bedload is trapped in upstream point-bar systems and the lower fresh-water lakes that are now almost dry. The River has been heavily dammed for the last 100 years. The relatively minor portion of the suspended load that is permitted into Lake Alexandrina no longer reaches the ocean. Conversely, fresh water no longer flows into the Coorong Lagoon, thereby leading to ever increasing

salinity. Prior to extensive barraging the River emptied directly onto the Lacepede Shelf and sediments were largely dispersed southeastward by strong longshore and tidal currents.

9.6.2   Local Oceanography

Waters that cover this eastern shelf are on the whole cooler than those in the west. General water move-ment is to the southeast throughout the year and is strongest in winter months, resulting in general down-welling and low nutrient levels. Physical structure of the waters is, however, strongly seasonal, similar to that on the Lincoln Shelf, being stratified over the summer period and more vertically homogeneous throughout the winter months. Stratification is due to upwelling of deep, cold, open ocean waters onto the shelf (Schahinger 1987). Important areas of cold water occur along the Bonney Sector of the Otway Shelf and south of Kangaroo Island. These areas of cold water incursion have alongshore velocities of 25–40 cm s^{-1} and can transport waters 215–430 km over a 10-day period. Thus, water can potentially be transported as far west as Eyre Peninsula, where local winds during subsequent upwelling bring it to the surface. Middle-ton and Bye (2007) suggested that a pool of upwelling water is likely to be maintained off Kangaroo Island (Figs. 9.2, 9.10).

Bonney Coast upwelling seems to be especially important because Cape Jaffa marks the eastern-most extent of many broadleaf sea grasses but more

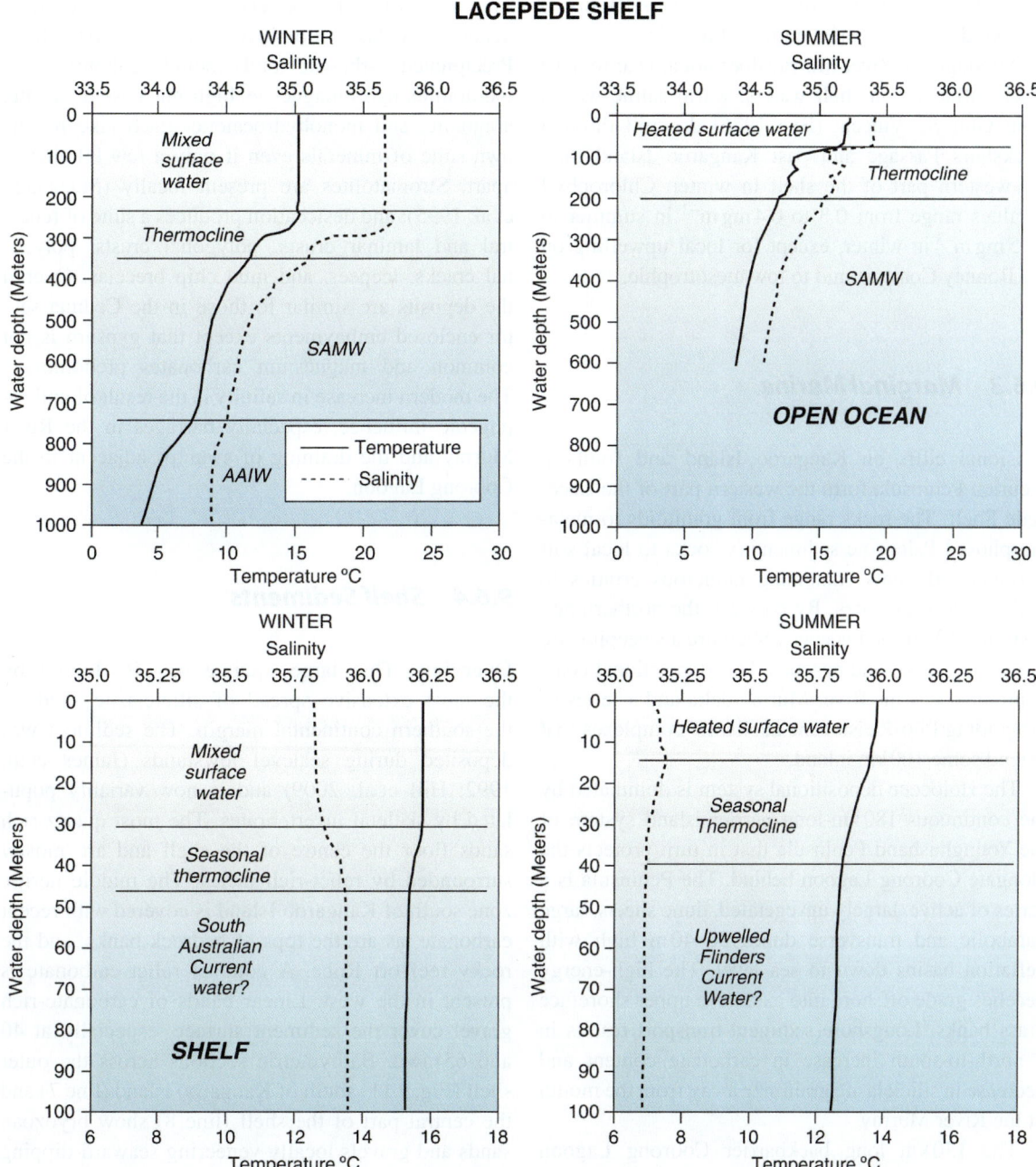

Fig. 9.10 Temperature and salinity profiles for the upper 1,000 m of the open ocean water column (*top*) and the upper 100 m of the shelf water column (*bottom*) during summer and winter seaward of the Lacepede Shelf. *AAIW* Antarctic Intermediate water, *SAMW* Subantarctic Mode Water. The thermocline and surface layers are composed of Sub-tropical

Surface Water modified by summer heating and winter cooling. Summer upwelling is thought to come from waters moved by the Flinders Current. (Compiled from CSIRO (2001) Data Trawler, Marine and Atmospheric Research (CMAR) Data Center, Commonwealth Scientific and Industrial Research Organization)

importantly, the northernmost extent of the large kelps *Durvillea* and *Macrocyctis*, phaeophytes that dominate all nearshore environments to the east and around all of Tasmania.

When modeled, upwelling should also be strong off the Coorong (Middleton and Platov 2003). As it upwells off the eastern Coorong the water is moved to the west and northwest. These winds also lower coastal sealevel (~25 cm), thus driving a weak (<10 cm s⁻¹) westward flowing coastal current (Fig. 9.2).

Although the River Murray does not appear to have much influence on shelf waters, warm saline waters from Gulf St. Vincent (Bye 1976) do spill through Backstairs Passage and past Kangaroo Island onto the western part of the shelf in winter. Chlorophyll a values range from 0.3 to 0.4 mg m⁻³ in summer to >0.5 mg m⁻³ in winter, except for local upwelling on the Bonney Coast, is mid to low mesotrophic.

9.6.3 Marginal Marine

Erosional cliffs on Kangaroo Island and southern Fleurieu Peninsula form the western part of the Lacepede Shelf. The rocks range from granitoids to metamorphosed Paleozoic sedimentary rocks to local soft Permian sediments containing numerous erratics to Pleistocene aeolianites. By contrast, the northern and eastern sides of the Lacepede Shelf are a sweeping arc of barrier islands and beaches. This depositional coast is produced by the River Murray delta and a series of prograding Plio-Pleistocene aeolianite complexes that extend some 100 km inland.

The Holocene depositional system is dominated by the continuous 180 km-long barrier island system of the Younghusband Peninsula that in turn protects the elongate Coorong Lagoon behind. The Peninsula is a series of active, largely unvegetated, dune sheets, large parabolic and transverse dunes 30–40 m high with deflation basins down to sea level. The high-energy beaches grade offshore into extensive upper shoreface grass banks. Longshore sediment transport results in a north-to-south increase in carbonate content and decrease in siliciclastic grain size away from the mouth of the River Murray.

The 130 km long backbarrier Coorong Lagoon is a series of depressions that become increasingly restricted and more saline southward (see Chap. 5 and Belperio 1995 for details). Waters are derived from winter rains and groundwater. Lagoon sediments are very fine bioclastic high-magnesium calcite and aragonite sand and pelleted mud derived from molluscs dominated by the gastropod *Coxiella* sp., ostracods, and benthic foraminifers. Strandline environments are characterized by a unique bryozoan-serpulid community (Bone and Wass 1990; Sprigg and Bone 1993).

The southern end is a series of isolated lacustrine depocenters (lakes) dominated by groundwater flow. Precipitated carbonate muds include dolomite, protodolomite, hydromagnesite, high-magnesium calcite, aragonite, and monohydrocalcite. Each lake has its own suite of minerals even if only a few kilometers apart. Stromatolites are present locally (Mazzoleni et al. 1995); and desiccation produces a suite of fenestral and laminar crusts, polygonal crusts, polygonal cracks, teepees, and mud chip breccias. Overall the deposits are similar to those in the Ceduna sector enclosed embayments except that gypsum is not common and magnesium carbonates predominate. The modern increase in salinity is the result of anthropogenic influence, especially barrages in the River Murray and the draining of swamps adjacent to the Coorong Lagoon.

9.6.4 Shelf Sediments

Overview: This large arcuate bay is floored by the most extensive spread of siliciclastic sand on the southern continental margin. The sediment was deposited during sealevel lowstands (James et al. 1992; Hill et al. 2009) and is now variably populated by skeletal invertebrates. The most quartz-rich sands floor the centre of the shelf and are mostly surrounded by relict-rich facies. The middle neritic zone south of Kangaroo Island is covered with recent carbonate, as are the tops of bedrock banks, and the rocky reef off Robe. A zone of relict carbonate is present in the west. Linear bands of carbonate-rich gravel cover the sediment surface, especially at 40 and 65 mwd. Bathymetric sections across the outer shelf (Fig. 9.11) south of Kangaroo island (line 7) and the central part of the shelf (line 8) show bryozoan sands and gravels locally veneering seaward-dipping submarine outcrops of Cenozoic limestone. Slumps and listric faults characterize the slope towards the east (Fig. 9.11—line 9, 9.8).

Shallow Neritic: The present outlet of the River Murray can be classified as a wave-dominated delta (cf. Galloway 1975) in a region of microtidal range, whose front is formed by beach-barrier shorelines of the Coorong Strand and whose delta plain is the strandplain of paleodune complexes to the north. The

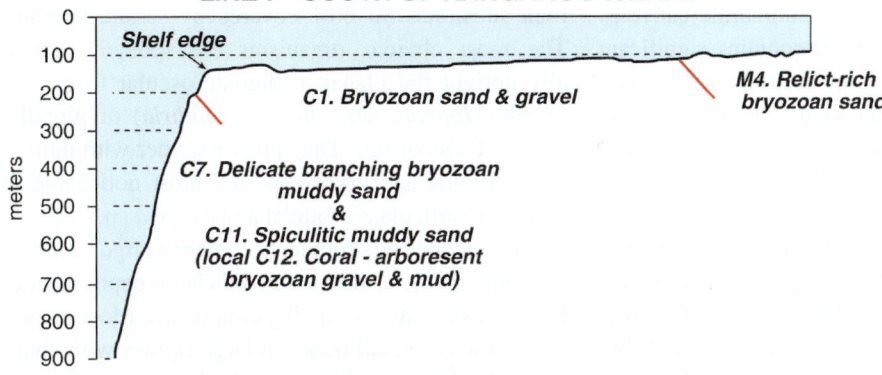

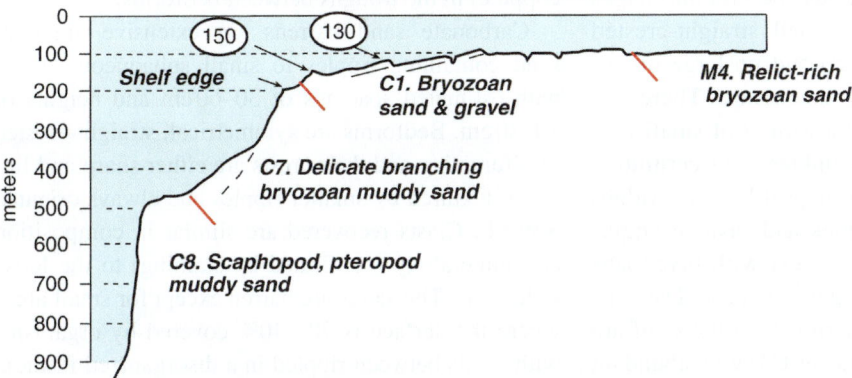

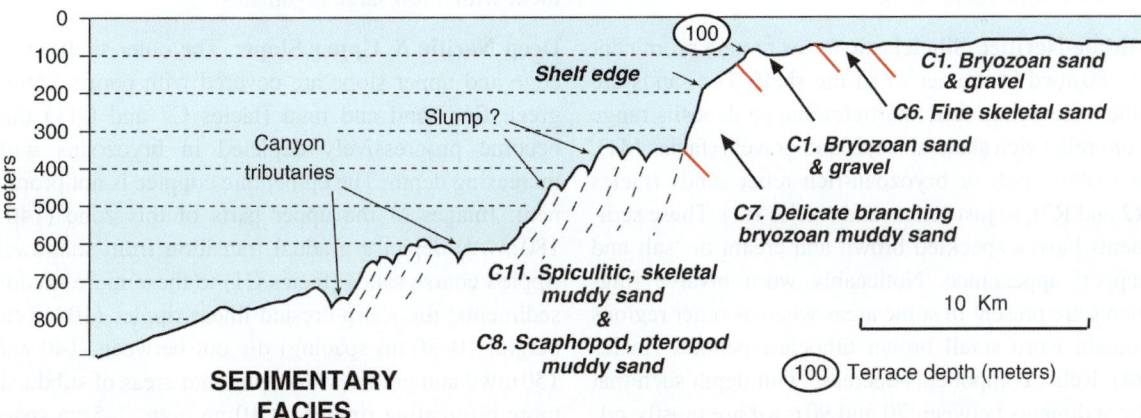

Fig. 9.11 Bathymetric profiles across outer part of the Lacepede Shelf (see Fig. 9.8 for locations) constructed from precision depth recorder records and illustrating distinctive bottom features, positions of prominent submarine terraces where possible, and the disposition of sedimentary facies. (see Fig. 9.8 for facies compositions)

barrier-beach system is as much constructed from offshore-derived carbonate as river-borne terrigenous sand.

The extensive terrigenous sand sheets on the shelf attest to the importance of the River Murray in delivering sediment to this region over the long term and confirm that during lower stands of sea level, siliciclastic sands were spread across the now-submerged delta plain (Hill et al. 2009). Even when sea level was low the same processes would be operative; terrigenous clastics would be largely trapped along the shoreline and only spill over the edge via submarine canyons. All sediment inboard of 80 mwd contains significant amounts of quartz, but the content drops off abruptly below these depths. Seafloor sands adjacent to Fleurieu Peninsula are more calcareous as facies R2 (mollusc-rich relict sand) and facies M1 (relict-rich skeletal sand and gravel).

The siliciclastic grains in this rippled sand barren environment are fine to medium grained and generally very well sorted (facies M2 and Q1). Bivalves (*Glycymeris*, *Katelysia*) are mostly infaunal but epifaunal *Chlamys* is locally conspicuous. The seafloor is smooth and monotonous and bottom images show clean sediments with small straight-crested wave ripples (10–15 cm apart) and no large organisms other than burrowing anemones. There are also scattered slit shells and a variety of small gastropods (olives, cones, periwinkles, and cerithids). Bottom images and bathymetric profiles show ridges to sheets of limestone pebbles and bivalve coquinas (facies R4–marooned beaches) with bryozoans growing on many of the larger particles. The tops of shallow banks in the western part of the shelf are encrusted rocky substrates (facies C4) with abundant macroalgae that are veneered with and surrounded by rhodoliths (facies C3).

Middle-Neritic: Siliciclastic facies remain prominent to ~80 mwd in the center of the shelf. The sands are otherwise rich in relict particles and so deposits range from relict-rich skeletal sands and gravels (facies M1), to mollusc-rich or bryozoan-rich relict sands (facies R2 and R3), to just relict sand (facies R1). These sediments have a speckled brown and cream or 'salt and pepper' appearance. Noticeably worn bivalve fragments are present in some areas whereas other regions contain worn small brown lithoclast pebbles (facies R4). Relict components decrease with depth such that the sediments between 70 and 90 mwd are mostly relict-rich bryozoan or skeletal sands and gravel (facies M4 and M1).

Diverse but robust and variably abraded bryozoans typify the bryozoan sand and gravel (facies C1) outboard. There are two end-member environments, epibenthic coppices and rippled carbonate sand barrens. Bottom images and bathymetric profiles (Fig. 9.11) show rock substrates covered by variable thicknesses of coarse sediment.

The epibenthic coppice environment comprises hard rock substrates 50–60% covered by sessile benthos. The most obvious organisms are sponges, generally upright flat planar to digitate oscular forms (cf. *Mycale*, *Iophon*, *Chondropsis*, *Clathria*) or globular types (cf. *Ancorina*). They grow together with numerous hydroids and bryozoans, the most noticeable of which are articulated zooidal and *Adeona* sp.

Intermediate environments comprise large subaqueous dune fields, as defined by precision depth profiles in a zone of seaward-dipping clinoforms. Most images depict starved small dunes to large ripples with abundant organisms growing on hard substrates (epibenthic coppice) in the troughs between bedforms.

Carbonate sand barrens are extensive areas of sand containing ripples to small subaqueous dunes with estimated spacings of 30–60 cm and heights of 10–30 cm. Bedforms are symmetrical, straight-crested to bifurcating and their crests are either sharp and linear or textured by smaller ripples and always oriented NW-SE. Clasts recovered are similar in composition and mineralogy (confirmed by staining) to the loose sediments. The sands are barren except for small areas where the surface is 20–40% covered by organisms with sands between rippled in a disorganized fashion. The seafloor below 120 mwd is mostly rippled sediment with a few large organisms.

Deep Neritic & Upper Slope: The outer shelf, shelf edge and upper slope are covered with poorly sorted green fine sand and mud (facies C7 and C11) that become progressively depleted in bryozoans with increasing depth. The epibenthic coppice is not prominent. Images of the upper parts of this zone (140–180 mwd) depict a gradual transition from shallower rippled coarse sands (facies C1) to these more muddy sediments; the sharp-crested linear ripples (10–15 cm height, 10–30 cm spacing) die out between -140 and 150 mwd and grade, with depth, into areas of subdued, more bifurcating ripples (5–10 cm high, 2–5 cm spacing) textured by smaller confused and irregular ripples interspaced with regions with no obvious physical sedimentary structures, but pockmarked by numerous burrows. Areas of the seafloor between 150 and 180 mwd, although mostly burrowed, are locally textured by subdued cm-scale diffuse ripples with ladder ripples in their troughs that have been modified by burrowing.

The shelf edge between 180 and 200 mwd is a relatively flat to gently undulating surface dimpled with cm-size depressions and bumps, pockmarked with vertical burrow openings and disturbed by rare trails of the mobile benthos (a burrowed mud barren). Numerous, small, cm-size or less, delicate, branching bryozoans grow either scattered across the seafloor or clumped in patches 10–30 cm across (epibenthic scrub). These clusters are dense intergrowths of bryozoans (especially adeonids and catenecellids), small sponges, and hydroids that form a "ground-cover" to heights of 10 cm or so. Sediments are either bryozoan sand and gravel or fine skeletal sand (facies C1 and C6)

The upper slope between 250 and 350 mwd is composed of facies C7 (delicate branching bryozoan muddy sand), facies C11 (spiculitic skeletal muddy sand), or facies C8 (scaphopod, pteropod muddy sand). The coarse fraction in facies C7 is almost entirely branching cyclostome bryozoans (15% of material). Other scattered bryozoans are *Celleporaria* sp. as well as erect rigid robust branching and fenestrate forms. *Celleporaria* sp. grows as irregular encrusting sheets, free small particles, hollow tubes, and subhemispheres up to 6 cm in diameter. The only bivalves are *Chlamys* sp. and *Venericardia* sp. whereas rare gastropods are the slit shell (*Tenagodus* sp.), turitellids, olives, and whelks. Other components are solitary corals, regular echinoid spines, siliceous sponge spicules, pteropods and serpulid tubes.

9.7 Synopsis

The South Australian Sea is a region characterized by complex geography wherein regional climate exerts a strong influence on seawater temperature and salinity and thus the seafloor biota. Water temperatures over the shelf decrease eastward. The two large gulfs that extend into the continental interior become progressively more affected by the arid continental climate northward, resulting in a warm-temperate environment much like the northern Great Australian Bight. Their semi-enclosed nature and greater intensity of evaporation compared to the Great Australian Bight, however, leads to higher evaporation rates. The shelf proper over most of the region is cool-temperate.

1. The two gulfs have many similar properties.

    (a) A clockwise circulation pattern wherein open ocean waters flow into the gulf from the west, northward up the western coast, are heated and evaporated at the head during summer, and then flow southward down the eastern coast.

    (b) Rhodolith pavements along the leeward southwestern sides of the gulf.

    (c) Extensive shallow subtidal seagrass banks, particularly toward the northern ends.

    (d) Locally important muddy tidal flats especially along the eastern coasts.

    (e) Locally important subaqueous tidal sand shoals.

    (f) Mollusc-dominated facies, both marginal marine and on the basin floor.

2. The large salients also have several differences.

    (a) Spencer Gulf has higher salinity because it narrows more at the head and extends much further into the continental interior.

    (b) Waters have a longer residence time in Spencer Gulf because there is a submarine ridge and a sharp salinity front at the gulf mouth that prevents continuous exchange with the open ocean during summer, thus enhancing the effects of heating and evaporation. High salinity is especially promoted at the northern end by the presence of an elongate sand spit (Ward Spit) that impedes water exchange. By contrast, such waters flow out of Gulf St. Vincent via Investigator Strait during most of the year with a secondary flow via Backstairs Passage in winter.

    (c) Rhodolith pavements are much more extensive in Spencer Gulf.

3. The Lacepede Shelf comprises the largest region of quartzose sand on the southern margin. This sand is, however, relict and stranded, originating when the River Murray flowed across the shelf during MIS 2 lowstand and earlier. The river also flowed more intensively during these stormy periods and delivered much more sediment to the ocean. The extensive rippled sands are either barren or populated by infaunal molluscs.

4. The mid- to outer neritic facies are similar to those west on the Lincoln Shelf; mollusc- and bryozoan-dominated or relict sediment and bryozoan-dominated facies respectively.

5. Whereas large benthic foraminifers, especially *Peneropolis* sp. are, because of the warm waters, present in the upper reaches of the gulfs there are none on the shelf because the water is too cold.

6. The upper slope is cut by numerous canyons and subject to local mass wasting. The reason for the numerous canyons is interpreted to be the numerous small and large rivers that flowed across the shelf during the wetter climates of the last MIS 2 glacial. The reason for the mass wasting is the continuing seismic activity in the region, unlike the relatively quiescent Great Australian Bight.

Chapter 10
The Southeastern Continental Margin

10.1 Introduction

The southeast part of the continental margin comprises the Otway Shelf and the Bonney Coast, Bass Strait, and the western shelf of Tasmania (Fig. 10.1). The 400 km long Otway Shelf lies between 37° and 43.5° S, and 139.5° E (Cape Jaffa) and 143.5° E (Cape Otway). The narrowest point is off Portland, where the shelf is less than 20 km wide. It broadens progressively westward, to 60 km off Robe, and eastward to 80 km off Warrnambool. Bass Strait, an intracontinental basin, has an area of 66,000 km², extends ~400 km east-west and 250 km north-south between 39° and 41° S and has a string of islands at each end. Depth ranges from 55 to 80 mwd with maximum water depth of 83 mwd near the geographic center. The Basin is essentially a bowl with a deep center that shallows toward the Victoria coast, the northern Tasmania coast, and sills along the eastern and western margins. The western Tasmania continental shelf that ranges from 20 to ~60 km wide, is the geomorphic continuation of the Otway Shelf but swings around into a meridonal orientation. The shelf is relatively deep, whereby the 50 m bathymetric contour lies near the coast. The continental slope has variable relief due to numerous submarine canyons and uplifted fault blocks and extends to the abyssal plain at about 5,000 mwd (Figs. 10.2, 10.3).

The shelf stretches from 37° to 43.5° S and so extends towards the sub-polar realm with attendant extreme weather, high sea states, and cooler seawater temperatures. The sea is a complex mixing zone between the Pacific Ocean to the east, the Indian Ocean to the west and the Southern Ocean to the south. There is an area of conspicuous summer upwelling along the Bonney Coast (Fig. 10.2). There is also upwelling accompanying severe storms off northern Tasmania between King Island and Cape Grim during winter; this is a potentially important source of nutrients (Middleton and Black 1994; Evans and Middleton 1998; Cresswell 2000). The subtropical convergence moves from south of Tasmania in winter northward to the central coast in summer, with nutrient levels and associated biota substantially higher south of the front (Fig. 10.3). Thus, nutrients are generally higher than around mainland Australia, but still lower than temperate latitudes worldwide. Highest nutrient values are associated with the south coast and the intrusion of subantarctic water. The Zeehan Current along the western margin and the East Australian Current along the eastern margin of Tasmania dominate circulation (Fig. 10.2). Both currents flow southward but they are out of phase; the East Australian Current is strongest and reaches furthest south in summer; the Zeehan Current is strongest and reaches furthest south in winter. Prevailing westerlies result in a west-to-east water flow through Bass Strait but Paleozoic granite ridges that locally rise above sea level as islands along the eastern and western margins restrict this oceanic flow somewhat. These islands, such as King Island and Tasmania proper also tend to protect most of Bass Strait from SW swells and waves.

Surficial sediments are mainly cool-temperate (cf. James and Lukasik 2010), commensurate with the lower seawater temperatures. There are no large, symbiont-bearing benthic foraminifers. Articulated corallines are not as common as in the west. The rocky inshore platforms are fringed with the bull kelp *Durvillea* sp. south and east of Cape Jaffa; this macroalga is not present in warmer waters to the west of Cape Jaffa.

Megafacies are areally dominated by carbonates (Fig. 10.4). Relict sediments are not as widespread as on other parts of the southern continental shelf with relict-rich deposits localized to shallow neritic environments along the mainland, around islands in Bass

N. P. James, Y. Bone, *Neritic Carbonate Sediments in a Temperate Realm*, DOI 10.1007/978-90-481-9289-2_10, © Springer Science+Business Media B.V. 2011

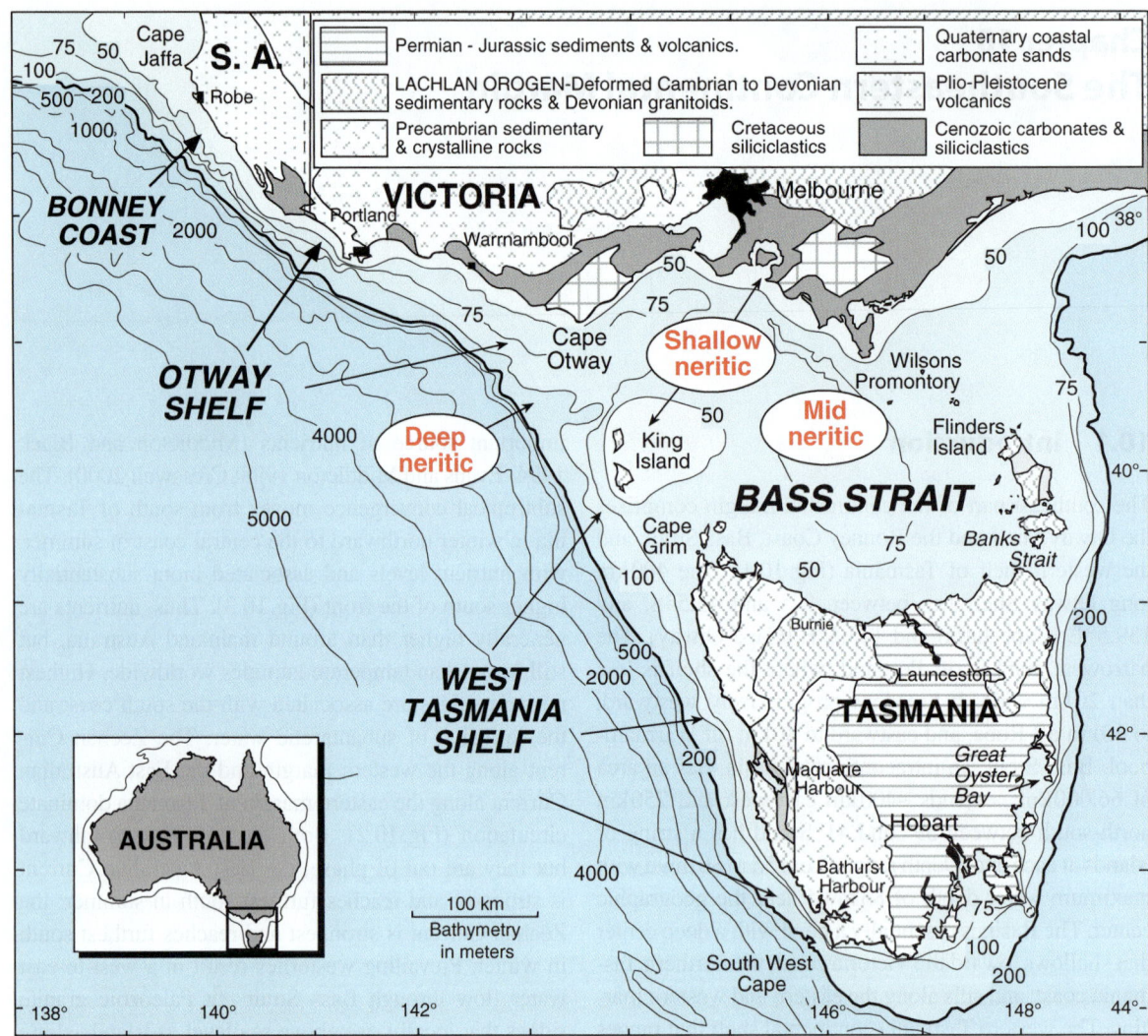

Fig. 10.1 Map and chart of the southeastern continental shelf, comprising the Otway Shelf and Bonney Coast, Bass Strait, and west Tasmanian shelf, the adjacent mainland, and the island of Tasmania, illustrating regional geology, bathymetry, and neritic zones

Strait and along the Tasmanian shelf. Quartzose sediments are most numerous along the coasts of Tasmania. A zone of complex facies lies along the outer part of the Bonney Coast and is detailed below.

10.2 Otway Shelf

10.2.1 General Attributes

This shelf extends from Cape Jaffa southeastward to Bass Strait (Fig. 10.1). It is geographically contiguous

but partitioned by oceanography. The northern part, called the Bonney Coast, is subject to strong seasonal upwelling. It borders a similar coastline to that of the Lacepede Shelf but with numerous small embayments. The eastern part of the shelf, however, lies adjacent to uplifted Cretaceous siliciclastic rocks of the Otway Ranges.

10.2.2 Local Oceanography

Shelf waters are cool temperate (<18°C) with open ocean salinities (35.1–35.6‰) (Schahinger 1987;

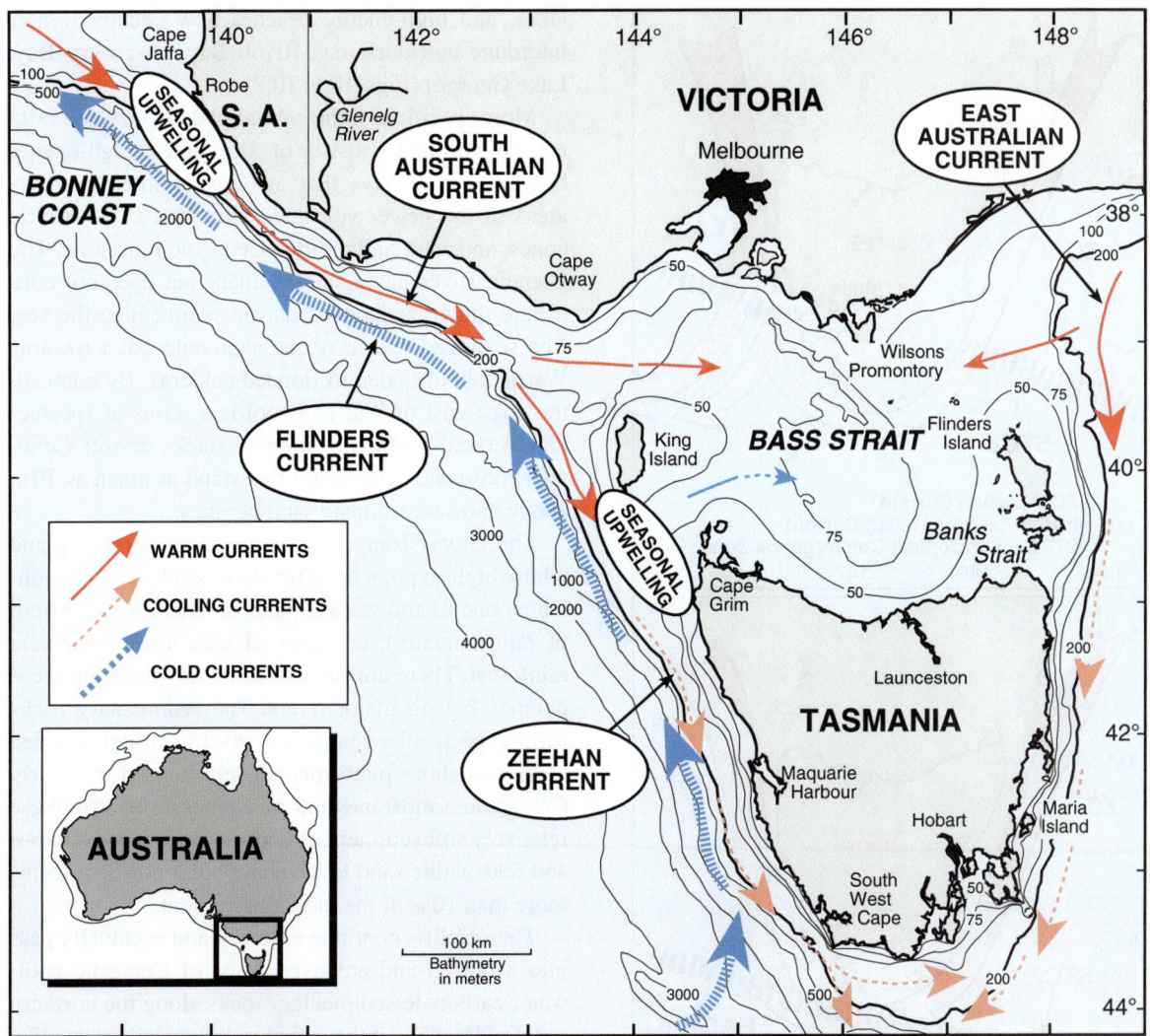

Fig. 10.2 Chart of the southeastern continental shelf illustrating major ocean currents and areas of upwelling

James et al. 1992) (Fig. 10.5). The water column is well mixed during winter months, with near uniform salinities of 35.6‰ and temperatures of 14°C across the shelf. Strong summer stratification results in two distinct water masses; a 30–40 m thick relatively warm (18°C) and saline (35.2‰), surface layer that overlies a cold (11–12°C), less saline, oxygen-depleted, nitrate-rich (6.0–7.0 μM) and dense bottom layer (Lewis 1981). This pronounced layering is produced by strong, wind-induced summer upwelling events that episodically move cold, nutrient-rich, offshore Antarctic Intermediate Water and Antarctic Mode Water from the Flinders Current onto the shelf platform south of Robe. This water reaches the surface as a band from just north of Portland to the middle part of

the Lacepede Shelf. Thus, surface waters in this region are 4°C cooler in summer than in winter. Seasonally high shelf edge nutrient levels have also been recorded in the Warrnambool sector (Fandry 1983; Levings and Gill, in press). Chlorophyll-a values are, however, generally low, usually between 0.3–0.4 mg m^{-3} in summer and 0.4–0.5 mg m^{-3} in winter (low mesotrophic) (Fig. 10.2).

10.2.3 Marginal Marine

The coastline between Cape Jaffa and Cape Otway, except for the Otway Ranges in the far east, is a relatively

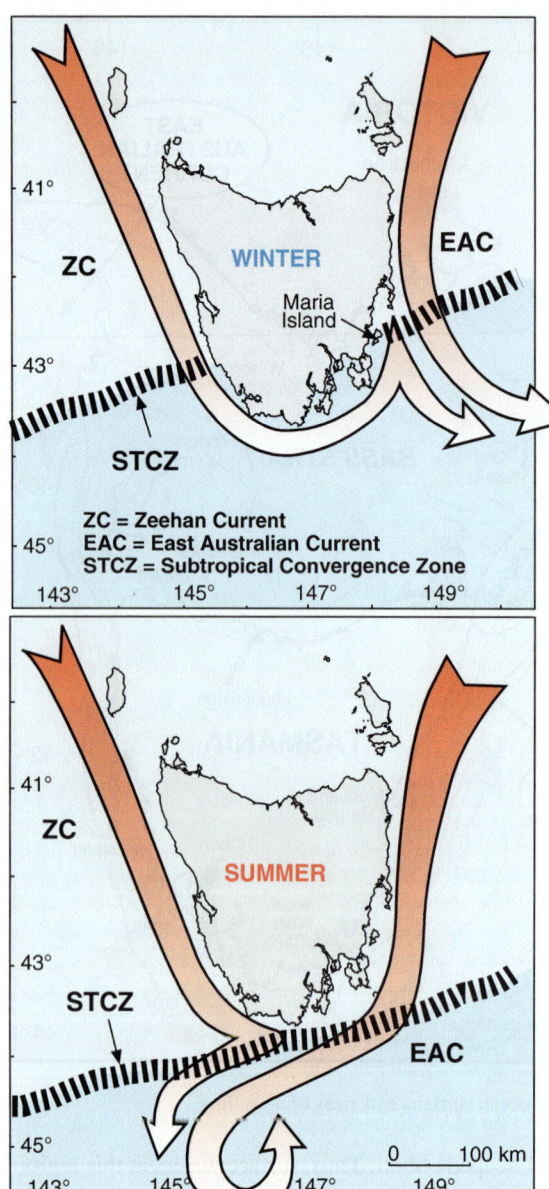

Fig. 10.3 Chart of the ocean around Tasmania illustrating seasonal variation in the position of the Subtropical Convergence Zone (*STCZ*), flow of the Zeehan Current (*ZC*), and flow of the east Australian Current (*EAC*). (After Cresswell 2000)

low coastal plain <200 m in elevation. Gradual Plio-Pleistocene uplift around Mount Gambier resulted in the seaward progradation of Bridgewater Formation aeolian dune complexes (see Chap. 3). Thus, the western part of the coast resembles the Ceduna Sector wherein the youngest seaward facing dune ridges have been breached and numerous coastal lagoons, saline

lakes, and high-energy beaches now occupy former interdune corridors (e.g. Rivoli Bay, Discovery Bay, Lake George) (Figs. 10.6, 10.7).

More specifically, the coastal plain from Cape Jaffa east to Portland consists of aeolianites, high-energy beaches, small bays that are locally interrupted by areas of the newer volcanics (thin lava flows, scoria cones, and ash), and uplifted Cenozoic limestone. The Glenelg River has a small estuary but does not contribute much terrigenous clastic sediment to the sea. The scalloped nature of the aeolianite coast towards Warrnambool is due to flooded calderas. By contrast, the coast east of Warrnambool is a series of spectacular vertical cliffs and offshore stacks of soft Cenozoic cool-water carbonates that stand as much as 70 m above narrow carbonate sand beaches.

The Otway Ranges are a deeply dissected upland whose highest point is 700 m above sea level. The highlands, one of the wettest parts of Victoria (1,200 mm of rain annually), are covered with a tall temperate rainforest. There are small streams flowing from these uplands but no major rivers. The sedimentary rocks form a spectacular coastline of 90–150 m-high wooded cliffs and shore platforms actively incised into early Cretaceous sandstones and mudstones. Erosion of these relatively soft sedimentary rocks contributes quartzose and feldspathic sand to the shore but it nowhere forms more than 10% of the modern sediment.

The seacliffs continue eastward and eventually pass into a cliffed and embayed coast of Cenozoic cool-water carbonate sedimentary rocks along the northern coast of Bass Strait that are morphologically much like the Cenozoic cliffs to the west in the Great Australian Bight. Holdgate et al. (2001) documented the marine geology of Port Philip Bay.

10.2.4   Shelf Sediments

Overview: Most of this relatively narrow shelf is covered with recent carbonate sediment (Fig. 10.7) (Boreen et al. 1993). Bathymetric profiles show distinct ridges of limestone gravel and an outer shelf with seaward-dipping limestones (Fig. 10.8). The slope off the Bonney Coast is a zone of disturbed bedding incised by tributaries leading to major submarine canyons.

Shallow Neritic: Much of the shallow neritic environment is a series of encrusted rocky reefs (facies C4).

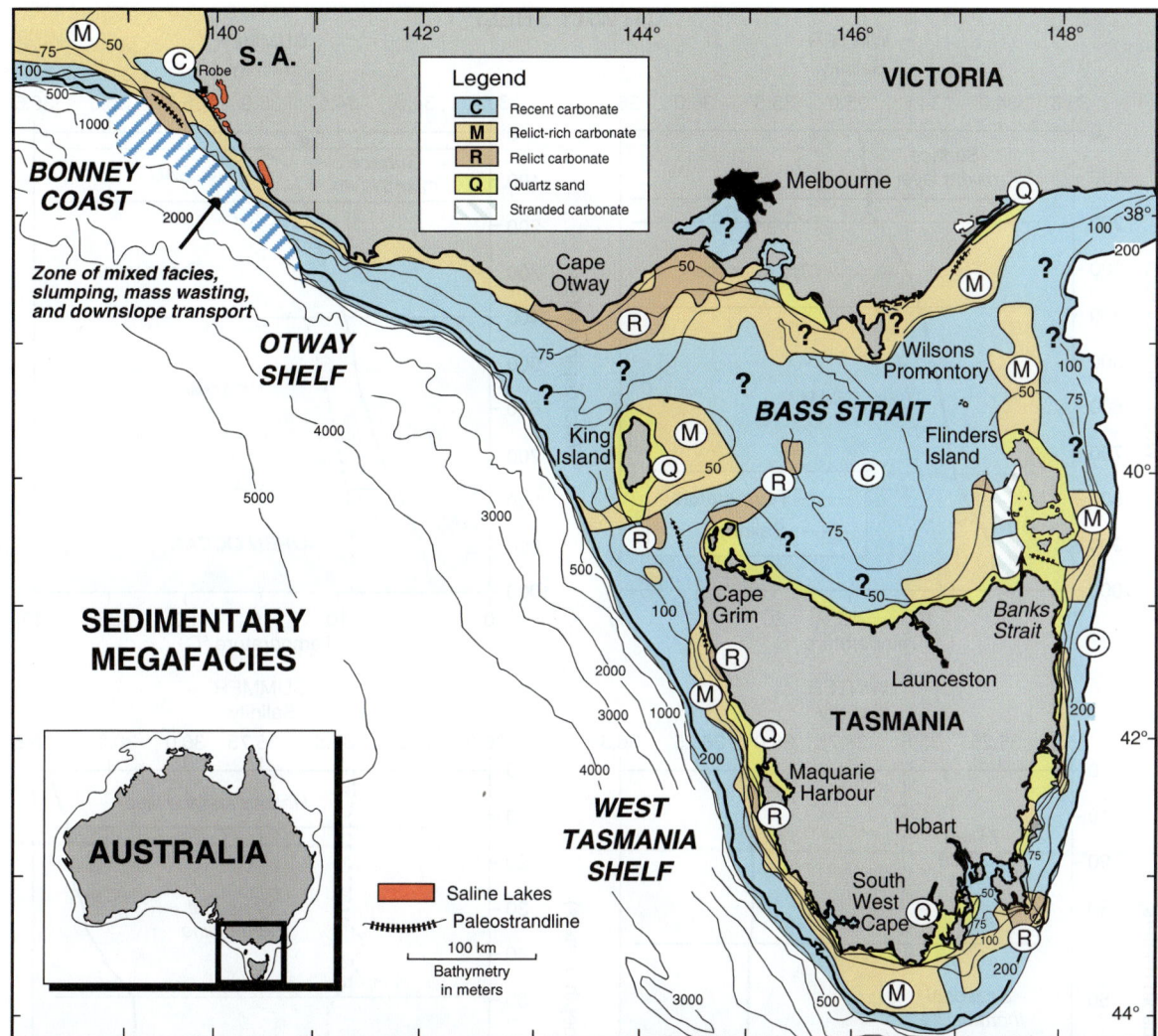

Fig. 10.4 A chart of surface sediment megafacies on the southeastern continental shelf, most of which are carbonates, with quartzose and relict facies restricted to the shallow neritic zone. The area off the Bonney Coast is a zone of rapidly changing facies, slumping, mass wasting and downslope transport making facies differentiation difficult

Relict-rich facies M1 and M2 occur as local sediment patches <50 cm thick and are a continuation of the similar sediment on the Lacepede Shelf. Bioclastic grains are extensively fragmented, abraded, polished, and often intensely bioeroded. Bryozoans are mainly encrusting and robust growth forms. Coralline algal fragments and irregular rhodoliths form up to 15% of the sediment. Cored sediments are uniformly homogeneous and lack identifiable physical and biological structures.

Mid-Neritic: The bryozoan sands (facies C1) of epibenthic coppice and sand barren environments are clean and poorly-sorted with up to 30% bioclastic gravel frag-

ments and no mud. Deposits have lower quartz (generally <10%) and relict grain contents than shallow neritic accumulations; skeletal grains are fragmented and variably abraded. The sediments are cross-stratified in core. Angles of dip vary from horizontal to 30° and scour-truncation surfaces are common. Cross-bed sets are 15–30 cm thick and typically fine-upward from gravelly-to medium grained-sand. Coarser layers (bottom sets) have a more open framework than the interbedded finer-grained sediments (foresets) and are dominated by erect rigid robust branching and arborescent bryozoans, with locally abundant mollusc fragments, and echinoid debris. Fine sand layers are homogenous and facies C6 in composition.

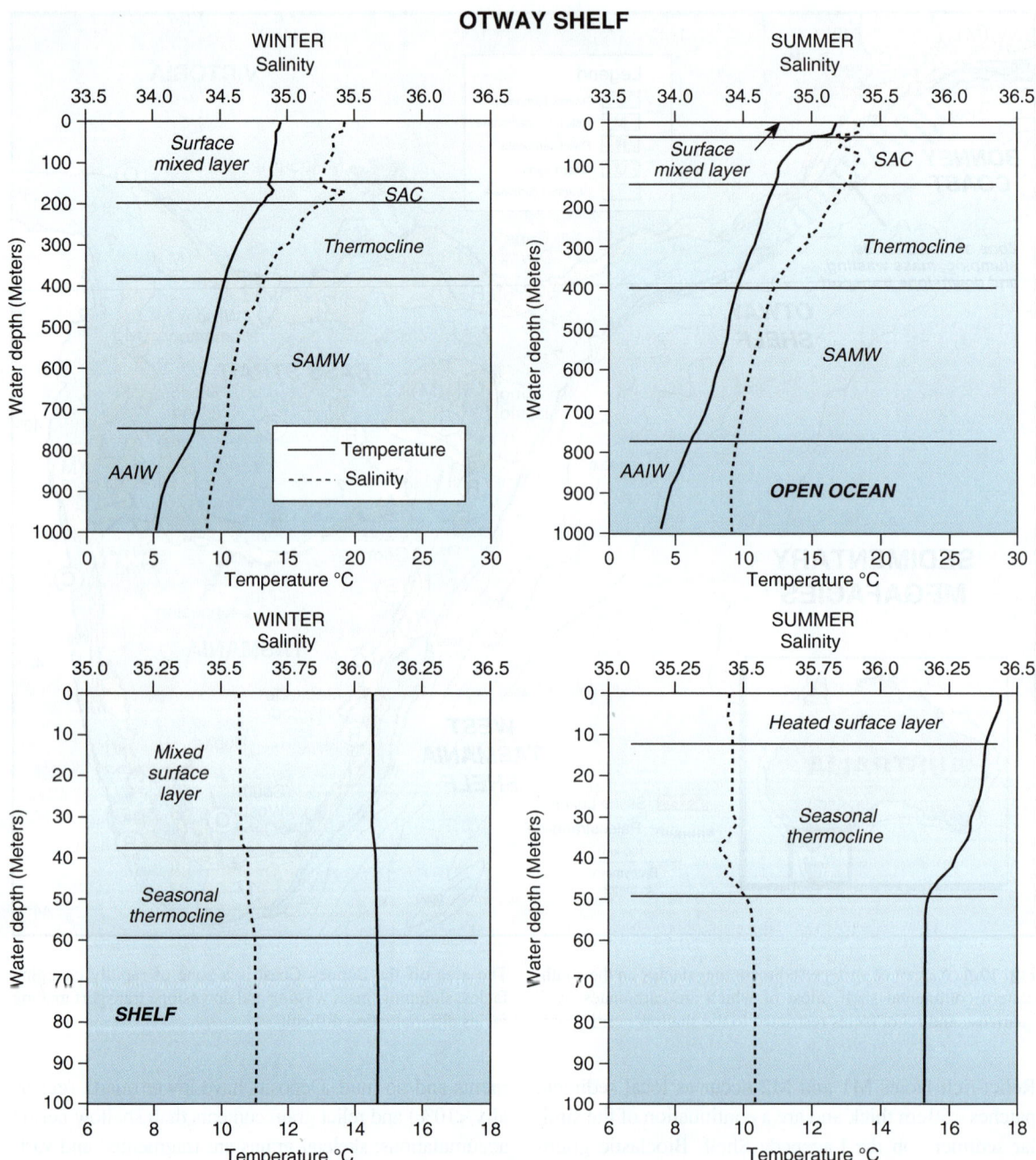

Fig. 10.5 Temperature and salinity profiles for the upper 1,000 m of the open ocean water column (*top*) and the upper 100 m of the shelf water column (*bottom*) during summer and winter on the Otway Shelf seaward of Portland (Fig. 10.1). *AAIW* Antarctic Intermediate water, *SAMW* Subantarctic Mode Water,

SAC South Australian Current. The surface layer is composed of Sub-tropical Surface Water modified by summer heating and winter cooling. (Compiled from CSIRO (2001) Data Trawler, Marine and Atmospheric Research (CMAR) Data Center, Commonwealth Scientific and Industrial Research Organization)

Fig. 10.6 Marginal marine environments and adjacent areas along the coast of the southeastern continental margin. (**a**) Carbonate beach and dune complex near Beachport 30 km south of Robe on the Bonney Coast (Fig. 10.1). (**b**) Cliffed coast and sea stacks of Miocene carbonates about 100 m high near Port Campbell ~40 km south of Warrnambool (Fig. 10.1). (**c**) Heavily wooded hills of the Otway Ranges that are ~500 m in elevation, near Cape Otway (Fig. 10.1). (**d**) Dense forest near the top of the Otway Ranges resulting from high annual rainfall. (**e**) A coast of Jurassic volcanic rocks in eastern Tasmania ~100 km north of Hobart (Fig. 10.1). (**f**) A dense forest with numerous ferns and dense stands of Gondwana flora such as Southern Beech and Huon Pine, on the western coast of Tasmania near Macquarie Harbour (Fig. 10.1)

Deep Neritic & Upper Slope: The deep neritic shelf and local intrashelf depressions are covered by homogeneous and bioturbated fine to very fine sand (facies C6). They form a continuous band that parallels the shelf edge.

Delicate branching bryozoans dominate olive-coloured muddy bryozoan sands in epibenthic scrub environments (facies C7) blanket the shelf edge and upper slope between 180 and 400 mwd and form a semi-continuous band paralleling the shelf-slope break

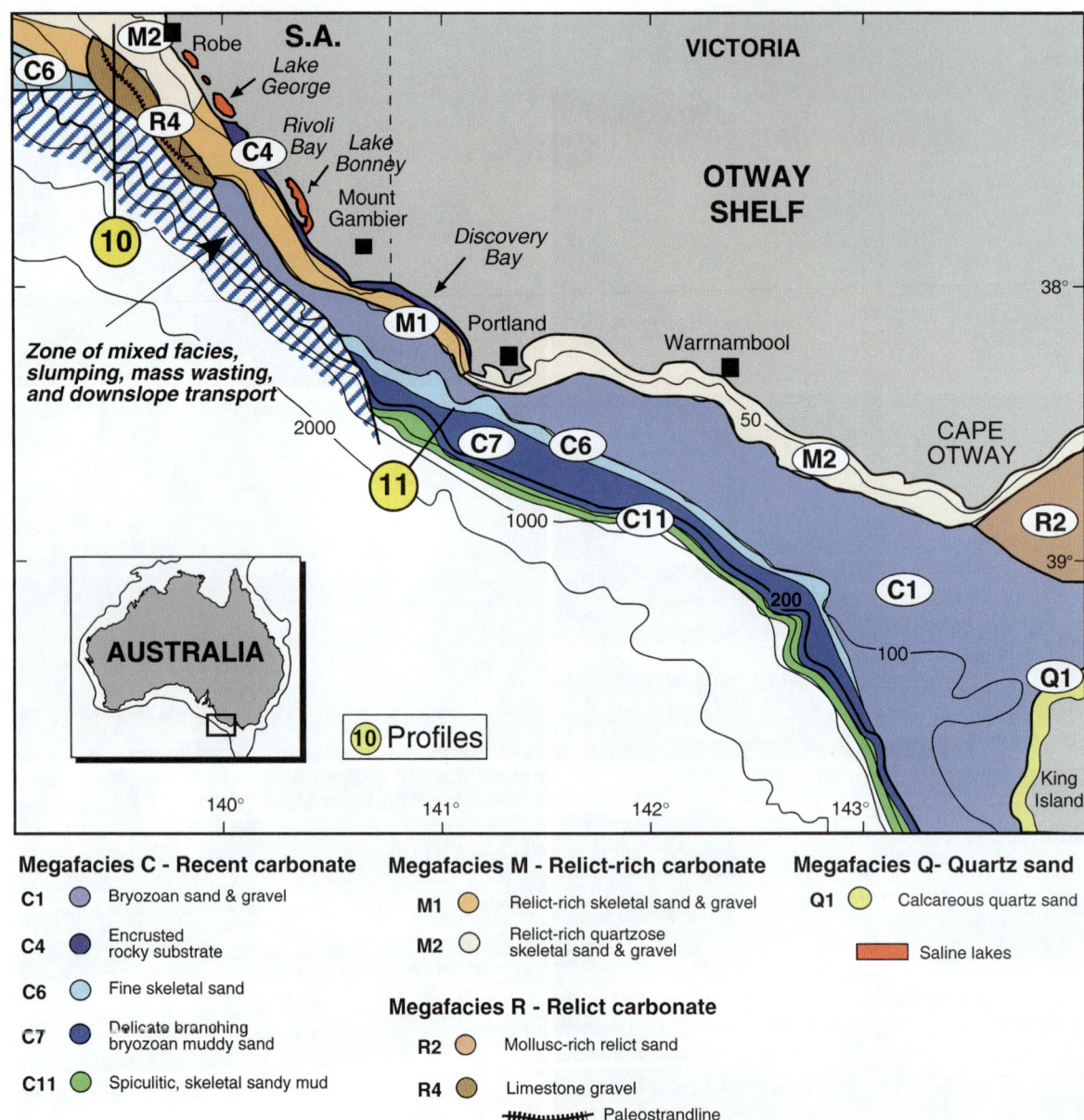

Fig. 10.7 Composition of surface sediment facies on the Otway Shelf. Lines of section in Fig. 10.8 are designated by number

along the entire length of the shelf. Fragmented bryozoans are not significantly worn and preservation of extremely delicate structures and whole fragile bryozoans are common.

The top of the slope also contains local Coral-*Celleporaria* gravels (facies C12—too local to depict on Fig. 10.7) composed of stranded and Holocene components. These sediments have a framework- to matrix-supported fabric (rudstone-floatstone) and consist of coarse bryozoan gravel with a fine-grained, bioclastic muddy sand matrix. Most of the coarse carbonate is arborescent bryozoans (*Celleporaria* sp.) wherein large cylindrical (5 cm max long axis) and sheet-like, nodular and arborescent forms are characteristic and locally dominate the facies. The corals, mostly large living and dead azooxanthallate *Caryophyllia planilamellata* occur as scattered cups among and on the bryozoans or as prolific, monospecific accumulations with few bryozoans. They are locally attached to one another.

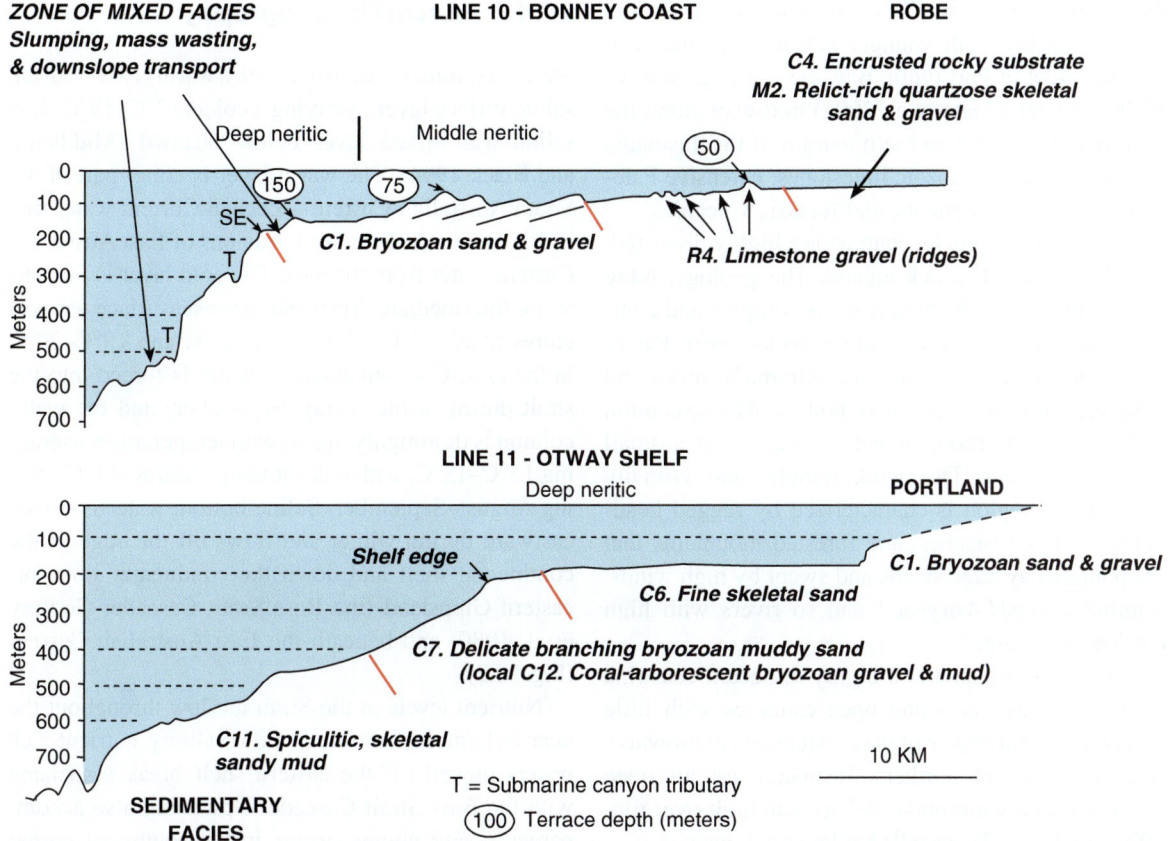

Fig. 10.8 Bathymetric profiles across the shelf off the Bonney Coast at Robe and off the Otway Shelf at Portland (see Fig. 10.7 for location) constructed from precision depth recorder records and illustrating distinctive bottom features, positions of prominent gravel ridges and submarine terraces where possible, and the disposition of sedimentary facies

Sediment between depths of 350 and 500 mwd in the burrowed mud barren environment is a mixture of pelleted foraminifer-nannofossil mud and skeletal sand (facies C11) wherein carbonate content varies between 25% and 80%. This thoroughly bioturbated sediment also contains articulated bryozoan singlets, pteropods, and sponge spicules.

10.2.5 Zone of Mixed Facies

The deep neritic and upper slope of the Bonney Coast, especially between Portland and Robe is a zone of mixed facies. There the facies are not easily separated into contiguous bands but instead samples of the seafloor sediment, commonly quite close to one another, are dramatically different. Furthermore, sediment otherwise identified as shallow water deposits elsewhere are present in relatively deep water. This is a region of ongoing seismicity, slope failure and mass wasting (von der Borch and Hughes-Clarke 1993). These deposits are either an intimate mixture of materials deposited at different sea level stands or allochthonous shallow water sediments that have been transported into deep water environments (Figs. 10.4, 10.7, 10.8).

10.3 Tasmania

10.3.1 Introduction

The two depocenters here are Bass Basin and the continental shelf. Bass Basin is bounded by shallow bedrock ridges on either side with local island complexes on the ridges themselves. The northern side of the basin is framed by Cenozoic limestone cliffs and narrow beaches with younger volcanics in the east, and the outlet of Port Philip Bay, the large harbour for Melbourne (Holdgate et al. 2001) mid-way along the Victorian coast. The northern margin of the Tasmania consists of local Cenozoic limestones, extensive Paleozoic glacigene sediments, and Jurassic volcanics.

Western Tasmania by contrast is a highly dissected, heavily forested bedrock upland. The geology, lying west of the Tamar Fracture zone, is complex and composed of Precambrian crystalline rocks, early Paleozoic sedimentary, volcanic, and ultramafic rocks and scattered Devonian granitoid bodies. The exception is Macquarie Harbour, which is located in a small Cenozoic graben. This wild, remote, and virtually unpopulated region is characterized by ragged headlands, deserted beaches, and forested mountains that are pounded by huge swells and swept by high winds. Rainfall exceeds $4\,\mathrm{m\,year^{-1}}$ and so rivers with high outflow dominate.

Southern Tasmania is a highly dissected coastline with numerous inlets and open estuaries with little freshwater input. Many of these estuaries are drowned. Such estuaries in northern Tasmania are all open because they are mesotidal (>2 m) with high river runoff (rainfall = 2–$3\,\mathrm{m\,year^{-1}}$) but low gradient.

The West Tasmania Shelf itself, between 40° and 43.5° S, and 143.5° E and 147° E, is relatively narrow and overlies the Mesozoic Sorell rift-basin. There are few rivers of importance other than the Franklin River which has its outlet in Macquarie Harbour. Most of the beaches are siliciclastic sediment.

10.4   The Bass Basin

10.4.1   General Attributes

The basin, now manifest as Bass Strait, is largely a protected depositional environment wherein King Island and the Hunter Group as well as the island of Tasmanian itself serve to protect the basin somewhat from southwesterly seas and swell. Surface sediment in this shallow depression, almost all of which is in the shallow neritic environment, ranges from coarse-grained across the eastern and western sills, to muddy in the basin center (Blom and Alsop 1988; Malikides et al. 1988; James et al. 2008).

10.4.2   Local Oceanography

Water in summer is stratified with a warm (13°C–20°C), saline surface layer overlying a colder (7°C–13°C) less saline well-mixed layer below 60 mwd (Middleton and Black 1994). The warm layer is either part of the South Australian Current—Zeehan Current water that is driven into the Strait or intrusions of East Australian Current water from the east. The cold layer is Subantarctic Intermediate Water. Summer seasurface temperatures reach 17°C–18°C in the west and 19°C–20°C in the east. Cool subantarctic water is forced into the strait during winter (May–September) and the water column is thoroughly mixed with temperatures averaging 13°C–15°C, with coldest temperatures (11°C) during August–September. Saline bottom water is driven eastward during winter and flows off the edge of the continental shelf and down the continental slope off eastern Gippsland (the Bass Strait Cascade; Godfrey et al. 1980) and beneath the East Australian Current (Fig. 10.2).

Nutrient levels in the Strait are low throughout the year (<1 μmol of nitrates). Low salinity nutrient rich waters upwell off the eastern shelf break associated with the Bass Strait Cascade. Upwelling also accompanies severe winter storms in the southwest corner between King Island and Cape Grim; this is a potentially important source of nutrients. Chlorophyll-a measurements are always slightly above $0.5\,\mathrm{mg\,m^{-3}}$, relatively high compared to the rest of the margin.

Strong oscillatory currents are present in Bass Strait. The current pattern is mostly tidal with residual currents being wind-driven. Strongest components are the semidiurnal tides across the sills, with flows up to $2.5\,\mathrm{m\,s^{-1}}$ (~5 knots). The flow in the west is opposite to the flow in the east and so currents in the middle are slow.

10.4.3   Neritic Sediments

Overview: This shallow bowl-like depression is open to the west and to the east with shallow confining rims dotted with islands. Sediments near these islands are siliciclastic- and relict-rich with complex facies patterns. Deposits along the mainland coast to the north and Tasmanian coast to the south are likewise siliciclastic and relict-rich. Coarse carbonates cover the shallow submerged ridges away from the islands and are a mixture of recent and relict-rich carbonate. The basin centre is floored by recent muddy carbonate (Figs. 10.4, 10.10).

Coastlines and Islands: Sediment in shallows around the islands and close to the Victoria coast is locally quartzose (facies Q1). The relict inshore sands (facies R1, R2, R3) that typify the Otway coast continue eastward to Melbourne but further east they contain more recent materials and are thus relict-rich (facies M1, M2, M5). The most areally complex facies patterns occur around and adjacent to Cape Barren Island, Flinders Island and other associated small islands in the east. The facies range from quartzose (facies M2, Q1) to stranded coralline (facies C5), to relict-rich (facies M1, M2) to recent carbonate (C2, C6), all in relatively close proximity. Such complexity doubtless reflects local irregular bathymetry, contributions from island erosion and the influence of mixed Bass Strait, East Australian Current and seasonally upwelled waters.

Subaqueous Ridges: The shallow ridges that run north from Cape Grim through King Island and northward across Flinders Island to the Victoria coast are mostly sand-covered except near the islands where the sediment is more gravelly. Ridge sediments are mostly relict-rich (facies M2, M5) except at the crests where they contain numerous molluscs (facies C2). The sill along the eastern margin of the basin is similar with the ridge covered with bryozoan-dominated carbonate sands and gravels (facies C1) that are locally relict-rich (facies M1).

Subaqueous tidal dunes are present at 40–46 mwd in eastern Bass Strait, between Flinders Island and the mainland (Malikides et al. 1989). Similar dunes occur on the western side between King Island and Cape Grim. These bedforms (wavelength=55–1,730 m, height=2–12 m) are textured with many superimposed ripples and megaripples. Local tides are non-rectilinear and bedforms are transverse to ebb flow but subparallel to flood flow. Maximum spring tidal range is 3.0 m and tidal currents may reach speeds of 100 cm s^{-1} (2 knots).

Basin Floor: Bathymetry deepens towards the basin floor bivalve mud environment and sediment in this becomes muddier with molluscs and bryozoans augmented by progressively more numerous small benthic foraminifers, echinoderms, calcareous worms, and sponge spicules (Blom and Alsop 1988; Nelson

1988b). The muddy sands differ slightly along the western and eastern sides of the bowl; along the western side they are rich in scaphopods and pteropods (facies C8) whereas those in the east contain numerous delicate branching bryozoans (facies C7). The highly bioturbated, pelleted carbonate muds on the basin floor are coccolith-rich, and contain benthic foraminifers, tunicate spicules, broken planktic foraminifers, and some sponge spicules (facies C10).

10.5 West Tasmania Shelf

10.5.1 General Attributes

The shelf is narrow, mostly <50 km wide. The shallow neritic zone is <20 km wide and so most of the shelf is middle neritic. It borders a steep, forested hinterland with high rainfall, numerous rivers, and several prominent estuaries, especially Macquarie Harbour and Bathurst Inlet.

10.5.2 Local Oceanography

Numerous small and a few large, high-gradient, westward-flowing rivers deliver large amounts of fresh water to the ocean, resulting in a seaward-protruding low salinity lens in the upper 50 m (particularly off Macquarie Harbour) and seasonal freshening of the Zeehan Current (Cresswell 2000). Riverine flow is, however, not a source of nutrients because it is rich in tannin from peat bogs and such waters possess lower nutrients than the open sea (Crawford et al. 2000) (Figs. 10.2, 10.3, 10.9).

The Zeehan Current, averaging 0.3–1.0 m s^{-1}, dominates nearsurface water flow (Lyne and Thresher 1994). The Current is relatively warm (~12°C) in winter and streams down the western shelf around to eastern Tasmania and then northward as far as Maria Island before encountering and being mixed with the Eastern Australian Current. In summer, the Zeehan Current reaches southern Tasmanian where it is entrained into the Eastern Australian Current that overshoots and extends 200 km past the island. There is marked summer water stratification on the shelf. Relatively warm (16°C) high salinity surface water derived from the Great Australian

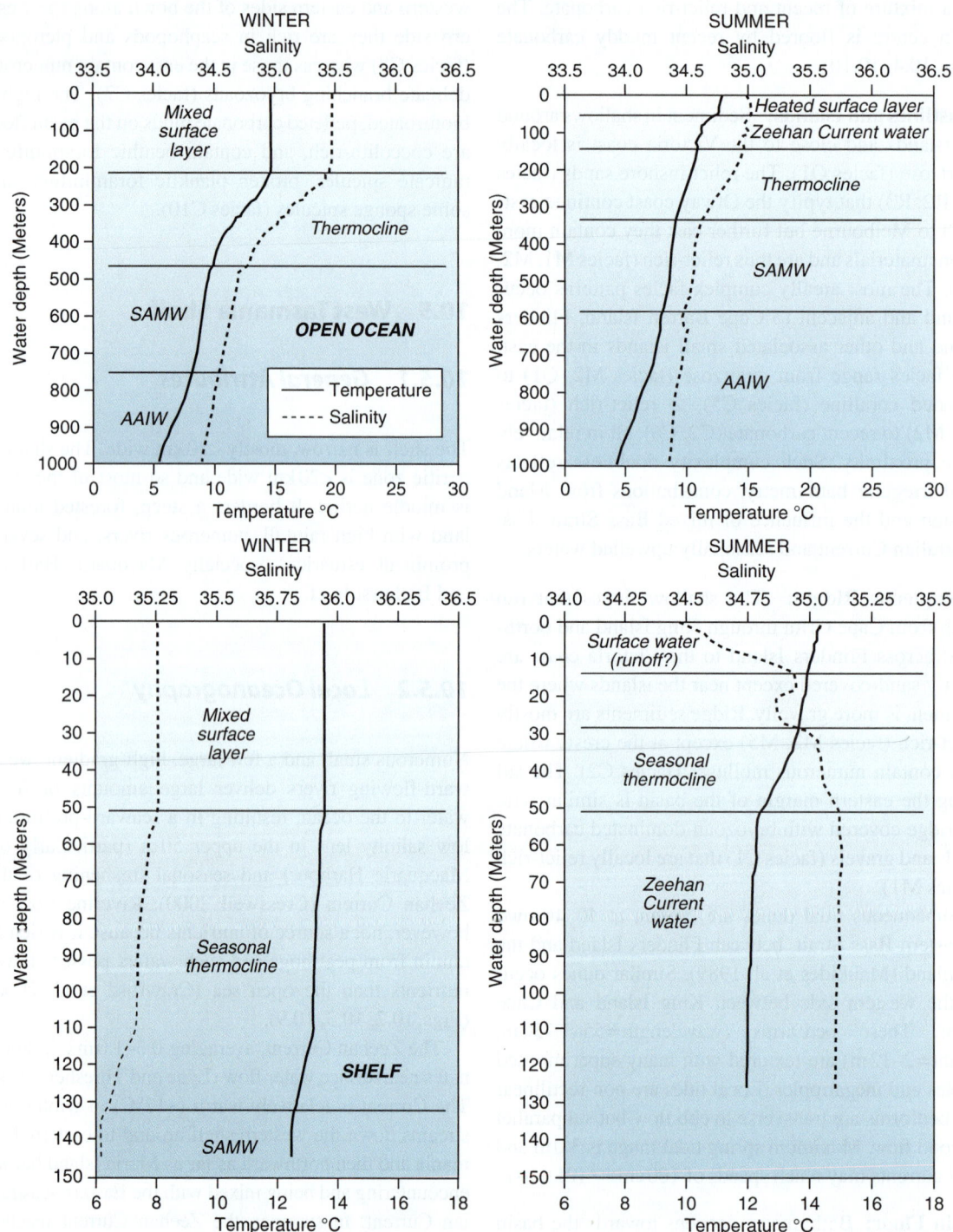

Fig. 10.9 Temperature and salinity profiles for the upper 1,000 m of the open ocean water column (*top*) and the upper 100 m of the shelf water column (*bottom*) during summer and winter seaward on the western Tasmania shelf. *AAIW* Antarctic Intermediate water, *SAMW* Subantarctic Mode Water. The thermocline and surface layers are composed of Sub-tropical Surface Water modified by summer heating and winter cooling. (Compiled from CSIRO (2001) Data Trawler, Marine and Atmospheric Research (CMAR) Data Center, Commonwealth Scientific and Industrial Research Organization)

Bight is underlain below the thermocline (50–80 mwd) by a layer of colder (12°C) high salinity water between 80 and 170 mwd (Cresswell 2000), likely Flinders Current water that is upwelled onto the shelf.

There is a 400 m deep surface mixed layer off the southern tip of Tasmania, coastal downwelling to 400 mwd and upwelling between 1,500 and 400 mwd. Chlorophyll-a measurements there are always >0.5 mg m^{-3}, again relatively high compared to most of the southern Australian continental margin.

10.5.3   Marginal Marine

The coastline of western Tasmania is a largely inaccessible, wild, and poorly studied marginal marine environment. It is not an important depositional system in areal extent. Precambrian and Paleozoic rocks dominate the region and form a mountainous coast that is covered by a dense, impenetrable temperate rainforest that includes relict Gondwana elements such as the Southern Beech. The erosional coast consists of rocky headlands and a few pocket beaches.

10.5.4   Marine Depositional Systems

Overview: Marine deposition is taking place in estuaries and on the open shelf. The estuaries are poorly known, except for Bathurst Harbour. The neritic zone has been sampled to depths of 200 m, the deep neritic zone (Jones and Davies 1983b). The shelf is covered by coarse-grained sediment either sand, gravelly sand, or sandy gravel. Megafacies are relatively straightforward; inner neritic quartzose sand, mid-neritic relict-rich sediment, and deep neritic-upper slope biogenic sands and gravels. This is a continuation of the general Otway Shelf facies trend although the sediments themselves are more quartzose overall. This relatively high siliciclastic content reflects outflow from the numerous rivers draining adjacent highlands all along the west coast of the island. Most of the shelf, however, is veneered with bryozoan-rich sands and gravels that are locally muddier outboard (Fig. 10.10).

Shallow Neritic-Estuarine: Bathurst Harbour, one of the two major estuarine systems along the west coast, has been recently studied and supports a unique depositional system (Reid et al. 2008). A sharp halocline at 2–3 mwd in this elongate estuary separates dark, tannin-rich, low-salinity surface water above from clear, normal marine bottom water. Tannins, while supplying few nutrients, substantially reduce light penetration to bottom environments and result in a profoundly thinned photic zone and a mixing of deeper-water sub-photic biotas comprising soft corals, bryozoans, and sponges with other organisms more typical of the regional temperate, shallow-water marine environment. The well-defined halocline allows a typically marine biota, including echinoderms and otherwise deep-water organisms, to live in the dark, lightless bottom waters of this estuary. The bioclastic factory, producing both carbonate and biosiliceous particles, exists in marine subphotic bottom waters along incised channel and shallow rocky shoreline environments in the outer parts and surrounding inner parts of the estuary. Organic-rich soft sediments in deeper (6–8 mwd) protected parts of the estuary, in contrast, generate few bioclasts, but contain high numbers of allochthonous sponge spicules that were transported from the adjacent bioclastic factory. Trapping of organic material within the inner estuary lowers sediment pH, thus promoting dissolution of carbonate biofragments, resulting in the preferential preservation of siliceous sponge spicules. The resulting sediment is a spicule-rich siliciclastic mud (facies B1).

Barrier estuaries and lagoons along the eastern and northeastern coast, which lie in a rainshadow and have rainfall amounts <100 cm, are hypersaline in the summer and so likewise do not deliver significant nutrients to the marine realm. Southern Tasmania possesses a highly dissected coastline with numerous inlets and open estuaries again with little freshwater input.

Shallow Neritic-Shelf: Quartzose sands on the shelf proper are especially prominent opposite Macquarie Harbour, Bathurst Inlet and D'Entrecasteau Channel. They are generally facies Q1, similar to the sands on the Lacepede Shelf except that the recent biogenic component is more diverse wherein small benthic foraminifers, echinoid fragments, and calcareous worm tubes are equal to the contribution from bryozoans and molluscs.

Middle Neritic: Images of the epibenthic coppices (see Chap. 7) show bryozoan growth that is conspicuously more extensive than on most of the Southwestern Continental Shelf (except off the eastern Albany Shelf) or the South Australian Sea shelf. Bryozoans

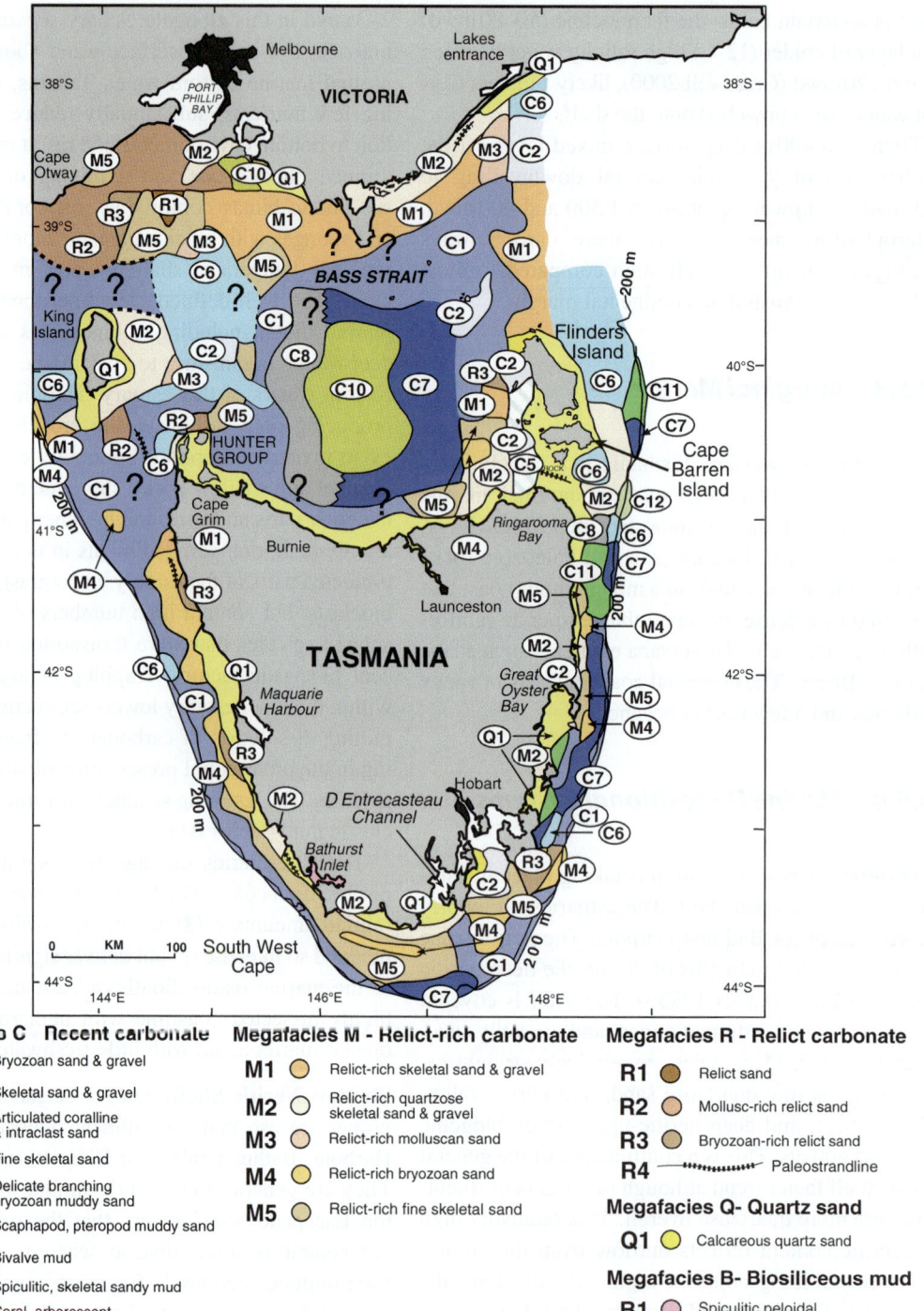

Fig. 10.10 Composition of surface sediment facies on the continental shelf around Tasmania. See Fig. 10.1 for additional nearshore bathymetric contours

are, furthermore, noticeably larger, more numerous, and prolific in surface sediment than in most other areas.

Sediment to roughly 100 mwd are relict-rich; with abundant bryozoans (facies M4) along most of the shelf that are replaced by fine skeletal sands in the south (facies M5). Deposits below ~100 mwd are bryozoan sands and gravels throughout (facies C1). Most bryozoan fragments are articulated zooidal, articulated branching and encrusting types inboard and robust rigid branching and fenestrate forms outboard. Coralline algal rods are absent although encrusting coralline fragments are present.

Deep Neritic: Only the shallow part of the deep neritic epibenthic scrub environment has been sampled (Jones and Davies 1983a) but it is either bryozoan sands and gravels or skeletal sands and gravels (facies C1 and C2). Sediment off the south coast at these depths is particularly rich in rigid delicate branching growth forms (facies C7). Living bryozoans are, as in the middle neritic zone, mostly more numerous than elsewhere.

10.6 Eastern Tasmania

10.6.1 Oceanography

The eastern shelf is protected by Tasmania from SW waves and swells and is a lower energy, somewhat warmer water environment than along the west coast. Summer sea surface temperatures range from 18°C in the north to 13°C in the south but are <13°C throughout winter. Shelf waters are stratified during summer with warm waters extending down to 75–100 mwd, but well mixed to depths of 300 m during winter. When flow of the oligotrophic East Australian Current is reduced in winter, the shelf is covered to depths of 300 m by mixed waters, mostly of subantarctic origin. Nutrient levels overall are low (~3 μmol nitrate) but interrupted locally during summer by shelfward incursions of slightly higher nutrient (up to 6 μmol nitrate) waters.

10.6.2 Sedimentology

Surficial sediments on the relatively narrow eastern shelf are conspicuously finer grained (generally muddy sands to locally muds) than those on the west coast, even though they display the same general facies disposition. Quartzose sands (facies Q1) occupy a narrow nearshore strip and floor Great Oyster Bay. Middle-neritic deposits contain fewer relict particles (facies M5, R3) and include a higher proportion of biogenic fragments other than bryozoans and molluscs (facies C7, C11). Bryozoans are mostly fenestrate and articulated branching types. The deep neritic environment is, however, similar to the west coast of Tasmania and dominated by bryozoan sands and gravels (facies C1) with many articulated branching, delicate branching, and fenestrate forms (Fig. 10.10).

10.7 Synopsis

The continental margin that extends from Cape Jaffa eastward to Cape Otway and further south and east across the entrance to Bass Strait and south along the western coast of Tasmania is bathed in waters that become progressively colder southward. The shelf borders the Victorian and Tasmanian coast but is open across the entrance to Bass Strait. The area has several particular characteristics, mainly the southward flow of the Leeuwin Current system, seasonal local upwelling and northward movement of the Subtropical Convergence Zone to intersect the Tasmanian coast, a mostly wet climate, attendant fluvial runoff, and a wave and swell climate that intensifies southward. The progressive decrease in seagrasses eastward and their virtual disappearance off western Tasmania coincides with their replacement by macroalgae in shallow neritic environments and has a profound affect on carbonate sedimentation.

The shelf overall is cool-temperate (cf. Betzler et al. 1997; James and Lukasik 2010) and grades southward into a marginal cold polar setting. The carbonate factory appears to be most important in the mid- to deep neritic zone, with shallow-neritic seafloor environments that border the coast rich in siliciclastic sediment or bare rock substrates that support a rich epifaunal benthic community.

1. The most important control on carbonate sedimentation appears to be seawater temperature, with upwelling and increased trophic resources augmenting carbonate production and accumulation. Decreasing water temperature precludes some of

the main sediment producers from the temperate carbonate factory, namely large benthic foraminifers, articulated coralline algae, and seagrasses. Summer upwelling in the Bonney Coast and winter upwelling near King Island are not reflected by any profound differences in carbonate sedimentation. This may be because the waters are sourced from the Flinders Current system and so are not especially nutrient-rich. It does appear, however, that overall increased trophic resources result in amplified benthic productivity, as illustrated by the relatively larger proportion of bryozoans and the reduced numbers of relict grains in sediment.

2. The marginal marine environment along the northern part of the Otway Shelf is similar to that in the South Australian Sea but comparatively higher rainfall across the Otway Ranges and over western Tasmania leads to greater fluvial runoff. Rivers that deliver sediment to the coast are only significant, however, in Tasmania where shallow neritic environments are characterized by siliciclastic sands.

3. The change from a seagrass-dominated to macroalgal-dominated shallow neritic zone eastward has several implications. Lack of extensive seagrass meadows means that:

   (a) The sediment is not bound by extensive root systems and so is more easily moved, leading to more grainy deposits with physical bedforms.

   (b) The epiphytes that are typically associated with the grasses, namely benthic foraminifers, articulate coralline algae, spirorbid worms, bryozoans, and encrusting corallines are less well represented in the sediment.

   (c) The diverse and abundant seagrass-associated mollusc community, especially gastropods, is low.

   (d) Mud produced by epiphyte breakdown and bioerosion is absent.

   (e) The highly productive seagrass carbonate factory is eliminated and so carbonate sedimenta-

tion rates are correspondingly reduced. This, in turn, results in part in the somewhat higher proportion of siliciclastic sand inboard and relict grains on the mid-shelf.

By contrast, macroalgae, that are localized to hard substrates, have an associated benthic biota that is dominated by crustose coralline algae, echinoids, epifaunal molluscs, and bryozoans. Barnacles, which are associated with such kelp forests elsewhere, are comparatively minor here, probably because of the relatively low trophic resource levels.

4. The shallow neritic environment at the entrance to Bass Strait is quite different from the open shelf. Unimpeded water movement amplified by tidal resonance has created a series of impressive subaqueous carbonate sand shoals. These dunes are mainly bryozoan-rich deposits but with sands along the crest rich in mollusc fragments and relict grains.

5. The marooned arborescent bryozoan (*Celleporaria* sp.) facies C12 that is today colonized by azooxanthellate corals is particularly prominent in the deep neritic zone of the Otway Shelf but does not appear to be present off western Tasmania. The facies may, however, occur at the eastern entrance to Bass Strait.

6. Bass Strait, an intrashelf basin like the Spencer Gulf and Gulf St. Vincent is, however, quite different from them. It is open to the ocean at either end, has currents flowing into the depression from both east and west, there is no significant evaporation, and there are only local seagrass facies. Like the gulfs it is mostly shallow neritic with deepest water facies generally muddy. These fine-grained, coccolith-rich sediments contain a significant small benthic foraminifer, echinoid, worm and sponge spicule component. It is not certain if these muddy deposits contain the same large bivalve (e.g. *Pinna*) as in the gulfs.

Chapter 11
Diagenesis

11.1 Introduction

Diagenetic change begins early in carbonate depositional environments (Bathurst 1975; James and Choquette 1990) and is usually manifest in the form of biogenic and chemical alteration (Fig. 11.1). The overall result of such processes in temperate water systems is loss of carbonate (Smith and Nelson 2003). Whereas this theme is true for the southern Australian open marine realm, large segments of the same marginal marine depositional system, because of semi-arid climate, are sites of intensive mineral precipitation.

11.2 Marginal Marine Environments

Diagenesis in marginal marine settings across southern Australia is most pronounced on muddy tidal flats, in restricted lagoons and estuaries, and in saline lakes. Sandy beaches are not usually prone to intertidal cementation and beachrock is rare.

Most diagenesis in tidal flats, lagoons, and lakes is in the form of carbonate and evaporite precipitates (see Chap. 5), particularly adjacent to the Albany Shelf, Great Australian Bight, the South Australian Sea and the Otway Shelf. These precipitates have been extensively documented with the most important and widespread being dolomite and other magnesium carbonates together with gypsum and minor halite (Belperio 1995). The dolomite and gypsum grow displacively in the muddy carbonate sediment whereas the gypsum also precipitates directly out of seawater. These environments have many similarities to those in tropical environments such as the Persian Gulf (Purser 1973; Alsharhan and Kendall 2003), and the Caribbean (Hardie 1977).

Such chemical precipitates are not present along the western Tasmania coast because of the relatively high rainfall. By contrast, however, at least one estuary along the humid Tasmanian west coast is an area of carbonate loss wherein the skeletal particles are dissolved and the remaining biosiliceous spicules form the bulk of the sediment on the estuary floor.

11.3 Neritic Environments

11.3.1 Biological Alteration

Biological breakdown of carbonate particles is everywhere accomplished by predation and boring (Fig. 1.1). Predation is important in two ways; (1) particle size is reduced as a byproduct of biogenic breakage, especially by durophagus fishes, and (2) rasping of algal-covered rocky substrates by various herbivorous invertebrates produces mud-sized sediment. Boring weakens the rock or skeleton making it more susceptible to mechanical breakage and produces silt-size particles. The principal bioeroders in the neritic realm are boring micro- and macro-endoliths, as they are in most carbonate systems (Tucker and Wright 1990). The major microborers are cyanobacteria, green and red algae, and fungi (Fig. 11.2) (cf. Golubic et al. 1975, 2000). The most important macroborers are bivalves, sponges, barnacles, and worms. This process leads to the production of progressively finer grains and eventually muds (Flügel 2004). The empty holes left by the endolith upon death also increases the surface area susceptible to chemical alteration as well as weakening the skeleton.

The effects of such alteration have not yet been quantified on the southern Australian shelf. Nevertheless, as a broad generalization, the processes documented

N. P. James, Y. Bone, *Neritic Carbonate Sediments in a Temperate Realm*,
DOI 10.1007/978-90-481-9289-2_11, © Springer Science+Business Media B.V. 2011

Fig. 11.1 A flow diagram illustrating the different types of biological and chemical alteration processes affecting carbonate particles on and just below the modern neritic seafloor off southern Australia and highlighting the major processes and products

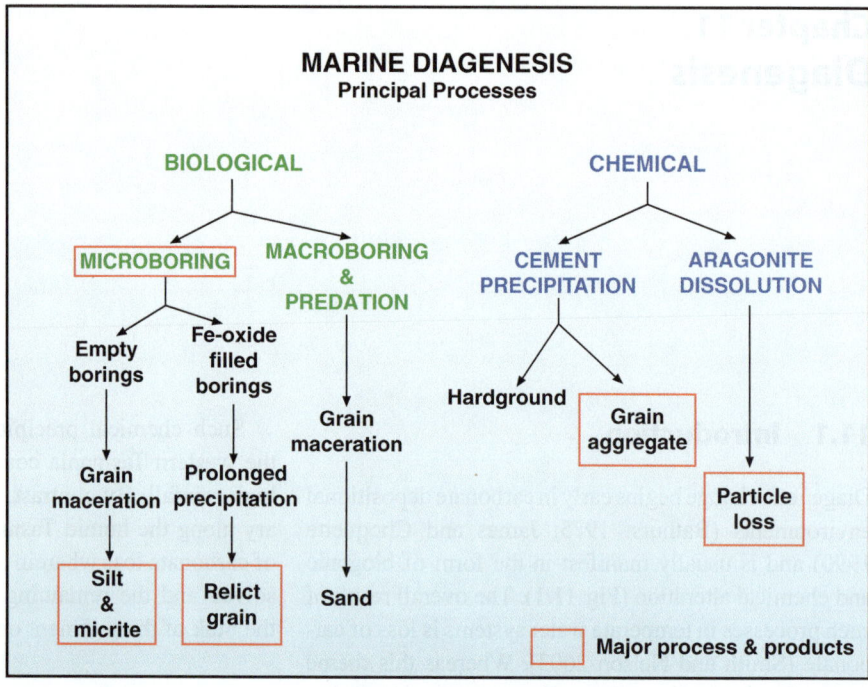

MARINE DIAGENESIS
Principal Processes

BIOLOGICAL   CHEMICAL

MICROBORING   MACROBORING   CEMENT   ARAGONITE
                    &   PRECIPITATION   DISSOLUTION
               PREDATION

Empty   Fe-oxide   Hardground   Grain
borings   filled   aggregate
          borings
                    Grain   Particle
          maceration   loss

Grain   Prolonged
maceration precipitation

Silt   Relict   Sand
&   grain
micrite

                                                  Major process & products

for other cool-water carbonate systems (Farrow and Fyfe 1988; Young and Nelson 1988; Smith and Nelson 2003) are also present here. In this context there are several pertinent observations concerning recent carbonate particles; (1) although microboring is present, it is not pervasive, particles are generally well preserved and, (2) microboring is much more prevalent in molluscs than in bryozoans.

11.3.2 Chemical Diagenesis

Synsedimentary diagenesis of neritic carbonates is generally thought of as the combined effect of chemical and biological processes that lead to micrite envelopes, total micritization, and facies-specific cementation in warm-water environments (Bathurst 1975; James and Choquette 1990; Tucker and Wright 1990), and maceration or seafloor dissolution in cool-water settings (Alexandersson 1978, 1979; Freiwald 1995; James 1997; Smith and Nelson 2003).

The prevailing view, until recently, has been that cool-water carbonates undergo little synsedimentary, chemical (constructive) diagenesis generally manifest as cementation (Nelson et al. 1988b; James and Bone 1989; James 1997). This notion is supported by several lines of evidence, in particular the absence of precipitated grains such as ooids, the relative paucity of non-biogenic carbonate mud

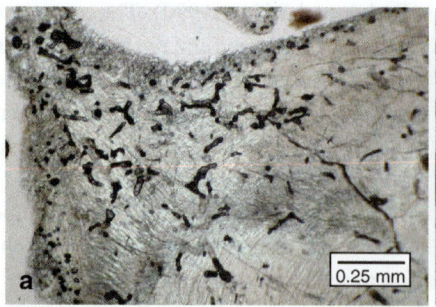

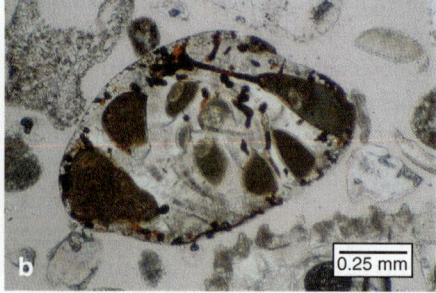

Fig. 11.2 Thin section images under plane polarized light of microborings. **(a)** A mollusc shell infested with empty microborings. **(b)** A relict grain composed of a foraminifer and Fe-stained microcrystalline carbonate that has been rounded by marine erosion; microborings in the foraminifer are filled with Fe-oxides

in neritic environments, the lack of micrite envelopes, and the apparent scarcity of synsedimentary cements. This perception seems to be supported by the fact that relatively low water temperatures result in slower reaction rates, higher ion activity products, and lower saturation levels for the carbonates. Whereas this is largely true, recent studies in southern Australia indicate not only that synsedimentary carbonate precipitation is taking place, but paradoxically that mineral-specific dissolution is also widespread (James et al. 2005; Rivers et al. 2008) in this cool-water system. These findings are in turn supported by observations of similar Cenozoic deposits in the region (James and Bone 1989, 1994; Kyser et al. 2002). Such processes are important because they dramatically change the character of the resultant deposits that enter the geological record.

11.3.3 Synsedimentary Cementation

Synsedimentary lithification in the form of extensive hardgrounds has not been demonstrated in neritic environments on the southern Australian shelf. This could be for two reasons; (1) they are not there, or (2) they are there but cannot be sampled. The relatively deep nature of the shelf and the high-energy sea state necessitates remote sampling and observation, as opposed to direct observation in shallow warm-water environments. By contrast, there is recurring evidence of particle cementation, largely in the form of grain aggregates. Perhaps the most convincing evidence, however, comes from the composition of relict grains, wherein early cement precipitation plays an important role.

11.3.3.1 Grain Aggregates

Sand-size particles of all types, calcareous skeletons, dolomite crystals, and siliciclastics, can be cemented by white to buff, microcrystalline Mg-calcite (Fig. 11.3) (Bone et al. 1992; Rivers et al. 2008). Such sand-size particle aggregates are found across the southern margin to ~100 mwd. They usually form a minor, <5% fraction of the sediment but are not present everywhere. The calcite cement crystals are platy to euhedral, <10 μm in size, and contain 10–14 mol% $MgCO_3$. These cements are identical to those found in the intraskeletal pores of various skeletal grains.

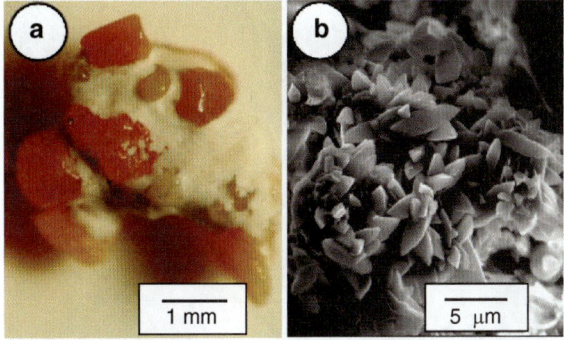

Fig. 11.3 Aggregate grains. (a) Skeletal particles and dolomite rhombs cemented together by microcrystalline magnesium calcite. (b) Scanning electron microscope image of the magnesium calcite crystals. (Images courtesy of J. Rivers)

11.3.3.2 Relict Particles

These iron-stained, abraded grains are either skeletons or intraclasts (Chap. 3). The material in intraskeletal pores and the red-brown microcrystalline material cementing intraclasts is Mg-calcite with minor amounts of clay. This calcite is identical to that in grain aggregates, <10 μm rhombs containing 10–14 mol% $MgCO_3$.

11.3.3.3 Hardgrounds

There are almost no recorded hardgrounds other than granule-size clasts. One exceptional example, however, comes from 130 mwd on the Ceduna sector (Nelson and James 2000). The limestone slabs have intra- and interparticle cements in the form of <10–70 μm-thick isopachous rinds of fibrous calcite spar and generations of interstitial micrite, especially microbioclastic and micropeloidal varieties (Fig. 11.4). Cement is predominantly Mg-calcite averaging 13–14 mol% $MgCO_3$. The hardground is likely stranded shallow neritic because the ^{14}C age of a *Glycymeris* cemented into the limestone is 18,400 ± 150 years BP.

11.3.3.4 Micrite Envelopes

In spite of evidence that cementation is occurring, albeit not extensive, grains are rarely if ever surrounded by a micrite envelope. Microborings in tropical settings are generally filled with microcrystalline Mg-calcite or aragonite, much of which is cement (Bathurst 1975).

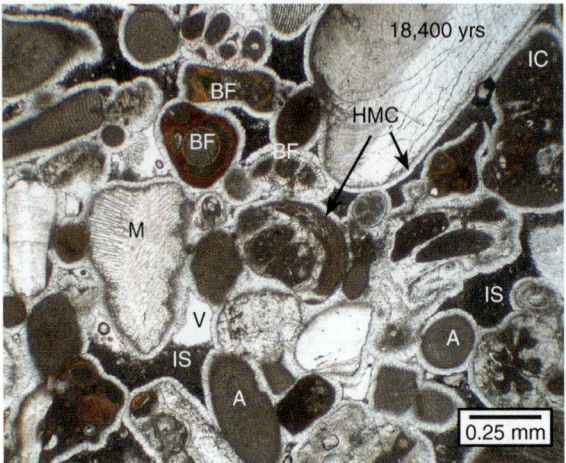

Fig. 11.4 Thin section image under crossed-polarized light of a marine hardground dredged from 130 mwd in the eastern Great Australian Bight south of Port Lincoln. The particles of coralline algae (A), benthic foraminifers (BF), molluscs (M), and intraclasts (IC) are cemented together by an isopachous rind of magnesium calcite (HMC–13 mol% $MgCO_3$) cement. The mollusc has a ^{14}C age of 18,400 years BP. Remaining spaces between the grains and postdating the cement is a fill of planktic foraminifer-rich carbonate mud (IS). The cement is interpreted to be shallow marine and coincide with the last glacial maximum sealevel lowstand whereas the muddy internal sediment records the subsequent sealevel rise; V = void

Microborings in these sediments are either empty or more often filled with amorphous Fe-oxides.

11.3.4 Fe-oxide Coating and Impregnation

Iron is a conspicuous component in these carbonates sediments. It is most obvious as the buff, yellow, and brown discoloration of stranded and relict grains. Rivers et al. (2007, 2008) have documented the nature of iron associated with particles in the neritic zone. No discoloration is apparent in some grains, whereas in others the staining is evident as a rim of brown impregnation with the central portion of the grain unaltered. Most discoloration follows borings, is blotchy, or is pervasive. Intraskeletal microcrystalline cement is generally more intensely stained than the skeleton itself. Attempts to identify the clay and Fe-bearing phases were undertaken using selective dissolution techniques and a Gandolfi X-ray camera, but were unsuccessful perhaps due to the minute portion of the phases. Nevertheless, under SEM it appears that the dull buff color of stranded grains and the orange-brown color of relict grains is due to clay particles and associated Fe-bearing phases.

This situation is superficially similar to that on the outer shelf (200–300 mwd) of eastern Australia where Marshall (1983) found that the dark reddish brown or yellow-green planktic foraminifer chamber fillings and mollusc shell impregnations were verdine (beretherine) and goethite. Such sediments were interpreted to be relict with verdine originally forming in shallow water and presently breaking down to goethite under the modern oxidizing conditions. Although such a process is viable for neritic sediments off southern Australia, no verdine phases have been observed.

Quartz grains from southeastern Australian beach sand are stained by Fe oxides coating clay particles (Sullivan and Koppi 1998) with Si/Al/K ratios similar to those detected by Rivers et al. (2008). The clay is a mixture of illite and kaolinite, whereas the Fe oxide phase is goethite.

Goethite is the most common Fe oxide in marine sediments (van der Zee et al. 2003). Precipitation of Fe oxides in marine pore waters is a product of redox cycling (Burdige 1993). Microbes may, however, play a more direct role in the Fe-staining of the sediments. Raghukumar et al. (1989) noted the presence of Fe oxide deposited by boring cyanobacteria in bivalve shells. The staining of borehole walls in relict skeletal grains demonstrates that a similar process might be partly responsible for dissemination of Fe oxide in the shelf sediments. With little fluvial input in the region, the ultimate source of iron in the sediments is likely aeolian (Fig. 11.5).

Fig. 11.5 Iron-rich siliciclastic dust from central Australia blowing south off the cliffs and out to sea along the northern shore of Investigator Strait, Yorke Peninsula; circled automobile for scale

11.3.5 Synsedimentary Dissolution

A recurring observation in neritic cool-water carbonates, particularly Cenozoic deposits, is that the original particles were composed mainly of calcite skeletons. This is because there is no evidence in many Cenozoic limestones, either in the form of calcitized shells or skeletal molds, that there were any aragonitic skeletal particles originally. This observation appears to be in direct contrast to the mineralogical composition of sediments in the southern Australian neritic realm.

Analysis of sediment samples from the Great Australian Bight and Lacepede shelves (James et al. 2005) indicates that the Holocene sediment fraction is a mixture of aragonite, LMC, IMC, and HMC. They are, however, overwhelmingly dominated by aragonite (Figs. 11.6, 11.7). Compositionally on the Great Australian Bight, the aragonite components range from 100% molluscs and 0% bryozoans to 15% molluscs and 85% bryozoans, averaging 63% molluscs, 35% bryozoans, and 2% other grains. On the Lacepede Shelf (as measured in 21 thin sections) the ranges are similar, from 100% molluscs and no bryo-

zoans to 14% molluscs and 86% bryozoans, averaging 58% molluscs and 39% bryozoans with 3% other particles.

Spatially on the Great Australian Bight most such sediments have >50% Holocene aragonite particles (Fig. 11.6), except for inshore facies west of the Roe Plain, that are dominated by coralline algal rhodoliths (HMC). Most inner and middle neritic environments contain >70% aragonite, with large areas having >80%. A similar pattern is evident on the Lacepede Shelf (Fig. 11.7), where areas with >70% aragonite extend across the middle and deep neritic zones and comprise the whole shelf off Robe. Like the Great Australian Bight, there are areas of the outer shelf that are conspicuously low in aragonite directly adjacent to regions of high aragonite content.

This apparent mineralogical contradiction is not new; there has always been a nagging suspicion that much of the aragonitic biota in Cenozoic cool-water limestones was somehow lost prior to initial lithification (Beu et al. 1972; Nelson 1978; Nelson and James 2000). The abundance of aragonite in southern Australian neritic sediments compared to similar Cenozoic deposits would seem to confirm this notion and

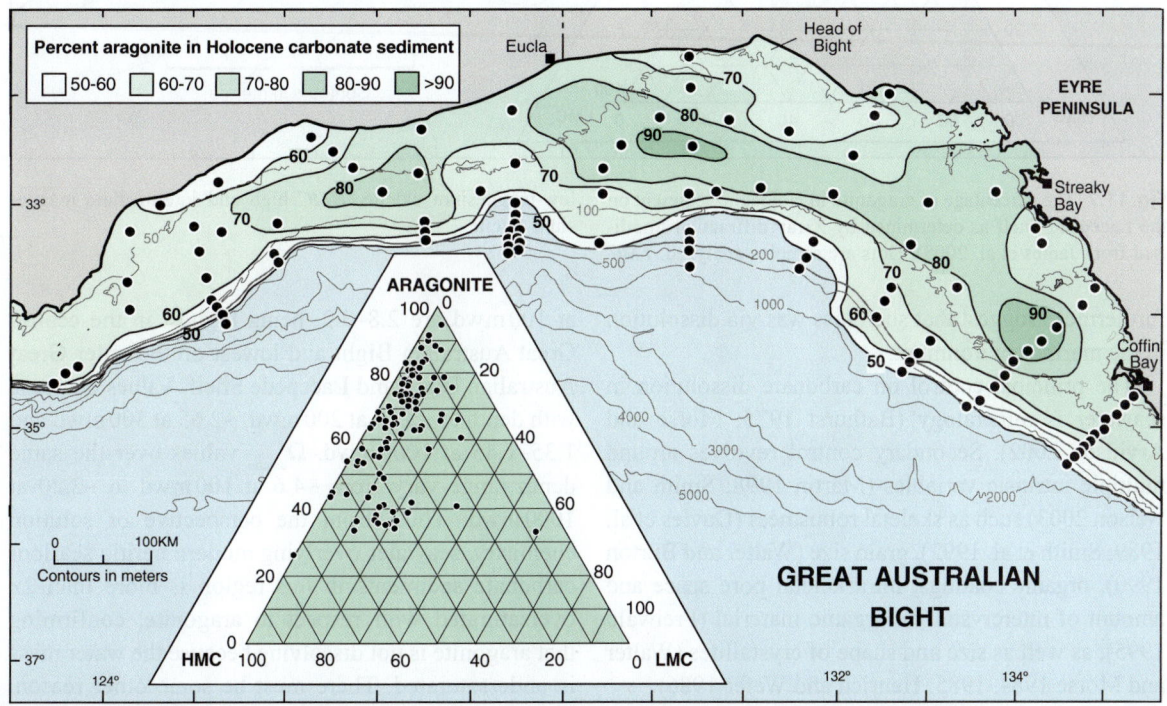

Fig. 11.6 The percentage of aragonite in Holocene sediment on the Great Australian Bight as determined by X-ray diffraction (modified from James et al. 2005). Dots are samples analyzed. *LMC* low-magnesium calcite, *HMC* high and intermediate magnesium calcite

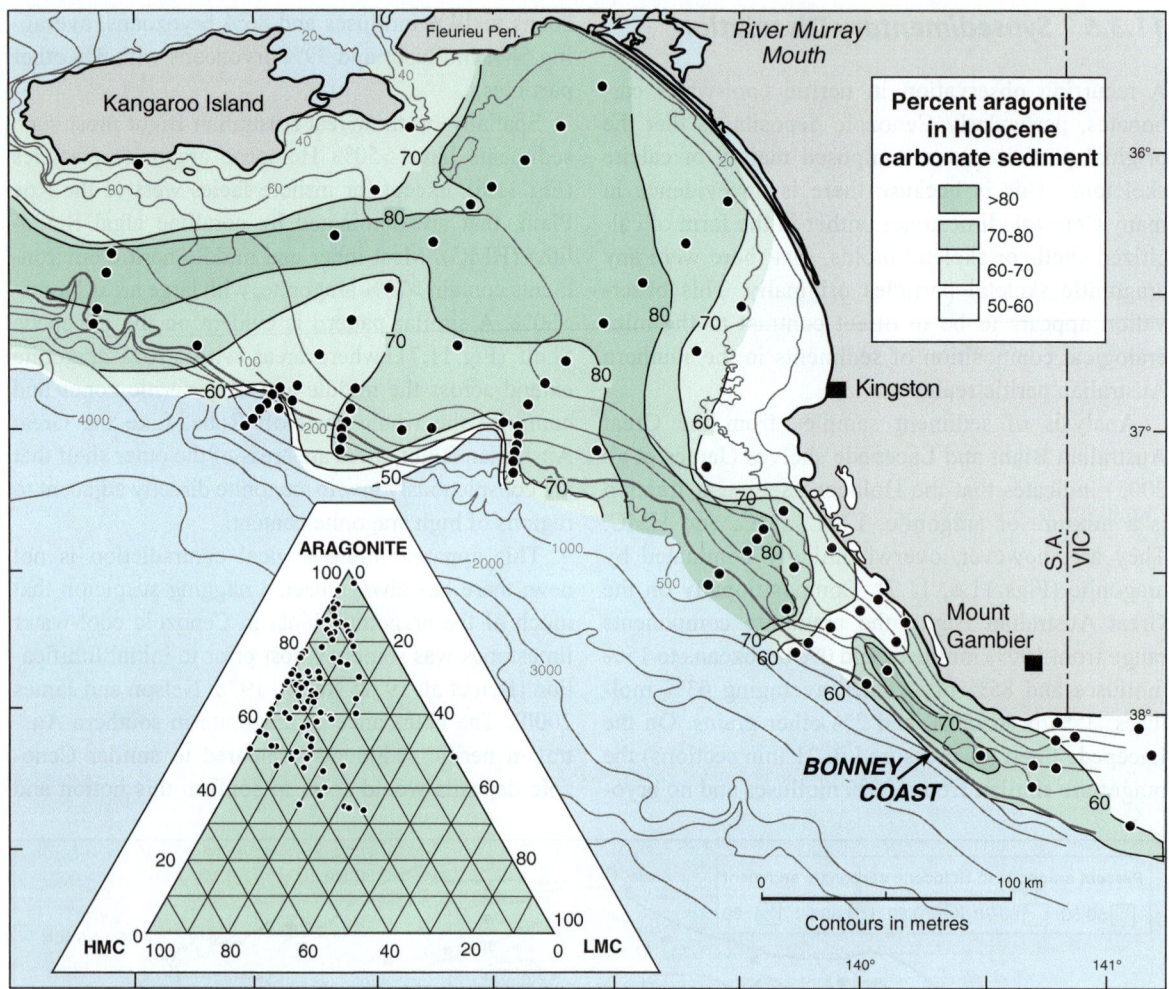

Fig. 11.7 The percentage of aragonite in Holocene sediment on the Lacepede Shelf as determined by X-ray diffraction (modified from James et al. 2005). Dots are samples analyzed. *LMC* low-magnesium calcite, *HMC* high and intermediate magnesium calcite

furthermore suggest that such loss was via dissolution in the marine environment.

The principal control on carbonate dissolution in seawater is mineralogy (Bathurst 1975; Morse and Arvidson 2002). Secondary control revolves around multiple intrinsic variables (Martin 1998; Smith and Nelson 2003) such as skeletal robustness (Davies et al. 1989; Smith et al. 1992), grain size (Walter and Burton 1990), organic coatings, intraskeletal pore space and amount of intercrystalline organic material (Freiwald 1995), as well as size and shape of crystallites (Walter and Morse 1984, 1985; Henrich and Wefer 1986).

James et al. (2005) determined that seawater in this region is oversaturated with respect to both aragonite and calcite to depths of 1000 m. $\Omega_{aragonite}$ values

at 100 mwd are 2.8–3.2, being highest in the central Great Australian Bight and lowest on the outer Great Australian Bight and Lacepede Shelf. Values decrease with depth to ~2.80 at 200 mwd, ~2.65 at 300 mwd and 1.35–1.45 at 1000 mwd. $\Omega_{calcite}$ values over the same depth range vary from ~4.6 at 100 mwd to ~2.20 at 1000 mwd. Thus, from the perspective of solution chemistry, seawater overlying modern neritic seafloor carbonate sediments in this region is more than 2× oversaturated with respect to aragonite, confirming that aragonite is not dissolving because the water mass is undersaturated. There must be some other reason. James et al. (2005) concluded that changes leading to undersaturation could have been brought about by the oxic to anoxic microbial degradation of sedimentary

Fig. 11.8 A diagram illustrating the major loss of aragonite in surface sediments just below the seafloor across much of the southern continental margin over an estimated period of ~20,000 years

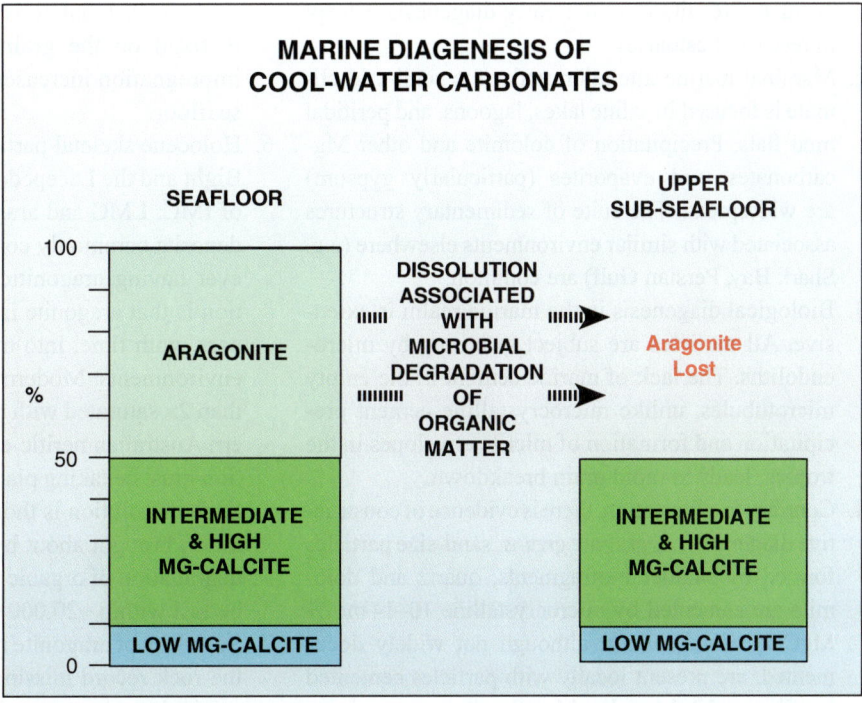

organic matter (Froelich et al. 1979) via several pathways, including: (1) aerobic oxidation of organic matter, (2) oxidation of reduced by-products and H₂S, (3) sulphate reduction and (4) anaerobic methane oxidation. Such processes are decoupled from overlying bottom waters, and have little relationship to the saturation state of the ambient seawater. Walter and Burton (1990), Walter et al. (1993), Patterson and Walter (1994), and Ku et al. (1999) demonstrated that in photozoan sediments from Florida and the Bahamas, where seawater is oversaturated with respect to aragonite yet pore waters are saturated or undersaturated, approximately 50% of the sediment has undergone dissolution.

Refining this argument, Rivers et al. (2008) observed that although aragonitic components make up 40–90% of Holocene carbonates on the Lacepede Shelf and the Great Australian Bight, there is ~40% less aragonite in stranded sediments compared to Holocene ones. Furthermore aragonite only comprises ~5% of relict grains and in the case of gastropods the grains are mainly steinkerns. That this loss of aragonite is a marine process is confirmed by the unaltered mineralogy of *Marginopora vertebralis* (10–13 mol% MgCO₃) in both stranded and relict sediments. This observation further indicates that there has been no Mg-calcite neomorphism over this time. Such results

imply that in the present shelf depositional system most aragonite should be lost via reactions associated with the aerobic oxidation of organic matter in the very shallow subsurface in ~20,000 years (Fig. 11.8).

A caveat to the foregoing discussion is that aragonitic skeletons, particularly gastropods, infaunal bivalves, and locally parts of bimineralic bryozoans, are preserved in some shallow neritic and marginal marine carbonates. The precise reasons for this situation are unclear. It could be related to the faster deposition rates in these environments and thus less time for shallow sub-seafloor alteration.

11.4 Synopsis

Early diagenesis is widespread in marginal and marine environments along the southern Australian continental margin.

1. Alteration and mineral precipitation in the marginal marine setting is climate controlled. These processes are most active where the climate is semi-arid or locally arid, particularly adjacent to the Albany, Great Australian Bight, and South Australian Sea sectors. The humid climate adjoining the eastern parts of the Otway Shelf and

Tasmania results in minor early diagenesis, except in restricted estuaries.

2. Marginal marine alteration under the semi-arid climate is focused in saline lakes, lagoons, and peritidal mud flats. Precipitation of dolomite and other Mg-carbonates, and evaporites (particularly gypsum) are widespread. The suite of sedimentary structures associated with similar environments elsewhere (e.g. Shark Bay, Persian Gulf) are common.

3. Biological diagenesis in the marine realm is extensive. All particles are subject to boring by micro-endoliths. The lack of marine cement in the empty microtubules, unlike microcrystalline cement precipitation and formation of micrite envelopes in the tropics, leads to rapid grain breakdown.

4. Contrary to expectation, there is evidence of constructive diagenesis. *Aggregate grains*, sand-size particles formed by smaller biofragments, quartz and dolomite are cemented by microcrystalline 10–14 mol% $MgCO_3$. *Hardgrounds*, although not widely documented, are present locally with particles cemented by fibrous 13–14 mol% $MgCO_3$. *Relict grains* are cemented by microcrystalline 10–14 mol% $MgCO_3$.

5. Fe-oxides coat relict grains and to a lesser extent stranded particles. The source of the iron is likely aeolian dust but it is as yet uncertain how this is fixed on the grain surface. The intensity of impregnation increases with length of time on the seafloor.

6. Holocene skeletal particles on the Great Australian Bight and the Lacepede Shelf are composed mainly of IMC, LMC and aragonite, yet similar Cenozoic deposits commonly contain little or no evidence of ever having aragonitic components. The implication is that aragonite is lost before these sediments pass, with time, into meteoric or burial diagenetic environments. Modern seawater is, however, more than 2× saturated with respect to aragonite in southern Australian neritic environments. Thus, dissolution must be taking place in the shallow subsurface. Such dissolution is thought to occur via undersaturation brought about by oxic and anoxic microbial degradation of organic matter. Aragonite appears to be lost within ~20,000 years.

7. Such loss of aragonite means that sediments go into the rock record missing a large proportion of their original components; thus the geological record of such deposits is biased. This loss may also account for the apparent low accumulation rates of many cool-water carbonates.

Chapter 12
Summary & Synthesis

12.1 Introduction

Carbonate sediment, generated by marine microbes, plants, and animals, covers many continental shelves of the world and has done so throughout much of geological history. Biofragments produced in the sea are unusually sensitive proxies that record the health of the oceans, their changing composition, and within their calcareous hard parts contain irrevocable evidence of global tectonic evolution. One of the ongoing quests of earth science is to ascertain how faithfully these calcareous particles encrypt environmental information. It has long been known, for example that modern tropical reefs, ooid shoals, and muddy tidal flats are powerful tools for unlocking past marine environments. Only recently has it become clear that carbonate sediments in temperate and polar oceans are widespread, distinctive, and important and thus can also be used to decipher the nature of ancient mid- and high-altitude oceans. The deposits that are documented herein, along the southern margin of Australia, are extremely important in this endeavour because they comprise the largest deposit of such carbonates in the modern world. Thus, this vast neritic realm contains information that can be used to unlock the nature of many Cenozoic and Mesozoic limestones, and to help formulate models that together can act as an intellectual bridge into the older world of ancient Paleozoic carbonates.

Each large modern depositional province is unique in the sense that sedimentation is controlled by an integrated set of variables (Fig. 12.1). The southern Australian continental margin is no exception. Such controls are both extrinsic or allogenic and intrinsic or autogenic. The most important extrinsic control is sea level fluctuation. In this region the sediments have accumulated during the late Quaternary, a period

of dramatic sea level change on the order of 100 m. The intrinsic controls are more numerous, specifically geography, tectonic history, climate, oceanography, and seawater nutrient levels. The southern Australian shelf is mostly zonal (latitude parallel) and has undergone relatively recent uplift. Whereas the climate is largely Mediterranean, the ocean is cold, stormy, and extremely energetic. Trophic resource levels are relatively low throughout. It is the partial or complete integral of all these controls that determines the spectrum of the modern surficial sediments.

12.2 Environmental Controls

12.2.1 Geography

This immense shelf stands out against other temperate depositional regions in the modern world. Most others, such as those around New Zealand, along the margin of continents in the Atlantic, adjacent to southern Africa in the Indian Ocean, and around the Pacific Rim are meridonal and so change rapidly from north to south. Only the roughly latitude parallel Mediterranean Sea is somewhat similar but it is a much lower energy system overall. Southern Australia is different because it is zonal and so variations in oceanography and climate affect sedimentation in more subtle ways than elsewhere.

12.2.2 Sea Level History

The vast open shelf has over the last 100,000 years ranged from being completely inundated and under

N. P. James, Y. Bone, *Neritic Carbonate Sediments in a Temperate Realm*,
DOI 10.1007/978-90-481-9289-2_12, © Springer Science+Business Media B.V. 2011

Fig. 12.1 A sketch illustrating the autogenic and
allogenic controls on carbonate sedimentation along the
southern continental margin
of Australia

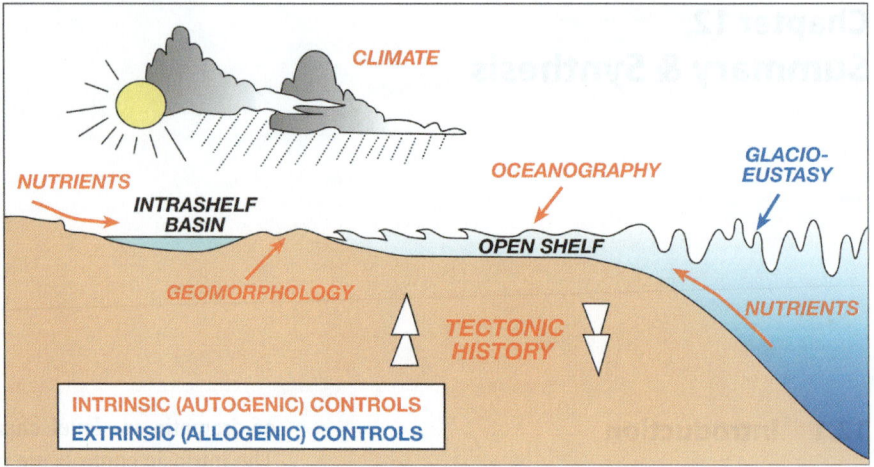

relatively deep water as it is today (MIS 5 and 1) to
being only partially covered by shallow water (<10 m)
with the shoreline moving back and forth across what
is now the middle neritic zone (MIS 3 and 4), to being
almost completely exposed with the coastline at modern depths of ~120 m (MIS 2) in the deep neritic zone
near the present shelf edge (Fig. 3.17). Thus, the sediments are first and foremost palimpsest, a mixture of
materials produced and diagenetically altered during all
of these phases. Relict grains are mixed with stranded
grains and these particles serve as a substrate for the
accumulation of Holocene biofragments (Fig. 12.2).

12.2.3   Geology and Tectonic History

The shelves and gulfs are the product of regional geology. The two Precambrian cratons of mostly hard
crystalline rocks are resistant to erosion and so have
relatively narrow shelves. The wide Great Australian
Bight is largely a product of northward moving marine
erosion of soft Cenozoic Eucla Basin carbonates. The
Otway and western Tasmania shelves are relatively narrow because of relatively recent uplift and hard coastal
lithologies. The two gulfs of the South Australian Sea
that pierce the hinterland are due to subsidence along
reactivated ancient north-south Precambrian faults.

Accommodation on the shelf is a function of the
interaction between sea level change and tectonic
movement. Rifting and passive margin subsidence
during the Cenozoic provided the framework for thick
and extensive Eocene to middle Miocene largely car-

bonate sediments, but this situation was interrupted by
late Miocene tectonic inversion. Such relatively recent
uplift, deformation, and tilting not only exposed the
inner parts of much of the shelf composed largely of
Paleogene and early Neogene sediment and sedimentary rocks, but also slowed or arrested ongoing shelf
subsidence, thus reducing sediment accumulation
space. The Great Australian Bight, for example, is a
bipartite shelf wherein the inner shelf is eroded into
Cenozoic limestone and dusted with Holocene carbonate sediment whereas the outer shelf and upper slope
is a late Neogene, prograding carbonate sediment
wedge (James and Bone 1994; Feary and James 1998)
(Fig. 3.16).

12.2.4   Climate

The semi-arid Mediterranean climate (Fig. 2.6) over
much of the area has several important consequences,
namely lack of terrestrial sediment and nutrient input,
seasonal seawater heating, and extensive marginal
marine evaporite systems.

The relative paucity of rain throughout the region,
except for Victoria and western Tasmania, means
that fluvial input is negligible. This is even true for
the River Murray, the largest river in Australia, that is
currently not delivering any sediment to the sea. This
situation means that in most areas carbonate sediments
can accumulate unimpeded by terrigenous clastic
sediment deposition, a prime prerequisite for carbonate sedimentation in general (James et al. 2009) and

PALIMPSEST SEDIMENTATION

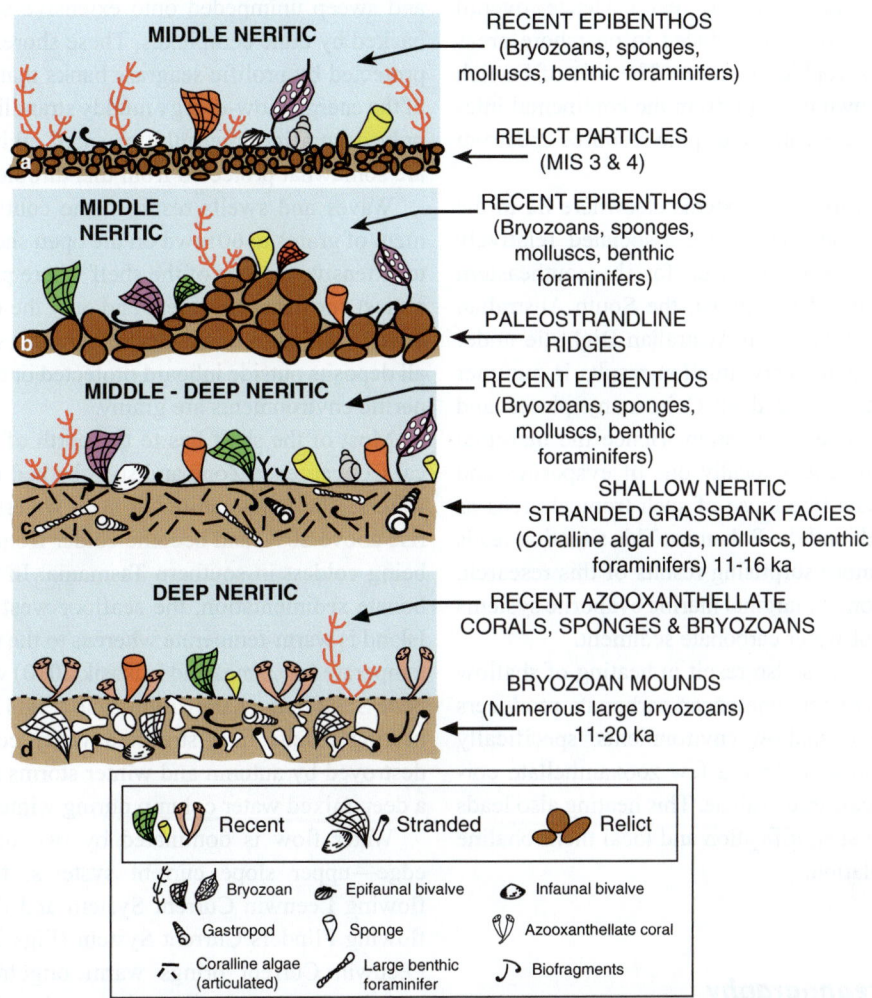

Fig. 12.2 Sketches depicting different types of palimpsest sedimentation on the southern Australian continental shelf. (**a**) A seafloor of relict sand particles, both lithoclasts and bioclasts, formed during marine isotope stages 3 and 4, that now serve as a substrate for the growth of a recent epibenthos dominated by bryozoans, sponges, molluscs, and benthic foraminifers that together produce a variety of carbonate sediment particles. (**b**) Paleostrandline ridges of relict boulders and cobbles together with shallow water bivalves that are today sites of recent rocky reef epibenthic growth dominated by bryozoans, sponges, molluscs, and benthic foraminifers. (**c**) A seafloor of carbonate sand dominated by stranded particles of articulated coralline algae and molluscs that formed in a seagrass environment during the early stages of the last sea level rise (16–11 ka) when the water depths were much shallower than today; the seafloor is now occupied by a middle to deep neritic epibenthic biota. (**d**) The exhumed surface of deep water sponge-bryozoan mounds (MIS 2) composed of large bryozoan skeletons that are now in the deep neritic environment and populated by living sponges, bryozoans, and azooxanthellate corals

cool-water carbonates in particular (Nelson 1988a). Nevertheless, the River Murray was an active fluvial depositional system during the LGM when it flowed across the shelf and into the Murray Canyon (von der Borch 1968; Hill et al. 2009). Much siliciclastic sand was blown out of the river and eastward across the coastal plain such that most of the shelf is now covered with this late Pleistocene quartzose sand (James et al. 1992; Hill et al. 2009). The modern biota is merely an addition to this siliciclastic sand blanket. By contrast, rivers along the wet coastline of Tasmania still deliver a considerable amount of terrigenous sediment to the sea and so shallow neritic sediments are largely siliciclastic.

Lack of fluvial input also results in little or no delivery of nutrients to the marine realm. Hence, overall biological productivity is reduced in nearshore areas because of low trophic resources. The role of Fe-rich aeolian dust blown offshore from the continental interior is as yet unclear in the trophic resource spectrum of this region.

Cool-water carbonate systems elsewhere lie in the stormy mid-latitude belts with associated relatively high rainfall. This is also true for the southeastern shelf of Australia. By contrast, the South Australian Sea and most of the Great Australian Bight lie under a semi-arid climate. This situation results in summer temperatures that exceed 40°C in many places and attendant extensive evaporation. Hence the marginal marine environment is locally one of evaporites and muddy tidal flats like those of such tropical areas as the Persian Gulf and the Bahamas. This situation leads to one of the more surprising results of this research, the juxtaposition of marginal marine evaporite systems and marine cool-water carbonate sediment.

Hot summer days also result in heating of shallow waters allowing some photozoan carbonate producers to exist in local shallow environments, specifically large benthic foraminifers, a few zooxanthellate corals, and green calcareous algae. This heating also leads to strong thermal stratification and local thermohaline seawater circulation.

12.2.5   Oceanography

The Southern Ocean has one of the most energetic sea states on the globe and this situation profoundly affects biology and sedimentation. Waves and swells from the south batter exposed rocky and cliffed coastlines and sweep unimpeded onto extensive sandy beaches backed by dune complexes. These shores are partially protected by prolific seagrass banks that absorb some of the energy. Low-energy muddy strandline deposition only occurs in large gulfs or small embayments that are somewhat protected from this turbulent ocean.

Waves and swells result in the continuous movement of grains to 60 mwd on the open shelf. This leads to extensive regions of the shelf where particles are in almost constant movement and only the most securely attached epifauna or adapted infauna can survive. Thus, all deposits outside inboard protected or outboard deep neritic environments are grainy.

Most of the shelf lies to the north of the Subtropical Convergence Zone and so is bathed in subtropical waters. The temperatures of such waters only locally rise above 20°C and become colder from west to east, being coldest in southern Tasmania. In terms of carbonate sedimentation, the seafloor west of Kangaroo Island is warm-temperate whereas to the east it is cool-temperate (cf. James and Lukasik 2010) with the Lacepede Shelf being a transition zone (Fig. 12.3). Summer heating results in a stratified water column that is destroyed by autumn and winter storms and results in a deep mixed water column during winter.

Water flow is dominated by two opposing shelf edge—upper slope current systems, the eastward-flowing Leeuwin Current System and the westward-flowing Flinders Current System (Figs. 2.7, 2.8). The Leeuwin Current brings warm oligotrophic waters into the Great Australian Bight region from the west during winter and the ongoing South Australian Current and Zeehan Current transport relatively warm and saline waters eastward along the outer shelf. The most important attribute of this system is that it blocks

Fig. 12.3 Southern Australian continental shelf illustrating the location of the warm temperate and cool temperate provinces with the southern Kangaroo Island and Lacepede Shelf being transitional regimens. Image courtesy of Geoscience Australia

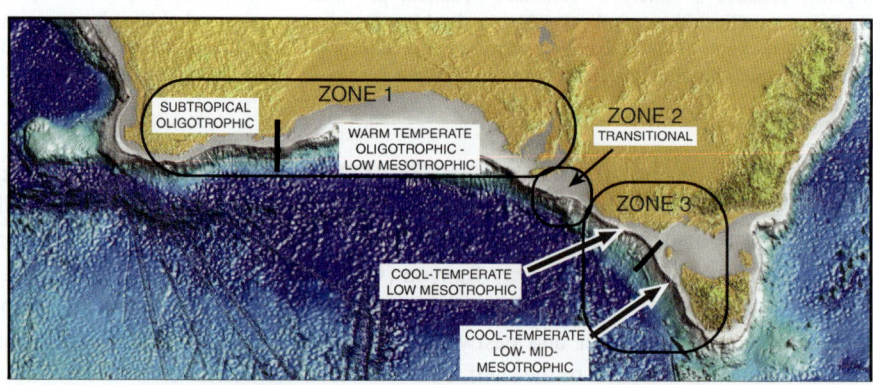

Fig. 12.4 A cross plot of minimum sea surface temperature and seawater chlorophyll content with appropriate environments wherein the different carbonate sediment associations are plotted; conditions on the southern Australian shelf are highlighted. (Modified from James and Lukasik 2010)

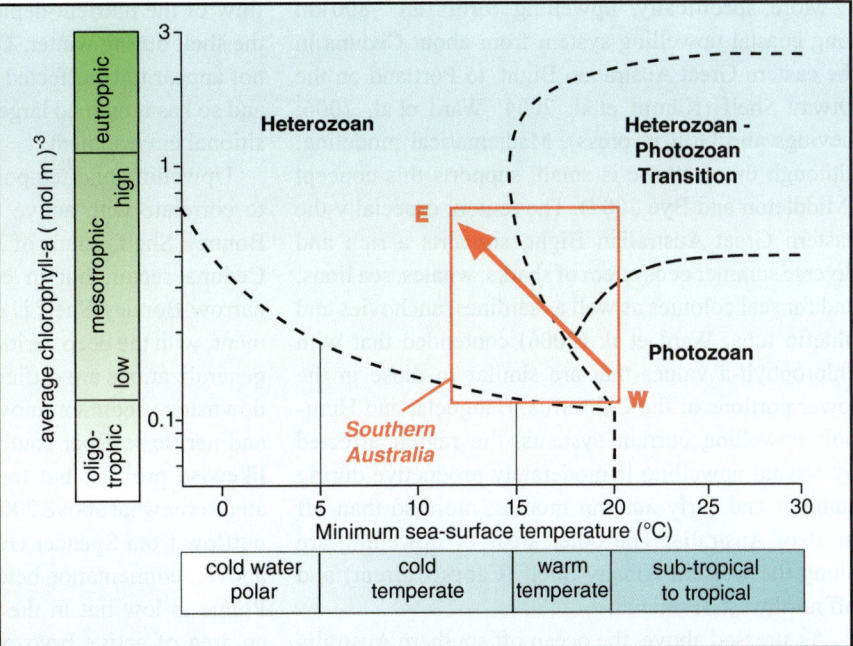

extensive upwelling despite favourable seasonal wind circulation. This overall downwelling situation, combined with the lack of terrestrial nutrient element supply, results in a generally low mesotrophic to marginal oligotrophic trophic resource system (see below). The Flinders Current system that flows below or outboard of the Leeuwin Current system advects Subantarctic Mode Water and Antarctic Intermediate Water with elevated nutrients westward parallel to the upper slope. Local conditions, especially on the Bonney Coast, south of Kangaroo Island, and in the eastern and possibly western Great Australian Bight allow this water to locally upwell onto to the shelf. Such an overall circulation pattern leads to a benthic and pelagic environment that is one of general low productivity, except in the discrete areas of upwelling.

12.2.6   Trophic Resources

Australian marine environments are generally low in nutrients and consequently have relatively low biological productivity (Zann 2000). The shelf waters are dominated by low-nutrient water masses (with especially low nitrates and phosphates), there are no major upwellings of nutrient-rich deep water, the runoff from ancient leached soils is naturally low in nutrients, and the rainfall is low. As a result, large areas of the ocean around Australia are virtual biological deserts. The low mesotrophic (0.1–0.5 chlorophyll-a $\mu mol\, m^{-3}$) nature of the southern continental margin (Fig. 12.4), with some waters borderline oligotrophic, contrasts with most other mid-latitude oceans that are mid to high mesotrophic (Longhurst 1998; Halfar et al. 2004).

The Subtropical Convergence Zone (Fig. 2.7), a biologically productive zone in itself, largely isolates the southern part of the continent from nutrient rich subantarctic waters. The continental margin is, nevertheless relatively close to these productive waters and they play an important role in overall bioproductivity. This is firstly manifest when the Subtropical Convergence Zone moves northward during summer months to lie at about the latitude of Macquarie Harbour, Tasmania (Fig. 10.3). Secondly, the Flinders Current is formed from Antarctic Intermediate water and Subantarctic Mode Water just southwest of Tasmania (Barker 2004) and it moves this water northward and westward along the upper slope across southern Australia to Cape Leeuwin and beyond. This water in turn upwells onto the shelf during summer and in this way elevated nutrients from Antarctica are delivered to a variety of neritic environments along the shelf.

More specifically, upwelling forms an ~800 km long coastal upwelling system from about Ceduna in the eastern Great Australian Bight, to Portland on the Otway Shelf (Kämpf et al. 2004; Ward et al. 2006; Levings and Gill, in press). Mathematical modeling, although the database is small, supports this concept (Middleton and Bye 2007). The region, especially the eastern Great Australian Bight, supports a rich and diverse summer ecosystem of sharks, whales, sea lions, and fur seal colonies as well as sardines, anchovies and bluefin tuna. Ward et al. (2006) contended that with chlorophyll-a values that are similar to those in the lower portions of the California, Benguela, and Humbolt upwelling current systems, the region affected by coastal upwelling is moderately productive during summer and early autumn months, more so than off most of Australia. The other areas of upwelling are along the western Albany Shelf (Capes Current) and off northwest Tasmania.

As stressed above, the ocean off southern Australia changes dramatically with the seasons; winter and spring are times of downwelling everywhere, summer and autumn are still characterized by overall downwelling but are also punctuated by local upwelling. There is a general correlation between nutrient levels and carbonate sedimentation at the broad scale on the open shelf. Relict sediments are particularly useful in this regard because these late Pleistocene particles can be used as proxies for relative rates of carbonate production. If the proportion of relict grains in the sediment between ~60 and ~140 mwd is high (>50%) then there has not been much Holocene sediment production, whereas if they are low then production has been somewhat greater. Although a qualitative measure, these assumptions are useful.

Relatively low nutrient levels are present over the Eyre Sector, the Lacepede Shelf, and the Lincoln Shelf. Summer heating and attendant evaporation, together with predominant downwelling in the Eyre sector produce the Great Australian Bight plume, a nutrient-poor, relatively saline water mass that drifts eastward during autumn (Fig. 8.6). Sediments beneath this plume contain the highest proportion of relict grains anywhere on the southern margin. This relationship is interpreted to reflect arrested Holocene sediment production. The perpetual downwelling leads to low rates of sedimentation in deep neritic environments here and attendant migration of slope facies up onto the shelf (Fig. 12.9). The Lincoln Shelf productivity may be reduced by

flow of the nutrient-depleted saline waters out across the shelf during winter. The wide Lacepede Shelf does not appear to be affected by any significant upwelling and so has remained largely a siliciclastic neritic depositional environment.

Upwelling and temporally elevated nutrients seem to correlate with active carbonate production on the Bonney Shelf, south of Kangaroo Island, and in the Ceduna sector, but in each case with caveats. The narrow Bonney Shelf is covered with carbonate sediment, with the deep neritic zone especially prolific, but generalizations are difficult to make because of active downslope sediment movement. The middle and deep and neritic seafloor south of Kangaroo Island may be likewise prolific, but there, sedimentation is attenuated somewhat above 200 mwd because of saline water outflow from Spencer Gulf during winter. As stressed above, sedimentation below the Great Australian Bight Plume is low but in the middle of this zone there is an area of active bryozoan and carbonate productivity (called the Ceduna Tongue by James et al. 2001). This area would appear to correlate with the observed upwelling in the eastern Great Australian Bight.

By contrast, our sampling has revealed that the deep neritic zone along the eastern part of the Albany Shelf around Cape Pasley and along much of the Baxter Sector is one of the most productive regions along the southern continental margin. This has not, as yet been identified as a region of elevated trophic resources. Finally, bryozoans in sediments around Tasmania, especially along the eastern and western margins are prolific with seafloor images showing high densities of epibenthic growth. These observations are tentatively correlated with higher nutrient levels associated with subantarctic waters within and south of the subtropical convergence zone.

Some inshore areas are, paradoxically, dominated by highly productive mangrove and seagrass communities. High productivity implies high nutrient demands. A large proportion of such nutrients in seagrass meadows probably come from efficient recycling (Hillman et al. 1989) since seagrasses can tap both the sediment and the water column for nutrients. As illustrated by the piles of grass on many beaches, there are at any given time, large amounts of leaf material in various stages of disintegration and dissolution. It is estimated that in situ remineralization within the sediment and uptake by roots may release 30–50% of all the nutrients required for new grass growth.

12.3 Zones

As a consequence of this research, it is now possible to partition modern neritic marine environments along the southern margin of Australia into zones. Utilizing the foregoing tectonic, oceanographic, sedimentological, and trophic resource continuum, the region can now be divided into three provinces (Fig. 12.3). Nutrient regimens are based on the combined attributes of foraminiferal biofacies (Li et al. 1996a, b, 1999) and measurements presented herein. The zones are defined using the terminology of James and Lukasic (2010).

Warm Temperate: This zone extends from the Albany Shelf and the Great Australian Bight onto the Lincoln Shelf and major Gulfs of the South Australian Sea. The area is nearly oligotrophic on the Albany Shelf in the west and is essentially a continuation of the environment that characterizes the neritic zone of southern Western Australia in the Indian Ocean (Collins 1988; James et al. 1994). Local seasonal upwelling in the deep neritic zone of the Albany Shelf, the Ceduna Sector of the Great Australian Bight, and south of Kangaroo Island increases available nutrient elements.

Transitional: An intermediate province across the wide Lacepede Shelf with fluctuating conditions that range from warm-temperate to cool-temperate but with overall low mesotrophic trophic resources.

Cool Temperate: The narrow Otway Shelf has overall low mesotrophic nutrient levels, but with important seasonal upwelling (Bonney Coast) and thus local elevated benthic productivity. Western Tasmania and Bass Basin are areas wherein seasonal influence of the subtropical convergence zone raises productivity levels and places a region in the low- to mid-mesotrophic zone.

12.4 Sedimentology

The shelf and adjacent environments can be viewed as a vast natural laboratory in which to study the nature of cool-water carbonate deposition. The processes of carbonate sediment formation and accumulation here are universally applicable whereas the products of this system are a function of local controls. Processes operative in these neritic and marginal marine environments are dominated by epibenthic growth, physical disturbance, and early diagenesis.

12.4.1 Epibenthic Sediment Production

Sessile, epibenthic invertebrates, the most important of which in terms of sediment production, are bryozoans, molluscs, and benthic foraminifers, dominate carbonate deposition in southern Australia. Whereas much is known about the role of molluscs and foraminifers, documentation of sedimentation in this region has dramatically clarified the way in which bryozoans produce sediment. Research documented herein has also revealed the importance of marine algae and plants in cool-water carbonate deposition.

12.4.1.1 Bryozoans

Our findings (Bone and James 1993; Hageman et al. 1995; Hageman et al. 1998) have built on the pioneer work of Nelson et al. (1988a) in New Zealand. The organisms, although present across the neritic realm, are most numerous and important in middle and deep neritic environments. They produce sediment across the grain-size spectrum (Fig. 4.3) from boulder size skeletons to mud-size particles.

- It is now clear that there is a fundamental difference in the way the particles are produced. Colonies that comprise a non-segmented skeletal element either go into the sediment as such or are fragmented by a combination of physical and biological processes; thus grain size is unpredictable. By contrast articulated colonies, in particular erect articulated branching and articulated zooidal growth forms, break down due to biological disintegration of the protenaceous connections between segments or zooids and hence generate grains of a predictable size. The former produce coarse sand- to small cobble- size rods whereas the latter generate individual silt-size grains of delicate individual zooids.
- The large (5–15 cm high) and common erect, rigid fenestrate bryozoan *Adeona* sp., typical of middle to outer neritic environments, is most numerous in warm-temperate environments. The colony is one of the few that has the ability to grow in shifting sediment substrates, because of its protenaceous root system that can extend tens of cm into the sand. This form, together with articulated zooidal types, is a pioneer.
- The erect, rigid multilaminar arborescent form *Celleporaria* sp. has been shown to be most numer-

ous in mesotrophic environments (Hageman et al. 2003). Whereas it can grow to large size (20 cm—the largest of all bryozoans in this area), it also grows as an encircling colony around oscular sponges.

- Encrusting types range from thin sheets only one zooid thick on grass blades to multilaminar forms that encrust ephemeral substrates to large encrusting masses growing on other invertebrate skeletons. Thus, they form sediment from mud to cobble size grains.
- Fenestrates and encrusters are the most numerous bryozoans in seagrass environments.
- There are depth-related differences in environmental space occupied by different bryozoans, for example, robust rigid branching forms (especially *Adeonellopsis*) grow to much deeper depths and the foliose form *Parmularia* is more numerous in cooler waters.
- The only significant bryozoans growing on macroalgae in southern Australia are encrusting types (especially *Membranipora membranacea*).

12.4.1.2 Marine Plants

The marine plants, macroalgae, coralline algae, and angiosperms, are fundamental components of the system. The photic zone is especially deep along this continental shelf, likely because of the low nutrient content of the water, and can extend to 100 mwd. The plants exhibit strong environmental partitioning. Marine grasses grow to 35 mwd throughout the warm temperate environments, but are much reduced in cool-temperate settings. Coralline algae, although present throughout, are most numerous in the warm temperate sector. Macroalgae, also present throughout, are more numerous, more robust, and of higher generic diversity in cool-temperate settings.

Marine grass beds are amongst the most prolific carbonate producing systems on the southern Australian margin. The accumulation rates of 20–100 cm ky^{-1} are, to date, the highest on the shelf. Carbonate comes from the epiphytes and is dominated by coralline algae, spirorbids, and encrusting foraminifers. This research and the work of others (see James et al. 2009 for references) has confirmed that these epiphytes produce copious amounts of carbonate that either stays in place or is swept onto the beach and blown into dunes. Dune systems are much reduced in areas where offshore grass is absent.

Coralline algae in grass beds are mostly geniculate (articulated) forms that produce great numbers of sand-size calcareous rods. This is in contrast to encrusting and rigid branching types that produce numerous broken branches or remain whole in the form of rhodoliths. Rhodoliths, although locally common, are nowhere as prolific as they are off western Australia, along the middle part of the eastern shelf and in the Mediterranean (Carannante et al. 1988; Betzler et al. 1997). The reasons for this are not clear but the cause may be low nutrient levels.

Macroalgae are particularly important on the rocky seafloor and are present throughout the photic zone. They are not prolific carbonate factories, however, because chemicals on the blades commonly prevent settling by epiphytic invertebrates. It is only on the holdfasts that corallines, bryozoans and, foraminifers are numerous. Barnacles and echinoids, which are common elsewhere in macroalgal environments, are not prolific in southern Australia, again likely because of relatively low seawater nutrient levels.

12.4.1.3 Substrate

The role of substrate has proven to be of fundamental importance for carbonate sedimentation. Nelson et al. (1988b) noted that New Zealand shelf sedimentary environments were puzzling because these were large areas of sand with little biota and rocky reefs with abundant growth and it appeared that these islands of growth were the only environments producing most of the sediment. This observation has been confirmed off southern Australia. The two main seafloor environments are open sands and rocky reefs. The rocky reefs are sites of epibenthic growth dominated by sponges and bryozoans throughout with macroalgae in shallow water. The change from coppice to scrub to turf with depth reflects a lessening of hydrodynamic energy together with a decrease in water temperature. These rocky substrates are mainly bedrock but some may be hardgrounds, an aspect that has yet to be assessed.

By contrast the sand barrens, although largely devoid of epibenthos, are locally host to numerous infaunal bivalves. Bryozoans that are specifically adapted to this environment of shifting sand are the fenestrate *Adeona*, various articulated zooidal growth forms, several vagrant (free-living) genera, and the brachiopod *Anakinetica cumingi*. Muddy substrates, such as those in the gulfs and in deep neritic to upper slope environments can be intensively burrowed, but any shell or rock is the site of epibenthic growth, especially in the gulfs.

Rocky reefs are, however, not the only hard surfaces. Epibenthic sediment producers also preferentially colonize old submerged beach ridges and large aggregations of large stranded skeletons such as those on the exhumed tops of bryozoan mounds (Fig. 12.2).

The nature of the sediment substrate also partitions the infauna. Sandy substrates of the shallow and middle neritic zone have few infaunal remains. Epifaunal echinoids are common but infaunal echinoid spines, where present, are not numerous. The sediments are not conspicuously burrowed and benthic foraminifers are mainly epifaunal (Li et al. 1996b). By contrast, muddy sediments in deep neritic and upper slope environments are burrowed, contain infaunal echinoid spines, and numerous infaunal benthic foraminifers, all of which attest to the sediment as a food source.

12.4.2 Physical Disturbance

As emphasized throughout this book, the hydrodynamic environment off southern Australia is one of the most energetic on the globe. Waves and swells disturb grains almost daily to 60 mwd whereas swells and exceptional storm waves intermittently promote seafloor sediment movement to 140 mwd. Such partitioning (Fig. 3.6) results in discrete zones of sedimentation (1) the shallow neritic is largely, except for grass beds, an environment of sediment production but little accumulation with most sediment transported landward, (2) extensive rippled sands and subaqueous dunes and rocky substrates characterize the middle neritic zone of sediment reworking, and (3) muddy sands and gravels mantle the deep neritic and upper slope, a setting of much sediment accumulation. Movement of grains in the middle neritic environment does not appear to result in extensive grain abrasion whereas it does on the shallow neritic seafloor.

12.4.3 Authigenesis and Diagenesis

12.4.3.1 Authigenesis

One of the most important findings of this research has been the recognition of the sediments as palimps-

est, and the origin of relict and stranded grains. The prevalence of these particles in middle and deep neritic environments further attests to the slow sedimentation rates across the region but also identifies the continental margin as a classic starved shelf. This is especially so for the relict particles that formed in the marine environment (cf. Rivers et al. 2005). They are a signal of low sedimentation rates and are useful as indicators of such conditions in the rock record. They are, for example, numerous in the Cenozoic Mannum Formation in the Murray Basin (Lukasik and James 2003).

Glauconite and phosphate should also be abundant in this situation (Nelson 1988a) but they are not. Glauconite is present in insignificant amounts in most shelf sediments in southern Australia and phosphate is virtually absent. Paucity of phosphate likely reflects the absence of deep water upwelling. Glauconite generally forms in low energy, calm water, confined microenvironments at the interface between oxidizing seawater and slightly reducing interstitial water between 50 and 500 mwd at water temperatures of <15°C on the outer shelf and slope (Odin and Fullagar 1988; Amorosi 2003). The origin of the mineral is poorly understood but the ions come from both seawater and the parent sediment. Importantly, glauconite is characteristic of transgressive and condensed sections (Amorosi 1995).

At first glance, the setting would seem to be ideal for glauconite formation. The sediments are transgressive, sedimentation rates are relatively low, and there is abundant iron in the system. Most environments are, however, high-energy and there is little fluvial input to provide K and Si. Thus, glauconite is not a common accessory authigenic mineral in this situation. This finding illustrates the capricious nature of glaucony facies and more importantly, highlights the fact that areas of low sediment accumulation are not necessarily sites of glauconite formation.

12.4.3.2 Diagenesis

A further consequence of marine diagenesis is the importance, in this area, of early dissolution of aragonite skeletal particles whereas the Mg-calcite grains appear to undergo little if any change on or just beneath the seafloor (Fig. 11.8). It is not yet clear how widespread such processes are but the implications are profound. The loss of aragonite in time frames of probably <20,000 years (Rivers et al. 2007) means that many, but not all, shelf sediments enter meteoric or burial diagenetic environments as calcite sediments with little if any carbonate available for dissolution and reprecipitation as cement. It also leads to a misconception as to the original composition of the sediments. Finally, it reduces the thickness of sediment accumulated over a given length of time and thus lowers sedimentation rates.

12.4.4   Sedimentary Facies

12.4.4.1   Shallow Neritic Facies

The shallow neritic environment (Figs. 12.5, 12.6) lies within the euphotic zone, is <60 mwd, and is most affected by summer seawater warming. It comprises two broad settings, low-energy intrashelf basins and the high energy open ocean. Climate is important because

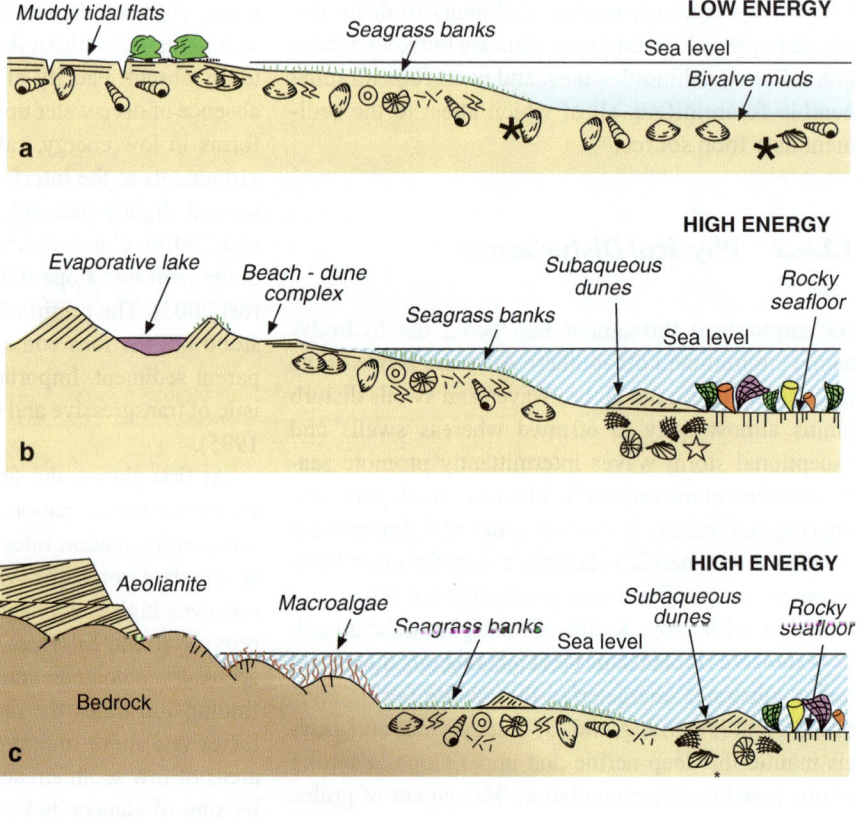

Fig. 12.5 A series of cross-sections depicting the different environments and resultant sediments in examples of semi-arid, warm temperate, shallow neritic environments in southern Australia

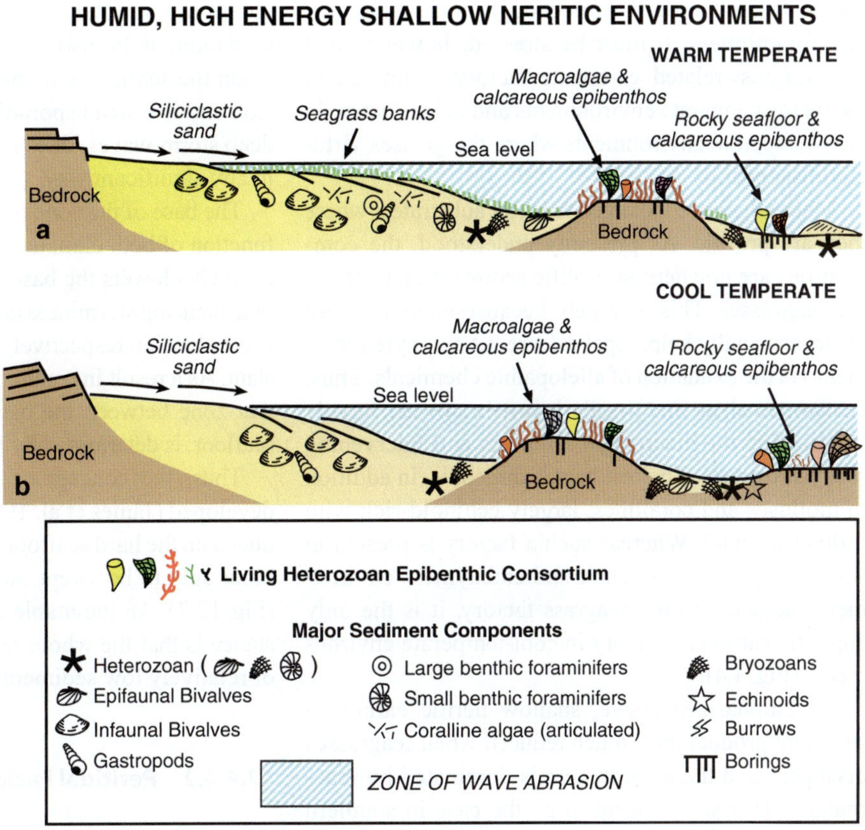

Fig. 12.6 Cross-sections illustrating the different environments and resultant sediments in selected examples of humid, high energy, shallow neritic environments in southern Australia

HUMID, HIGH ENERGY SHALLOW NERITIC ENVIRONMENTS

it controls, to some degree, the amount of siliciclastic sediment input and the water temperature. Shallow seafloor sediments in low energy settings, such as those exemplified by muddy tidal flats and prolific seagrass meadows, are dominated by a combination of molluscs, benthic foraminifers (including large forms) and coralline algal pieces. Hydrodynamics is not a key factor except for sweeping muds onto the tidal flats. High-energy shallow neritic sediments are likewise mollusc, benthic foraminifer, coralline-dominated in shallowest water. These sediments are the source of beach and dune sands with linear depression between dune sets, locations for lagoons and evaporative lakes. Deeper parts of the shallow neritic zone are constantly swept by waves such that sediment produced there is swept away and or locally concentrated into subaqueous dunes. Bryozoans, molluscs, and benthic foraminifers dominate these sediments. The seafloor adjacent to bedrock coasts is typically the site of macroalgal growth with grasses populating the adjacent sediment substrates.

Siliciclastic sand beaches and shore attached sediment wedges populated by seagrasses dominate warm-temperate nearshore environments in areas of higher rainfall. The grasses are missing in cool-temperate, humid settings and sediment production is focused on bedrock substrates and shed onto the surrounding seafloor.

High biological productivity in this area of low trophic resources suggests that nutrients must be effectively recycled. The most critical factories here are seagrasses in sediment substrates and macroalgae on rocky substrates. Grasses dominate in the sub-tropical warm water environments whereas macroalgae, especially large kelp, prevail in cooler water regions.

As in tropical settings, seagrasses both stabilize the sediment substrate and act as hosts for a cornucopia of epiphytes, but the sheer volume and abundance of the grasses in this area exceeds those in most tropical settings, the reasons for which are as yet unclear. The volume of calcareous epiphytes is directly proportional to the volume of seagrasses (Brown 2005; James et al. 2009). Sediments are like those in the tropics, muddy and mollusc-rich, but these environments also contain high numbers of coralline algae and if the water temperatures are warm enough, some large benthic foraminifers. It must be stressed, however, that this seagrass-related carbonate factory is limited to the warm-temperate environments and is ineffective in cool-temperate environments where the grasses virtually disappear.

Macroalgae are localized to rocky substrates, where they are prolific. As presently understood, the communities are nowhere as prolific sediment factories as the seagrasses. This is largely because many of these algae, especially kelp, repel calcareous epiphyte settlement via the exudation of allelopathic chemicals. Thus, calcareous algae and sessile invertebrates are largely localized to the macroalgal holdfasts and intervening rocky substrates. The resultant sediment is, in addition to molluscs and corallines, largely echinoid-rich with little or no mud. Whereas such a factory is present in warm-temperate inner shelf and embayment environments adjacent to the seagrass factory, it is the only euphotic carbonate factory in cool-temperate environments (Fig. 4.4).

As emphasized above, shallow neritic carbonate sediment production is much reduced when seagrasses disappear and the environment is dominated by macroalgae. This would seem to be the case in southern Australia where the Otway Shelf marginal marine environment contains aeolianites but they are not common along the western coast of Tasmania. The paucity of this facies in Tasmania probably corresponds to the virtual absence of seagrasses, with their place being taken by extensive macrophytes. Thus, it would appear that the cool-temperate inner shelf environment is not a region of high carbonate production and consequently few tidal flats or aeolianites should be expected. The consequence is that peritidal facies in such colder water settings are siliciclastic-dominated, even in semi-arid climatic settings, and not geographically extensive. This could be one of the reasons why peritidal and aeolianite deposits are so rare in the Tertiary of southern Australia.

The preceding situation, however, cannot be divorced from sea state and tectonics. The whole shelf has been uplifted relatively recently in the late Cenozoic such that much of the sea floor is a hard rocky, generally Cenozoic, limestone surface. This surface lies mostly within the zone of wave abrasion and so what sediment does form is swept away. In other words, accommodation is low partly because the hard seafloor is shallow relative to the deep zone of wave abrasion. It is not simply sea level that controls accommodation, it is also oceanography (Fig. 12.1). Only when the seafloor falls below the zone of wave abrasion and is instead sporadically affected by swells and deep storm waves, does sediment begin to accumulate in any significant way.

The base of the zone of wave abrasion, however, is a function of both climate and sea level. Lowering of sea level also lowers the base of wave abrasion. Increasing or decreasing storminess also lowers or raises the base of wave abrasion respectively while sea level remains constant. As a result in southern Australia, accommodation, that zone between the base of wave abrasion and the seafloor, is determined by both sealevel and climate.

Thus, the concept of a "shaved shelf" has been developed (James et al. 1994) wherein sediment is produced on the hard seafloor by an upright and encrusting biota only to be swept away and deposited elsewhere (Fig. 12.7). An inevitable consequence of this circumstance is that the whole region can be an environment of relatively low sediment accumulation.

12.4.4.2   Peritidal Facies

Peritidal mud flats (Fig. 12.5), with or without mangroves, in this semi-arid climatic realm are, because of the high summer air temperatures, virtually identical to those in more tropical settings such as the southern Bahamas and the Persian Gulf (cf. Pratt 2010). They contain the whole suite of sedimentary features that characterize such facies worldwide and in much of the geological record. Thus, muddy tidal flat facies can be expected adjacent to cool-water, low-energy subtropical marine environments.

Beach-dune complexes (Fig. 12.5) that face the southwest prevailing winds are some of the most extensive and spectacular on the globe. They are especially well developed in the eastern Great Australian Bight and the South Australian Sea as well as along the coast of Western Australia. The aeolianites are forming today and have done so throughout the Pleistocene. Sediment in these complexes is almost wholly derived from the adjacent shallow offshore sediment factory (Wilson 1991) and is locally mixed with variable amounts of siliciclastic particles. The extent of such aeolianites is likely due to the combined effects of the prolific shallow-water seagrass factories and the high-energy marine environment.

Fig. 12.7 A sketch of the continental shelf depicting the concept of a shaved shelf. The seafloor lies within the zone of wave abrasion with Cenozoic bedrock acting as a substrate for epibenthic growth. Carbonate produced by this biota is swept away to either lodge in local depressions or more typically to be moved landward or seaward and form significant accumulations in marginal marine and slope environments respectively. (Modified from James et al. 1994)

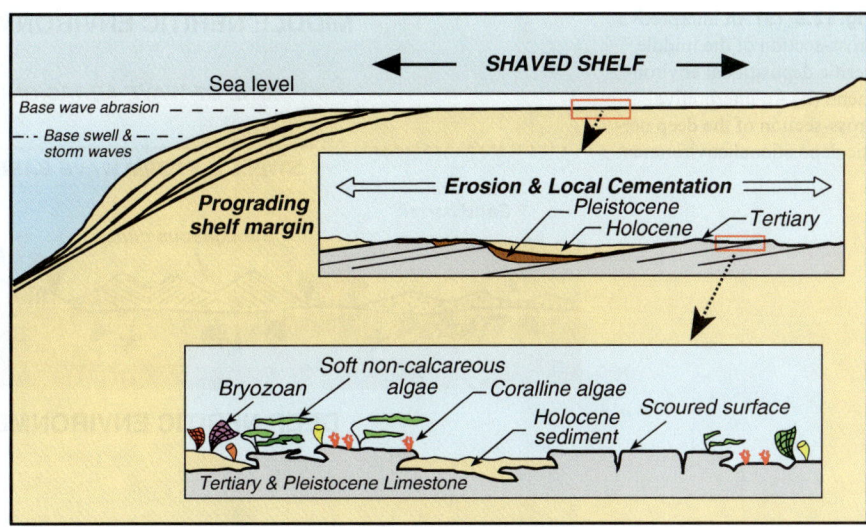

When viewed at the large scale, disregarding the carbonate components, this whole region in terms of peritidal deposition strongly resembles the Bahamas; i.e. aeolianites and muddy tidal flats. This similarity is doubtless in large part due to the virtual absence of terrigenous clastic material in the system wherein the adjacent marine carbonate factory controls all marginal marine deposition. It is also a function of the mid-latitude location of each system, between ~25° and 40° latitude, where climate is dominated by high-pressure systems and attendant onshore winds.

12.4.4.3 Middle Neritic Facies

The middle neritic shelf (60–140 mwd) is a region of rocky outcrops, shifting sand and subaqueous dunes (Fig. 12.8a). Macroalgae and seagrasses are not prolific but corallines can be locally numerous. The most striking aspect of this environment is the contrast between rocky reefs that are profusely decorated with a wide variety of sessile invertebrates, especially bryozoans and sponges, (the epibenthic coppice environment) and bare rippled sands with few obvious animals or plants other than opportunistic species such as rooted or unattached bryozoans, and uncemented brachiopods. This is in part due to the high-energy environment wherein the sediments are constantly or semi-constantly moving such that only specific animals can successfully inhabit the mobile seafloor. As a result, the rocky substrate is the only serious carbonate sediment factory. The seafloor can be viewed as rocky islands of sediment producers separated by areas of carbonate sand derived from such factories. The inevitable consequence of this situation seems to be that as the sediment is produced the islands get buried in their own carbonate debris and the factory shuts down. This does not, however, seem to be the case everywhere because storm waves and swells periodically impact the seafloor and transport much of this sediment away, towards the land or towards the shelf edge and slope (see below).

A prominent and recurring feature is the paleoshoreline formed during the LGM, usually a seacliff several tens of meters high whose base lies at ~120–130 mwd. The crest of this sharp to degraded escarpment is a seafloor of subaqueous dunes whereas the foot and seafloor extending out from it into deeper water is covered by the more muddy burrowed facies.

12.4.4.4 Deep Neritic and Upper Slope Facies

Deep neritic and shelf edge environments (140–230 mwd) usually lie below swell and storm wave base (Fig. 12.8b) but the bathymetry is complicated by relatively recent Pleistocene history. The seafloor is an irregular carpet of diminutive sponges and bryozoans or burrowed muddy sediment. The shelf edge is not generally a sharp declivity but is instead a gradual increase in depth that leads to the upper slope. It is usually a region of rock outcrops along southern Australia, locally in the form of seaward-dipping limestone beds, populated by a community of calcareous

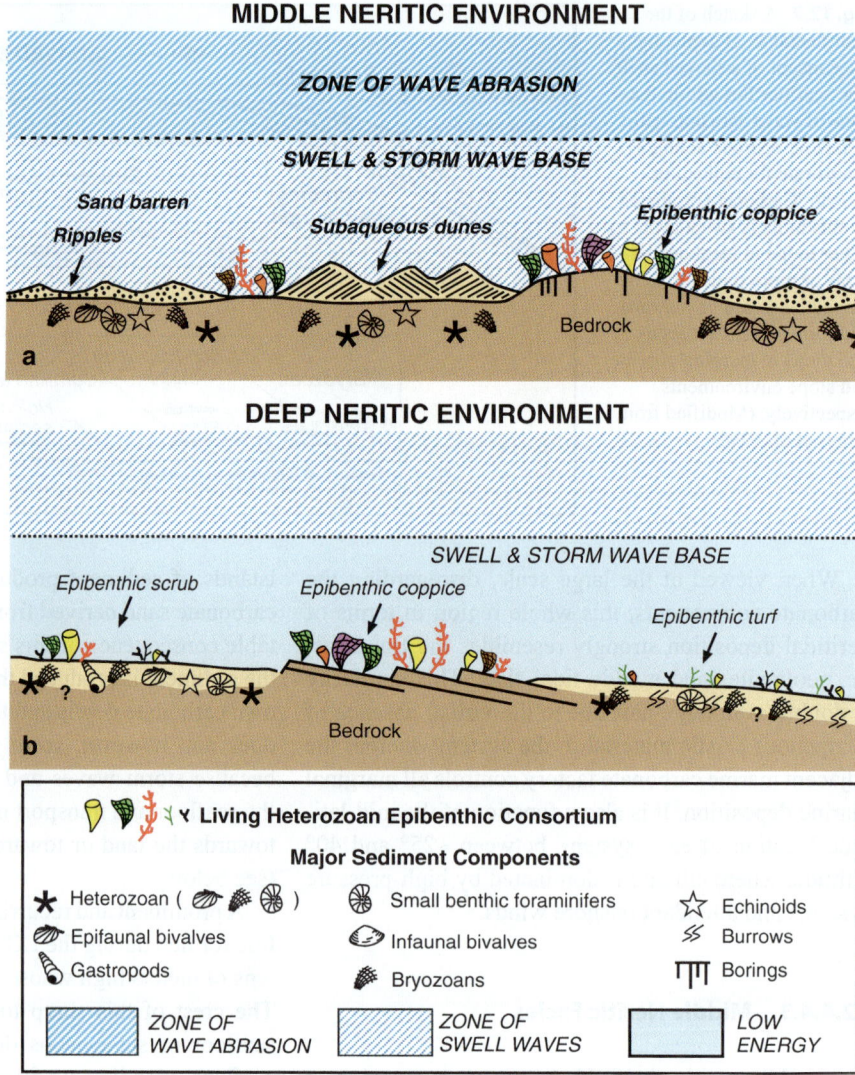

Fig. 12.8 (a) An interpretive cross-section of the middle neritic depositional environment. (b) An interpretive cross-section of the deep neritic depositional environment

invertebrates, particularly bryozoans, sponges, and molluscs (Fig. 12.8b). In regions of seasonal upwelling, these communities are prolific and extend into shallower water well onto the outer shelf (Fig. 12.9). By contrast, in areas of downwelling, muddy slope facies dominated by sparse delicate bryozoans can extend up onto the shelf edge. This relationship illustrates the importance of even slight nutrient enhancement on the nature of the carbonate-producing biota.

The upper slope is either a steep incline, a region of mass wasting, or a prograding wedge of sediment, locally incised by submarine canyons. Prograding margins, as exemplified by those documented in the central Great Australian Bight (Feary and James 1998) attest to the massive transport of sediment off the shelf

in this high-energy regimen. The prograding wedge of almost wholly carbonate sediment in the Eyre Sector is >600 m in thickness and all Quaternary in age (Feary et al. 2000a). This situation is the opposite of that inshore wherein sediment is transported landward. The outboard sponge- and bryozoan-dominated biota becomes less numerous with depth and below ~400 mwd the seafloor is burrowed muds and fine sands. The thickness and extent of this sediment wedge indicates that a similar oceanography-dominated sedimentary system has been operating here throughout much of the Pleistocene and probably longer.

It has long been known that off-shelf sediment transport is an important process along the southern Australian margin, with neritic sediment discovered on

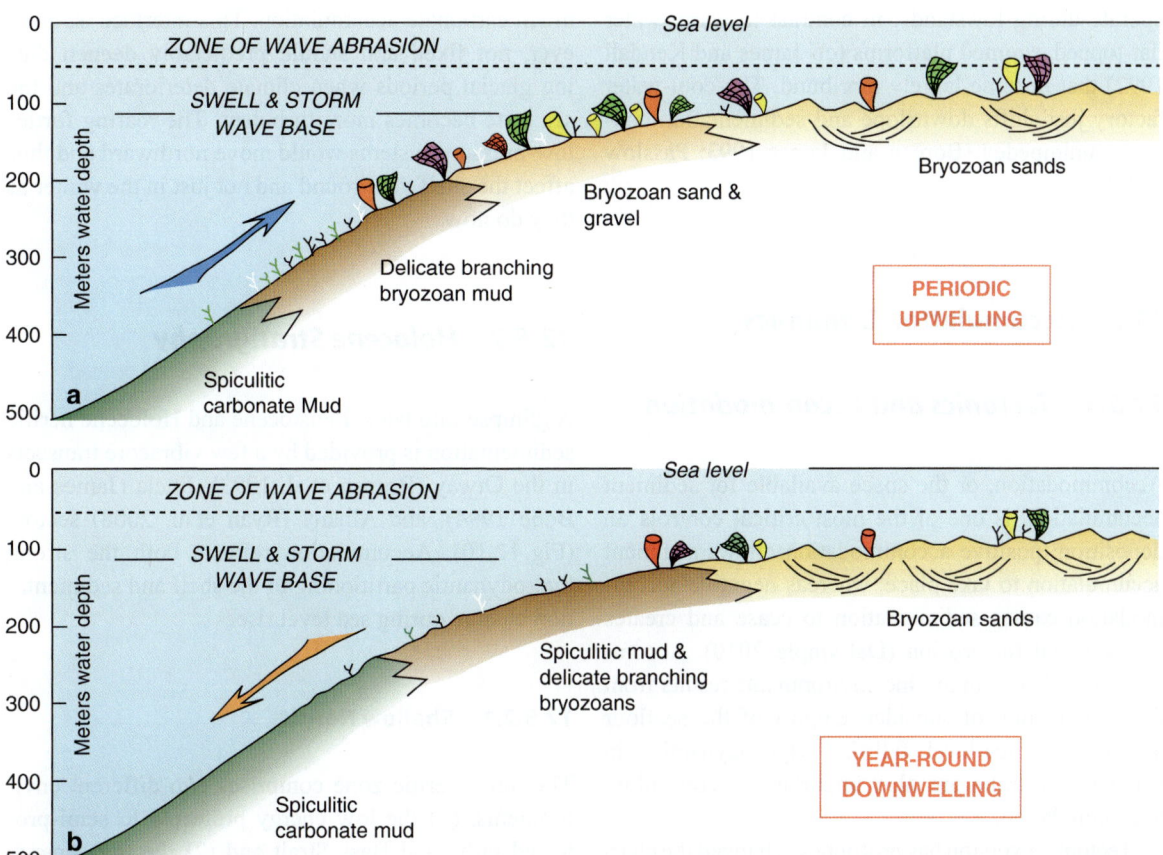

Fig. 12.9 A sketch illustrating the response of the seafloor epibenthos to upwelling and downwelling. (**a**) Seasonal upwelled waters deliver somewhat increased nutrients to the shelf leading to active growth and carbonate sediment production. (**b**) Down-welling waters with relatively low nutrients result in reduced growth of the carbonate sediment producing biota and a shift of slope or deep neritic facies into shallower waters. (Modified from James et al. 2001)

the abyssal plain (Conolly and von der Borch 1967). Recent studies of slope and submarine canyon processes have illuminated this depositional realm (Passlow 1997; Exon et al. 2005; Hill et al. 2005). Some of these canyons are the largest on earth. It appears that most canyons are ancient, having been initially cut in the late Cretaceous and episodically re-eroded during periods of eustatic sea level lowstand (especially the Oligocene). Their location is the topic of some controversy, whereas they seem to be most numerous off bedrock promontories (von der Borch 1968) and adjacent steep slopes, they are also conspicuous offshore from major river systems (Hill et al. 2005). Alternatively they may be related to erosion along zones of weakness such as Mesozoic basin-forming faults (Hill et al. 2009).

Consensus seems to be that erosion coincides with sea level lowstand and is mostly in the form of head-

ward erosion promulgated by active off-shelf sediment movement. Shelf-to -basin transport appears to be most active during sea level lowstand and early transgression, with highstand coinciding with periods of finer grained deposition. This relationship is also true for slope deposition in general wherein the Eyre sector wedge, for example is characterized by grainy sediment (packstone) during lowstands and muddy sediment (wackestone) during highstands (Feary et al. 2004). Thus, today LGM grainy sediments are covered by fine-grained sediment implying little downslope sediment transport (Feary et al. 1992; Passlow 1997). Much of this sediment, however, appears to be down slope mudflows, not pelagic fallout. Alternatively, mass wasting and downslope movement seems to be locally ongoing (Hill et al. 2005).

An important conclusion from the foregoing is that the cool-water carbonate factory continues to operate during lowstands, in contrast to warm-water flat-topped, rimmed platforms (cf. James and Kendall 1992) that become largely moribund. The cool-water factory just shifts downslope and sedimentation continues unimpeded (Boreen and James 1993; Passlow 1997).

much carbonate accumulation. This interface is, however, not fixed and should predictably deepen during glacial periods when climate deteriorates and the sea state becomes more turbulent. The roaring forties low-pressure systems would move northward and thus affect the shelf year-round and not just in the winter as they do now.

12.5   Accumulation Dynamics

12.5.1   *Tectonics and Accommodation*

Accommodation, or the space available for sediment accumulation, is one of the most critical controls on deposition: positive accommodation allows sediment accumulation to take place, whereas negative accommodation causes sedimentation to cease and creates the potential for erosion (Dalrymple 2010). Accommodation change in marine environments results from the combination of subsidence/uplift of the seafloor and eustatic sea-level fall/rise. Hydrodynamics in southern Australia are also important in controlling accommodation.

Tectonic inversion has profoundly changed the character of the continental margin over the last 30 my. The combination of a high-energy hydrodynamic system wherein normal storm wave base is 60 mwd and there is little subsidence means that most of the shelf is one of low or negative accommodation. Only the outer part is a region of positive accommodation. Sediment is not actively accumulating over large parts of the shelf; it is an area of sediment production but non-accumulation. Onshore transport is demonstrated by the extensive aeolianite complexes composed of marine sediment, at >200 m. Profound offshore transport is highlighted by the extensive upper slope progradational wedges with accumulation rates (up to 300 m my^{-1}) that are higher than those of most tropical photozoan systems.

Accommodation in carbonate depositional systems, especially those in temperate latitudes, is also determined by hydrodynamics and this is especially true for the high-energy shelf of southern Australia. The critical hydrodynamic interface is storm wave base, above which all loose sediment is moved almost continuously because of the combined action of swell and storm waves. Thus, in southern Australia, the base of wave abrasion is also practically the upper limit for

12.5.2   *Holocene Stratigraphy*

A glimpse into latest Pleistocene and Holocene neritic sedimentation is provided by a few vibracore transects in the Otway (Boreen et al. 1993), Eucla (James and Bone 1994), and Albany (Ryan et al. 2008) sectors (Fig. 12.10). Accumulation reflects both the strong hydrodynamic partitioning of the shelf and sedimentation change during sea level rise.

12.5.2.1   Shallow Neritic

The inner neritic zone comprises two different environments, (1) the low energy protected to semi-protected gulfs and Bass Strait and (2) the high-energy open shelf in the zone of wave abrasion. The gulfs have a wide spectrum of facies as outlined in Chap. 7 with basin floors covered by a 0.5–2.0 m thick, muddy sediment (facies C10) over bedrock or calcrete-covered Pleistocene MIS 5 deposits. Basin margins <10 mwd are sites of extensive, 4–8 m thick, seagrass banks. The stratigraphic succession under the banks ranges from 7.5–2.0 ka.

The best record of shallow open ocean but somewhat protected neritic sedimentation is in Esperance Bay in the lee of the Recherche Archipelago on the eastern Albany Shelf (Ryan et al. 2008). A basal lag of lithoclast—mollusc (especially *Pecten*) gravels (facies M3) in the innermost parts of Esperance Bay are overlain by quartzose sand with seagrass fibers in the upper parts (facies C9). The same basal lag in the outer bay (<50 mwd) is covered by very poorly sorted coarse biofragmental sand containing numerous rhodolith horizons (facies C3). The basal lag records sea level rise and ranges from 7–12.5 ka in the open bay but from 5.6–8.7 ka, inboard reflecting gradual sea level rise. The seagrass beds can be up to 7 m thick whereas open bay deposits are generally <3 m thick.

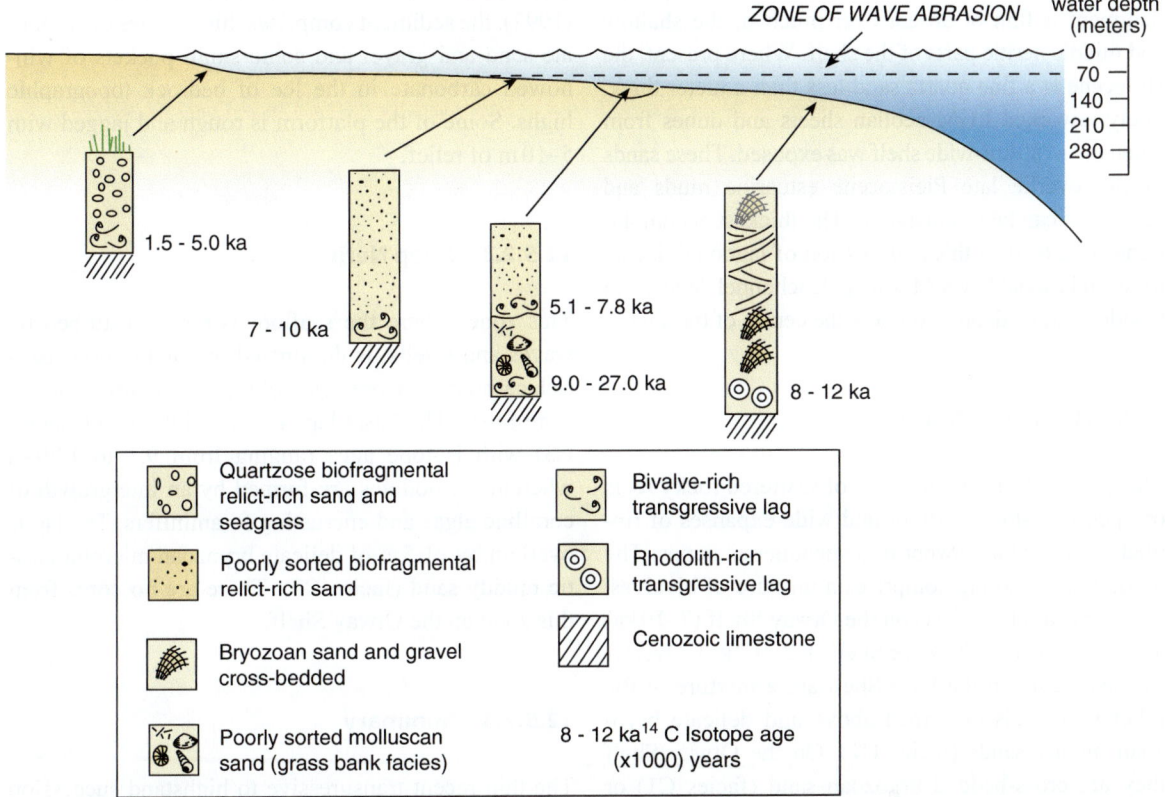

Fig. 12.10 A sketch summarizing the thin Holocene sediment stratigraphy on the southern continental shelf, mainly from the western sector and the Otway Shelf, synthesized from Boreen and James (1993); James et al. (1994); and Ryan et al (2008). The outer shelf has a basal rhodolith and epibenthic lag (8–12 ka) overlain by a succession of bryozoan-rich sediments that are locally cross-bedded. The middle-shelf has the same lag but is succeeded upward by more poorly sorted relict-rich sands.

The inner shelf is complex and can be a mixture of quartzose sediment and carbonate seagrass facies. The basal lag becomes progressively younger landward, i.e. it is a transgressive succession. The stacked succession in the depression from the Warrnambool Transect (Otway Shelf) comprises a basal seagrass facies overlain by an open shelf facies; they are separated by a hiatus and molluscan lag that is interpreted to represent arrested sedimentation in the zone of wave abrasion

There is a general theme of a shelf veneered by a thin ragged blanket of sediment over mostly Cenozoic limestone on the open, unprotected shelf within the zone of wave abrasion (>60 mwd) on both the Otway Shelf and the Great Australian Bight (Eyre Sector). The irregular limestone surface with large shallow depressions and small pockets filled by Holocene sediment alternates with large areas of rocky seafloor that are swept clean or populated by an active macrophyte and filter feeding invertebrate biota.

Sediment pockets are generally a meter or less thick and comprise a mollusc-rich lag overlying a bored and encrusted surface of Tertiary or locally Pleistocene limestone. *Glycymeris* and *Katelysia* together with

many other worn shells including surf clams (*Donax*) dominate bivalve gravels. Overlying sediments on the Eyre Shelf are relict-rich quartzose skeletal sand wherein about 1/3 of the sediment is bivalve fragments (facies M2). The numerous recent articulated corallines, benthic foraminifers, and molluscs together with scattered *Marginopora* attest to a seagrass environment (facies C9). Ages at the base of the cores are <3 ka. Similar sediments overlying the basal lag (1.5–5 ka) in shallow depressions on the Otway Shelf are medium to coarse, quartzose, mollusc, and bryozoan sands with abundant relict particles (facies M2, M3, and M4).

These findings have been confirmed by seismic, coring, and bottom sampling of the Lacepede Shelf

(Hill et al. 2009). The Pleistocene succession there lies on top of a relatively flat karst unconformity with up to 10 m of relief eroded into Miocene neritic carbonates. Sediment is thin or absent over much of the shallow and middle neritic parts of the shelf. Where present, the Holocene is a fine quartz sand less than a meter thick, likely reworked LGM aeolian sheets and dunes from when the ~180 km-wide shelf was exposed. These sands locally overlie late Pleistocene estuarine muds and sands that can be mollusc-rich. The thickest accumulations are a 6–10 m thick succession of lagoonal, lacustrine, and fluvial River Murray paleochannel deposits in a wide shallow depression near the centre of the shelf.

12.5.2.2   Middle Neritic

This part of the seafloor is one of scattered rocky reefs or open limestone seafloor and wide expanses of rippled sands that are swept into subaqueous dunes. The ubiquitous basal lag comprises a mixture of bivalves, molluscs, and lithoclasts on the Otway Shelf (7–10 ka) and bryozoans on the Eyre Shelf (8.2–9.2 ka). Overlying sediments on the Eyre Shelf are a mixture of the relict rich sands described above and delicate bryozoan muddy sands (facies C7). On the Otway Shelf they are cross-bedded bryozoan sand (facies C1) or fine bioclastic sand (facies C6) up to 1 m thick.

A particularly important succession comprising two stacked 30–50 cm thick units was cored between 60 and 80 mwd in the Warrnambool sector of the Otway Shelf (Boreen et al. 1993; Boreen and James 1993). The lower unit is dark-grey carbonate sand with a poorly sorted mix of bivalve fragments, shell fragments, minor quartz sand and mud, and a distinctive seagrass assemblage (facies C9). The basal bivalve lag has ages of 27, 11, and 9 ka. The upper unit is a red-brown fine-grained bioclastic sand with an open shelf biota (facies M1). The basal bivalve lag has an isotope age of 7.8–5.1 ka. Basal contacts of both units are abrupt, burrowed, and marked by a bivalve lag.

Similar sediment packaging is also present in cores taken from >60 mwd in western parts of the Bass Basin (Blom and Alsop 1988). A lower embayment facies up to 30 cm thick, rests upon earlier lacustrine sediments, and contains an assemblage of estuarine to lacustrine molluscs with ^{14}C ages ranging from 10.3 to 11.7 ka. A single sample of the overlying modern Bass Strait facies has an isotope age of 8.7 ka.

The middle neritic environment on the Lacepede Shelf is a Miocene hardground largely devoid of surface sediment cover. As documented by James et al. (1992), the sediment comprises thin patches of carbonate sand and gravel and some small pockets of winnowed carbonate in the lee of bedrock topographic highs. Some of the platform is rough and jagged with 5–10 m of relief.

12.5.2.3   Deep Neritic

This zone, where the seafloor is rarely disturbed by waves and swells, is dominated by muddy bryozoan-rich carbonates. Cores did not intersect bedrock on the Eyre Shelf. The basal lag is a rhodolith gravel (facies C3) with isotope ages ranging from 9.1 to 12.9 ka wherein the nodules are formed by an intergrowth of coralline algae and encrusting foraminifers. The lag is overlain by <1.5 m of delicate bryozoan microbioclastic muddy sand (facies C7). There are no cores from this zone on the Otway Shelf.

12.5.2.4   Summary

The thin recent transgressive to highstand succession everywhere sampled so far comprises a basal lag that becomes younger inboard. The bored and encrusted rock substrate on the middle and shallow neritic zone suggests ravinement and arrested sedimentation accompanied and immediately followed marine inundation. Although data is limited the lag appears to be rhodolith gravels outboard and quartzose, relict-rich mollusc sand inboard. The stacked succession on the Warrnambool sector records seagrass deposition in an embayment environment not unlike the gulfs today followed by a hiatus and then middle neritic facies. This hiatus is interpreted to be the record of the passage of the zone of wave abrasion across this area.

12.6   Forward Modelling

The main controls on neritic, cool-water carbonate sedimentation, as outlined in Chap. 4, are water temperature, hydrodynamic energy, seawater salinity, siliciclastic sediment input from land, and trophic resource levels. The southern Australian neritic shelf environment illustrates the sedimentological response to most of these controls except variations in hydrodynamic energy trophic resources and changing accommodation. The system present is one of high hydrodynamic energy, low mesotropic to marginal oligotrophic conditions, and minor accommodation. Using the processes of sediment formation determined from this environment, it is useful to vary the parameters and then predict what the differences in sedimentation would be.

12.6.1 Hydrodynamic Energy

Decreasing the hydrodynamic energy (Fig. 12.11) across the shelf, but maintaining the same water depths, should result in:

1. Migration of upper slope and outer neritic facies into shallower mid-neritic and shallow neritic environments.
2. An overall muddier facies spectrum.
3. Decreased off-shelf sediment transport, enhancing retention of muds in the neritic environment and reduction in the size of the seaward prograding slope wedge.

4. Co-occurrence of phototrophic taxa with invertebrate taxa that occur in deeper water settings on the present shelf in overall shallower environments.
5. Reduction in the number of carbonate producing taxa capable of living in mobile substrates and overall increase in benthic organism diversity in the mid-shelf region.

12.6.2 Trophic Resources

Increasing trophic resource (Fig. 12.12) leading to overall high mesotrophic conditions should have several consequences:

1. Higher planktic biomass, that would in turn result in:

   (a) Increased food sources for filter-feeding benthic invertebrates and thus increased overall biomass, carbonate production, and sedimentation rate.
   (b) Decreased water clarity resulting in the shallowing of phototrophs such as seagrasses, macroalgae, and calcareous algae.

2. Increased numbers of those organisms that require higher trophic resources such as barnacles, oysters, and bryozoans.

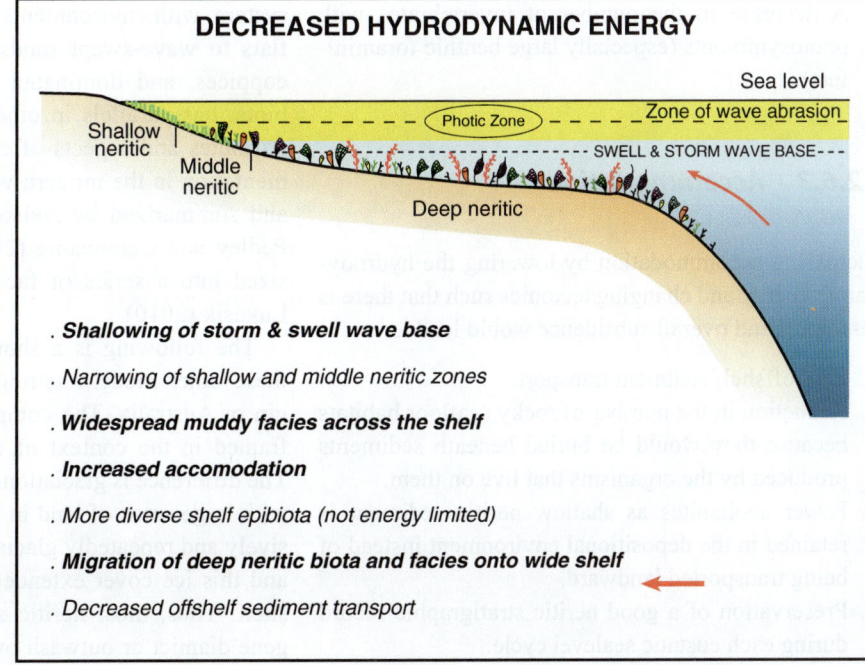

Fig. 12.11 An interpretive sketch of the shelf profile illustrating the effects of decreased hydrodynamic energy on sedimentation across the southern Australian continental margin

DECREASED HYDRODYNAMIC ENERGY

Sea level
Zone of wave abrasion
Photic Zone
Shallow neritic
Middle neritic
SWELL & STORM WAVE BASE
Deep neritic

. *Shallowing of storm & swell wave base*
. *Narrowing of shallow and middle neritic zones*
. **Widespread muddy facies across the shelf**
. **Increased accomodation**
. *More diverse shelf epibiota (not energy limited)*
. **Migration of deep neritic biota and facies onto wide shelf**
. *Decreased offshelf sediment transport*

Fig. 12.12 An interpretive sketch of the shelf profile postulating the effects of increased trophic resources on sedimentation across the southern Australian continental margin

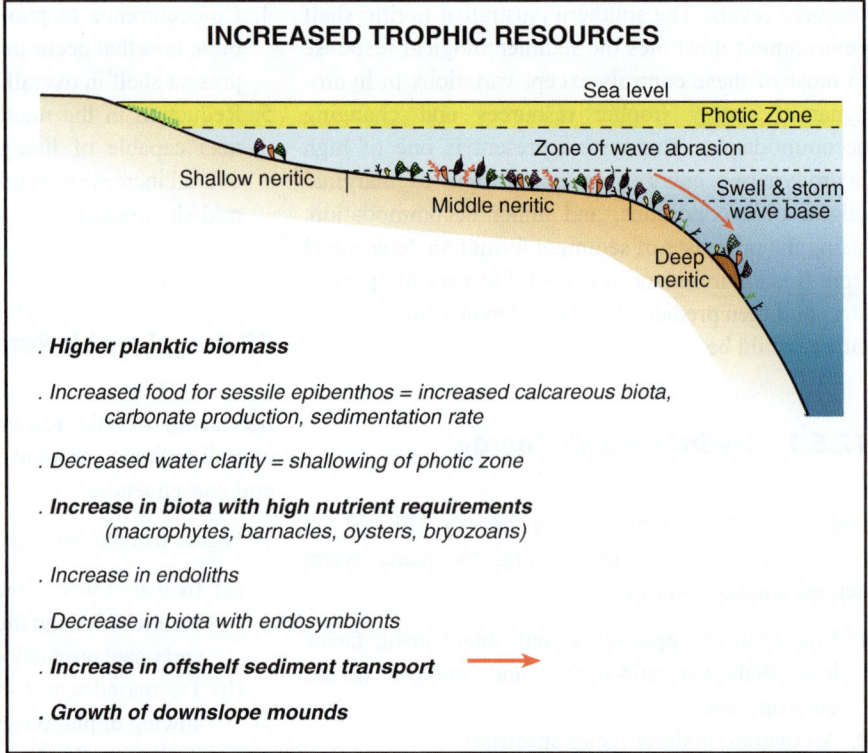

3. An increase in the number of endolithic organisms, resulting in higher rates of bioerosion and increased carbonate sand and silt production.
4. A decrease in the number of invertebrates with photosymbionts (especially large benthic foraminifers).

12.6.3 Accommodation

Increasing accommodation by lowering the hydrodynamic energy and changing tectonics such that there is less uplift and overall subsidence would lead to:

1. Less offshelf sediment transport.
2. Reduction in the number of rocky seafloor habitats because they would be buried beneath sediments produced by the organisms that live on them.
3. Fewer aeolianites as shallow neritic sediment is retained in the depositional environment instead of being transported landward.
4. Preservation of a good neritic stratigraphic record during each eustatic sealevel cycle.

12.7 Global Comparison

This high energy, low nutrient, cool-water carbonate system, with environments ranging from muddy tidal flats to wave-swept sands to deep water bryozoan coppices, and dominated by a heterozoan benthic biota, has parallels in other parts of the globe. The attributes and aspects of cool-water carbonate sedimentation in the modern world have been organized and summarized by Nelson (1988a), James (1997), Pedley and Carannante (2006) and recently synthesized into a series of facies models by James and Lukasik (2010).

The following is a short comparison of some of these other modern settings with the southern margin of Australia. The comparison is most practically framed in the context of the different hemispheres. The difference is glaciation. The northern hemisphere, with large areas of land in polar latitudes, was extensively and repeatedly glaciated during the Pleistocene and this ice cover extended out onto the continental shelf. Thus, most neritic sediment comprises glacigene diamict or outwash overlain by a thin veneer of

Holocene carbonate. The southern hemisphere, with comparatively little land near the pole, except for Antarctica itself, has continents whose shelves were exposed during the Pleistocene but not necessarily affected by glaciation.

12.7.1 Southern Hemisphere

12.7.1.1 Australia

The continental shelf extends northward and equatorward along the eastern and western margins of Australia and in both instances it is mainly a carbonate depositional province. A southward-flowing, relatively warm, low-nutrient current in both cases (Leeuwin Current in the west and East Australian Current in the east) dominates large-scale water movement (Fig. 2.7). The shelves extend into a tropical, mainly reef-dominated environment north of ~25° S latitude (Fig. 1.2).

The eastern Australian shelf is relatively narrow compared to the southern margin and south of ~25° S is composed of a shore-attached siliciclastic sediment wedge (Boyd et al. 2004) that passes outboard into cool-water carbonates. The shelf is blanketed with siliciclastic sediment inboard because of higher rainfall than in the south and the proximity of the Great Dividing Range, a source of such sediment. The region has only been documented in a reconnaissance fashion (Marshall and Davies 1978). The carbonates are similar in aspect to those on the Otway Shelf. An extensive region of coralline algal deposits marks the northward transition into truly tropical Great Barrier Reef carbonates (Marshall and Davies 1978; Marshall et al. 1998).

The shelf off Western Australia south of the Abrolhos Reef Complex at 30° S (called the Rottnest Shelf) is likewise comparatively narrow. The sediments are all oligotrophic, warm temperate (Collins et al. 1993; James et al. 1999) with conspicuous large benthic foraminifers and small coral accumulations as far south as Perth. The whole shelf has an elevated rim of unknown composition that is covered by encrusting corallines and rhodoliths along its entire length. Otherwise the deposits are most similar to those in the very western Great Australian Bight and Albany Shelf.

12.7.1.2 New Zealand

Neritic carbonates around New Zealand are particularly well studied (Nelson et al. 1988b). The deposits are cold temperate and high-energy throughout and most resemble those around western and southern Tasmania. Siliciclastic sediments predominate on the inner shelf next to the mainland, largely because of the wet climate. The outer shelf and the vast Three Kings platform to the north (34° S) and the Snares Platform to the south (47–49° S) are, however, sites of extensive carbonate deposition. Barnacles, macroalgae, echinoids, coralline algae characterize intertidal to very shallow, rocky neritic environments (<10 mwd) in the north with kelp forests deeper. Otherwise the composition and distribution of neritic carbonate sediments are similar to those described herein from southern Australia. The major differences are that there are no large benthic foraminifers or grass banks off New Zealand whereas barnacles comprise a significant proportion of all sediments and glauconite is common in intraskeletal chambers, neither of which are important in southern Australia.

12.7.1.3 South America

The meridonal continental shelf off Brazil extends from the tropical north to the temperate south. Outside the Amazon and San Francisco rivers there is little terrigenous sediment input and so the shelf is blanketed with carbonate sediment (Carannante et al. 1988). Corals are rare in the tropical environments north of 15° S. The transition zone between 15° and 25° S has all of the attributes of the warm-temperate system in southern Australia. South of 25° S, however, the environment is cold-temperate with bryozoans becoming significantly reduced as the seawater temperature decreases. The warm-temperate facies are similar to those in southern Australia but the apparent cold-temperate setting without bryozoans has no counterpart.

12.7.2 Northern Hemisphere

12.7.2.1 North Atlantic & North Pacific

These facies along the NW coast of Europe, both coasts of Canada, the South China Sea and Japan are cold-temperate. Of particular importance on these shelves is the presence of a hard substrate formed by boulder pavements resulting from winnowing of mud and sand from diamict during the Holocene transgression. Such a relationship is lacking in southern Australia although the low accommodation and extensive exposed bedrock in the form of rocky reefs creates a similar factory. Shallow neritic facies are kelp forests and coralline algal banks. Kelp forests with innumerable barnacles, common epilithic bivalves and echinoderms, generate copious amounts of barnacle-rich sediment. This material covers the adjacent seafloor in the form of rippled sands that are generally barren but in less energetic settings are burrowed and populated by numerous infaunal and epifaunal bivalves. The maerl facies occupies somewhat protected environments behind islands. The rigid branching corallines grow on hard surfaces or form banks (reefs) of interlocking branches and algal crusts. These particles are swept into banks and shoals of cross-bedded sand and gravel. Sediments in middle and deep neritic environments are likewise generated on hard substrates, either bedrock or winnowed glacigene boulder lags. Bivalves, bryozoans, corals, and serpulids, as well as echinoids, and benthic foraminifers densely colonize such substrates.

12.7.2.2   Mediterranean

The Mediterranean Sea is a large enclosed basin with limited water exchange with the Atlantic Ocean at its western end. It is a wave-dominated, almost tideless, comparatively low energy depositional system wherein the main sediment producers are red algae, bryozoans, molluscs, benthic foraminifers, and barnacles (Carannante et al. 1995; Betzler et al. 1997). Fairweather wave base is 5–15 mwd whereas storm wave base is 30– 40 mwd. The waters are near-oligotrophic and there is a west to east, cool to warm seawater temperature gradient. As pointed out by Betzler et al. (1997), the system, except for the size difference, is most similar to the large gulfs in southern Australia described herein. The shallow depositional environments are all warm-temperate with extensive seagrass banks and associated mollusc-benthic foraminifer facies (Betzler et al. 1997; Fornos and Ahr 1997). By contrast, there are much more extensive coralline algal facies both in the form of broken branching corallines (maerl) in 30–65 mwd and rhodoliths (pralines) at 55–75 mwd.

The Adriatic (McKinney 2007) has an east-to-west trophic resources gradient with relatively warm oligotrophic–mesotrophic waters adjacent to the Croatian coast and eutrophic waters near the Italian coast due to high nutrients and counterclockwise circulation. The seafloor under oligotrophic conditions is dominated by an upright filter-feeding calcareous biota whereas the more terrigenous clastic muddy seafloor immersed in eutrophic waters is dominated by an infaunal biota. Bryozoans are particularly evident in oligotrophic waters where local meadows of upright articulated bryozoans (*Cellaria* sp.) form extensive muddy floatstones. This setting strongly emphasizes the importance of terrestrial sediment and nutrient input, a feature lacking in southern Australia.

12.7.2.3   Gulf of California

The Gulf of California (Halfar et al. 2006a, b) is an elongate meridonal gulf wherein there is year-round upwelling with extensive tidal mixing at the northern end. A strong nutrient gradient is present from oligotrophic at the gulf entrance in the south to eutrophic in the north. Molluscs are ubiquitous and occur in all facies but zooxanthellate corals are only present in oligotrophic settings in the south; coralline algae dominate in the mesotrophic central part; bryozoans are most abundant in the more eutrophic head of the gulf. It would seem that only the northern part has parallels in southern Australia.

12.8   Conclusion

The continental margin of southern Australia has long been recognized as one of the great carbonate depositional systems in the modern world and now with this book it has been appropriately documented. The shelf and adjacent environments stand with the Great Barrier Reef, the Persian Gulf, and Florida-the Bahamas as fundamental sites of carbonate sediment formation and deposition. The sweep of the environments in southern Australia is astonishing; saline lakes, isolated lagoons, aeolianite ranges and cliffs, perpetually dark estuaries, formidable rocky coasts, turbulent surf beaches, quiet muddy tidal flats, luxuriant mangrove woodlands, illuminated seagrass banks and kelp forests, moving tidal sand shoals, rhodolith pavements, luxuriant bryozoan-sponge coppices, continuously shifting subaqueous dune fields, submerged fossil beaches, moribund biogenic mounds, and deep burrowed muddy slopes. Impressive biogenic and abiotic processes have produced an extensive suite of deposits that range from evaporites to muddy and grainy carbonates. The sedimentary processes and products here are part of the global temperate water depositional realm and as outlined above, contain elements of most similar systems worldwide. We have herein deliberately synthesized attributes over a vast area in an attempt to distill the essence of the system. In some cases the theme is emphasized at the expense of the detail but this is deliberate in order to highlight the pervasive themes of sedimentation.

This marine realm lies in an icehouse world of dramatic sea level, seawater temperature, and climate change. The shelf surface over the last 50 ky has at various times been one of large lagoons (some of which were evaporitic), high-energy beaches that swept back and forth across the shelf surface, abrading, sorting, and altering the sediment grains, a terrestrial landscape of locally vegetated dunes that were destroyed by the rising sea, and finally a deeply submerged shelf with a seafloor ranging from sponge-bryozoan meadows to shifting subaqueous dune fields to wave swept bedrock. All of this history has been deciphered from a ragged blanket of sediments that now cover the shelf. The sediment once generated is not, however, all retained and it now appears that much of it disappears via physical and chemical processes. Active hydrodynamics sweep much of the shelf sediment both inboard and outboard. Most aragonite skeletal components appear to vanish via dissolution soon after deposition. The fortuitous west-to-east temperature gradient led to the opportunity to contrast the warm-temperate seagrass environment in the west with the cool-temperate macrophyte regimen in the east. The warm-temperate seagrasses are the most vigorous carbonate factory in the system with production and accumulation rates that rival those of the tropics.

When we began to study this area we assumed that deposition would follow the rules and take place in a region where an anticlockwise southern hemisphere oceanic gyre dominated circulation and thus controlled carbonate production and accumulation. Nothing could be further from the truth! The region operates under conditions that are at odds with predicted global

mid-latitude oceanography. The sea state and hydrodynamic background are similar or perhaps more energetic than elsewhere. By contrast, the unforeseen near-oligotrophic nutrient regimen, locally harsh semi-arid climate, zonal orientation of the shelf, inverse tectonic regimen, and backwards clockwise oceanic

Fig. 12.13 Ancient cool-water limestones. (**a**) Bryozoan-bivalve grainstone-rudstone, Oligocene, western Yorke Peninsula, South Australia. (**b**) Bryozoan-brachiopod rudstone, Permian, Maria Island, Tasmania. (**c**) Core of Mississippian bryozoan rudstone-floatstone, Pekisko Formation, Alberta, Canada

circulation patterns impart an unexpected series of controls on sedimentation. These unique attributes have not hampered carbonate deposition but serve to highlight the fact that not all such sedimentation is controlled by predictable oceanographic attributes.

One of the current challenges of carbonate sedimentology is how to apply fundamental concepts of modern cool-water carbonate deposition deduced from regions such as this to the ancient rock record. Seafloor deposits provide a window into the ancient world but their direct application comes with many caveats, the major ones being biotic evolution and changing oceanography. There are now numerous accounts of interpreted cool-water limestones especially in the volumes edited by Nelson (1988a, b), James and Clarke (1997), and Pedley and Carannante (2006). An additional source of references outside these compilations can be found in James and Lukasik (2010).

This vast area contains many depositional environments from which to pluck information about sediment-organism and sediment-oceanographic relationships that can be used when interpreting the origin of carbonate rocks throughout geologic history. Our findings are most useful when deciphering Cenozoic rocks (Fig. 12.13), when the cool-water marine biota was largely similar to that in the modern ocean. Our discoveries are also applicable, with evolutionary caveats, to the Mesozoic but with trepidation to the Paleozoic.

The epicontinental seas of southern Australia are natural laboratories wherein we can clearly see the processes of sedimentation, even though the products are largely the result of time-limited interactions. We are, today, on the threshold of a new era in sedimentology where ancient carbonates can be interpreted in a much more sophisticated fashion than ever before. To make this achievable, more studies like this one are needed so that we can use as much information from modern neritic seas as possible to wring out all the information encrypted in ancient carbonates (Fig. 12.13) and thereby interpret them in an ever more intelligent fashion.

There is a wealth of information about the evolution of our globe entombed in carbonate sedimentary rocks. Although it seems that we have, by documenting some modern systems, acquired all the information that we need to decipher the rock record, this is an illusion. There is more information in these rocks than we have even begun to extract. To unlock this information we must return to the ocean and document or re-investigate modern carbonate depositional systems. New tools and innovative concepts will refine and perhaps even overturn previous interpretations of modern depositional systems. This book is but one example of such a study.

Appendices

Appendix A   Seafloor Sample Sites–Albany Sector

Sample #	Depth (m)	Latitude	Longitude	Facies
GAB 001	42	34°02.18′ S	114°57.14′ E	C3 Coralline gravel
GAB 086	90	34°26.19′	123°48.54′	C5 Articulated coralline and intraclast sand
GAB 087	90	34°32.54′	123°40.17′	C5 Articulated coralline and intraclast sand
GAB 088	98	34°34.81′	123°37.74′	C5 Articulated coralline and intraclast sand
GAB 089	210	34°37.14′	123°32.20′	C5 Articulated coralline and intraclast sand
GAB 090	403	34°38.45′	123°31.67′	C11 Spiculitic skeletal sandy mud
GAB 091	293	34°37.87′	123°27.62′	C11 Spiculitic skeletal sandy mud
GAB 092	154	34°37.24′	123°25.07′	C1 Bryozoan sand & gravel
GAB 093	95	34°31.59′	122°58.42′	M1 Relict-rich skeletal sand & gravel
GAB 095	430	34°40.75′	122°26.53′	C12 Coral, arborescent bryozoan gravel & mud
GAB 096	292	34°39.36′	122°26.03′	C12 Coral, arborescent bryozoan gravel & mud
GAB 097	190	34°38.99′	122°25.90′	C5 Articulated coralline and intraclast sand
GAB 098	156	34°38.71′	122°25.83′	C5 Articulated coralline and intraclast sand
GAB 099	105	34°33.72′	122°24.98′	C6 Fine skeletal sand
GAB 100	335	33°57.18′	122°01.87′	C12 Coral, arborescent bryozoan gravel & mud
GAB 101	236	33°57.22′	122°01.86′	C1 Bryozoan sand & gravel
GAB 102	35.5	33°57.18′	122°01.88′	C2 Skeletal sand & gravel
GAB 103	55	33°58.27′	121°55.17′	M2 Relict-rich, quartzose skeletal sand & gravel
GAB 104	67	34°00.06′	121°51.20′	M2 Relict-rich, quartzose skeletal sand & gravel
GAB 105	78	34°04.13′	121°44.01′	R1 Relict sand
GAB 106	87	34°12.00′	121°31.87′	R1 Relict sand
GAB 107	90	34°19.99′	121°31.80′	R1 Relict sand
GAB 108	101	34°28.66′	121°32.16′	M1 Relict-rich skeletal sand & gravel
GAB 109	123	34°32.36′	121°31.92′	C1 Bryozoan sand & gravel
GAB 110	154	34°32.72′	121°31.66′	C1 Bryozoan sand & gravel
GAB 111	95	34°29.50′	120°38.86′	C5 Articulated coralline and intraclast sand
GAB 112	65	34°20.16′	119°55.11′	M1 Relict-rich skeletal sand & gravel
GAB 113	106	34°36.03′	119°55.17′	C1 Bryozoan sand & gravel
GAB 114	190	34°36.74′	119°55.13′	C1 Bryozoan sand & gravel
GAB 115	350	34°37.37′	119°54.85′	C11 Spiculitic skeletal sandy mud
GAB 116	66	34°37.01′	119°21.77′	M1 Relict-rich skeletal sand & gravel
GAB 117	65	34°35.12′	119°00.04′	Q1 Calcareous quartz sand
GAB 118	87	34°59.40′	119°00.01′	C1 Bryozoan sand & gravel
GAB 119	149	35°00.20′	118°59.84′	C1 Bryozoan sand & gravel

N. P. James, Y. Bone, *Neritic Carbonate Sediments in a Temperate Realm*,
DOI 10.1007/978-90-481-9289-2, © Springer Science+Business Media B.V. 2011

Sample #	Depth (m)	Latitude	Longitude	Facies
GAB 120	325	35°00.52′	119°00.72′	C1 Bryozoan sand & gravel
GAB 121	82	35°17.80′	118°19.98′	M1 Relict-rich skeletal sand & gravel
GAB 122	109.5	35°24.80′	117°58.01′	M4 Relict-rich bryozoan sand
GAB 123	147	35°27.31′	117°48.15′	C1 Bryozoan sand & gravel
GAB 124	303	35°27.97′	117°45.78′	C1 Bryozoan sand & gravel
GAB 125	230	35°26.41′	117°40.06′	C1 Bryozoan sand & gravel
GAB 126	86.5	35°21.12′	117°26.61′	M4 Relict-rich bryozoan sand
GAB 127	86	35°12.92′	117°05.08′	M2 Relict-rich, quartzose skeletal sand & gravel
GAB 128	59	35°07.27′	116°52.44′	Q1 Calcareous quartz sand
GAB 129	70	35°06.97′	116°20.39′	Q1 Calcareous quartz sand
GAB 130	100	35°07.41′	115°58.56′	C5 Articulated coralline and intraclast sand
GAB 131	160	35°07.09′	115°51.07′	C1 Bryozoan sand & gravel
GAB 133	416–476	34°46.97′	114°45.66′	C6 Fine skeletal sand

Appendix B   Seafloor Sample Sites–
Great Australian Bight Sector

Sample #	Depth (m)	Latitude	Longitude	Facies
GAB 002	49.5	32°4.12′	127°30.79′	R3 Bryozoan-rich relict sand
GAB 003	30	32°19.10′	127°30.00′	C2 Skeletal sand & gravel
GAB 004	36	32°14.94′	128°00.26′	R1 Relict sand
GAB 005	40	32°06.16′	128°29.09′	R1 Relict sand
GAB 006	40	32°0.40′	128°29.12′	R1 Relict sand
GAB 007	42	31°50.49′	129°00.32′	C2 Skeletal sand & gravel
GAB 008	43	31°43.00′	129°30.59′	C2 Skeletal sand & gravel
GAB 009	47.5	31°59.94′	130°59.97′	M1 Relict-rich skeletal sand & gravel
GAB 010	54	31°59.92′	129°59.66′	M1 Relict-rich skeletal sand & gravel
GAB 011	60	32°25.01′	129°59.77′	M1 Relict-rich skeletal sand & gravel
GAB 012	70	32°42.04′	129°59.58′	C2 Skeletal sand & gravel
GAB 013	101	33°05.75′	129°59.90′	C11 Spiculitic skeletal sandy mud
GAB 014	153	33°17′	130°00′	C11 Spiculitic skeletal sandy mud
GAB 015	200	33°20′	130°00′	C11 Spiculitic skeletal sandy mud
GAB 016	308.5	33°21.31′	129°58.79′	C11 Spiculitic skeletal sandy mud
GAB 017	400.5	33°22.80′	129°59.02′	C11 Spiculitic skeletal sandy mud
GAB 018	498	33°24.47′	129°59.23′	C11 Spiculitic skeletal sandy mud
GAB 019	301	33°21.63′	129°19.19′	C11 Spiculitic skeletal sandy mud
GAB 020	156.5	33°19.98′	129°17.67′	C6 Fine skeletal sand
GAB 021	492.5	33°21.59′	128°28.29′	C11 Spiculitic skeletal sandy mud
GAB 022	347	33°19.99′	128°28.07′	C11 Spiculitic skeletal sandy mud
GAB 023	343	33°19.79′	128°27.88′	C11 Spiculitic skeletal sandy mud
GAB 024	305	33°19.31′	128°29.08′	C11 Spiculitic skeletal sandy mud
GAB 025	295	33°19.17′	128°29.13′	C11 Spiculitic skeletal sandy mud
GAB 026	309	33°19.28′	128°29.26′	C11 Spiculitic skeletal sandy mud
GAB 027	358	33°20.00′	128°29.00′	C11 Spiculitic skeletal sandy mud
GAB 028	205	33°19.79′	128°29.06′	C7 Delicate branching bryozoan muddy sand
GAB 029	148	33°16.04′	128°31.74′	C1 Bryozoan sand & gravel
GAB 030	137	33°13.08′	128°28.67′	C6 Fine skeletal sand

Sample #	Depth (m)	Latitude	Longitude	Facies
GAB 031	90.5	33°02.58'	128°28.71'	C1 Bryozoan sand & gravel
GAB 032	54.5	32°40.05'	128°28.90'	C2 Skeletal sand & gravel
GAB 033	87	32°59.65'	128°00.39'	C5 Articulated coralline and intraclast sand
GAB 034	200	33°18.10'	127°31.53'	C1 Bryozoan sand & gravel
GAB 035	153.5	33°16.40'	127°29.77'	C7 Delicate branching bryozoan muddy sand
GAB 036	100	33°10.16'	127°29.69'	C5 Articulated coralline and intraclast sand
GAB 037	76.5	32°59.03'	127°11.19'	C5 Articulated coralline and intraclast sand
GAB 038	51	32°40.48'	126°38.92'	R1 Relict sand
GAB 039	42.5	32°28.02'	126°22.79'	R1 Relict sand
GAB 040	41	32°24.00'	126°13.98'	C3 Coralline gravel
GAB 041	36	32°21.83'	125°57.55'	C3 Coralline gravel
GAB 042	234	33°25.94'	125°58.05'	C1 Bryozoan sand & gravel
GAB 043	227	33°25.77'	125°58.02'	C1 Bryozoan sand & gravel
GAB 044	163	33°25.20'	125°58.09'	C12 Coral, arborescent bryozoan gravel & mud
GAB 045	143.5	33°24.84'	125°58.04'	C1 Bryozoan sand & gravel
GAB 046	136.5	33°24.59'	125°58.00'	C1 Bryozoan sand & gravel
GAB 047	305	33°53.68'	125°21.88'	C12 Coral, arborescent bryozoan gravel & mud
GAB 048	182	33°53.57'	125°21.81'	C1 Bryozoan sand & gravel
GAB 049	156	33°52.55'	125°21.79'	C1 Bryozoan sand & gravel
GAB 050	118.5	33°52.47'	125°20.92'	C1 Bryozoan sand & gravel
GAB 052	71.5	33°36.00'	125°11.00'	R3 Bryozoan-rich relict sand
GAB 053	62	33°26.30'	125°04.98'	M1 Relict-rich skeletal sand & gravel
GAB 054	54	33°11.59'	124°55.20'	M1 Relict-rich skeletal sand & gravel
GAB 055	61.5	33°16.76'	125°18.16'	C6 Fine skeletal sand
GAB 056	72.5	33°19.94'	125°43.92'	C6 Fine skeletal sand
GAB 057	112	33°21.37'	125°59.25'	C1 Bryozoan sand & gravel
GAB 058	65	33°08.22'	125°57.73'	C1 Bryozoan sand & gravel
GAB 059	56	32°50.08'	125°58.31'	R1 Relict sand
GAB 060	49.5	32°36.00'	125°58.00'	R1 Relict sand
GAB 061	46	32°42.01'	125°42.84'	C3 Coralline gravel
GAB 062	50	32°46.97'	125°25.23'	C3 Coralline gravel
GAB 063	46	32°52.11'	125°05.22'	C3 Coralline gravel
GAB 064	42	32°57.55'	124°46.61'	C3 Coralline gravel
GAB 065	42.5	33°03.50'	124°22.92'	M1 Relict-rich skeletal sand & gravel
GAB 066	46	33°12.92'	124°22.94'	C2 Skeletal sand and gravel
GAB 067	50	33°22.00'	124°22.95'	M1 Relict-rich skeletal sand & gravel
GAB 068	59	33°37.90'	124°22.99'	R1 Relict sand
GAB 069	64	33°43.01'	124°23.03'	R2 Mollusc-rich relict sand
GAB 070	73.5	33°54.87'	124°23.07'	M1 Relict-rich skeletal sand & gravel
GAB 071	79	34°07.12'	124°22.91'	R1 Relict sand
GAB 072	87.5	34°13.79'	124°23.15'	C1 Bryozoan sand & gravel
GAB 074	117–125	34°14.66'	124°23.88'	C6 Fine skeletal sand
GAB 075	140–180	34°15.07'	124°24.79'	C1 Bryozoan sand & gravel
GAB 076	173	34°15.25'	124°24.55'	C1 Bryozoan sand & gravel
GAB 077	311	34°16.67'	124°22.63'	C11 Spiculitic skeletal sandy mud
GAB 078	380	34°16.98'	124°23.04'	C11 Spiculitic skeletal sandy mud
GAB 081	403	34°22.15'	124°11.16'	C11 Spiculitic skeletal sandy mud
GAB 083	180	34°20.92'	124°08.49'	R4 Limestone gravel
GAB 084	96	34°20.47'	124°08.04'	C1 Bryozoan sand & gravel
GAB 085	80	34°16.38'	124°00.12'	M1 Relict-rich skeletal sand & gravel
GAB 086	90	34°26.19'	123°48.54'	C5 Articulated coralline and intraclast sand
GAB 087	90	34°32.54'	123°40.17'	C5 Articulated coralline and intraclast sand

Sample #	Depth (m)	Latitude	Longitude	Facies
GAB 088	98	34°34.81'	123°37.74'	C5 Articulated coralline and intraclast sand
GAB 089	210	34°37.14'	123°32.20'	C5 Articulated coralline and intraclast sand
GAB 090	403	34°38.45'	123°31.67'	C11 Spiculitic skeletal sandy mud
GAB 091	293	34°37.87'	123°27.62'	C11 Spiculitic skeletal sandy mud
GAB 092	154	34°37.24'	123°25.07'	C1 Bryozoan sand & gravel
ACM 037	380	35°22.75'	134°42.02'	C12 Coral, arborescent bryozoan gravel & mud
ACM 038	200	35°21.68'	134°42.11'	C7 Delicate branching bryozoan muddy sand
ACM 039	124	35°09.11'	134°46.26'	C5 Articulated coralline and intraclast sand
ACM 041	83	34°40.00'	134°54.08'	R3 Bryozoan-rich relict sand
ACM 041A	75	34°12.13'	134°52.22'	R2 Mollusc-rich relict sand
ACM 042	55–45	33°47.09'	134°50.15'	R2 Mollusc-rich relict sand
ACM 043	48	33°40.01'	134°49.97'	R2 Mollusc-rich relict sand
ACM 044	500–480	35°00.86'	133°54.21'	C11 Spiculitic skeletal sandy mud
ACM 045	310–340	35°00.57'	133°55.46'	C7 Delicate branching bryozoan muddy sand
ACM 046	200–180	34°58.99'	133°54.96'	C1 Bryozoan sand & gravel
ACM 047	121	34°52.09'	133°55.54'	C1 Bryozoan sand & gravel
ACM 048	96	34°26.03'	134°00.16'	C5 Articulated coralline and intraclast sand
ACM 049	84	34°05.07'	134°05.02'	R2 Mollusc-rich relict sand
ACM 050	71.5	33°38.17'	134°10.49'	R2 Mollusc-rich relict sand
ACM 051	71.5	33°19.98'	134°15.04'	C6 Fine skeletal sand
ACM 052	65	33°12.88'	134°16.46'	M2 Relict-rich quartzose skeletal sand & gravel
ACM 053	500	34°27.34'	132°48.54'	C12 Coral, arborescent bryozoan gravel & mud
ACM 054	254–214	34°24.99'	132°51.25'	C7 Delicate branching bryozoan muddy sand
ACM 055	169–182	34°23.82'	132°51.31'	C1 Bryozoan sand & gravel
ACM 056	121	34°13.55'	132°59.01'	C5 Articulated coralline and intraclast sand
ACM 057	94	33°44.09'	133°15.97'	C1 Bryozoan sand & gravel
ACM 058	72.5	33°18.02'	133°32.05'	R1 Relict sand
ACM 059	62	32°53.39'	133°48.25'	R1 Relict sand
ACM 060	31	32°40.09'	133°59.21'	M2 Relict-rich quartzose skeletal sand & gravel
ACM 061	31	32°37.00'	133°58.55'	M2 Relict-rich quartzose skeletal sand & gravel
ACM 062	26	32°34.58'	133°57.08'	C4 Encrusted rocky substrate
ACM 063	515	34°02.61'	132°23.21'	C8 Scaphopod, pteropod sand & mud
ACM 064	261–315	34°00.05'	132°25.32'	C1 Bryozoan sand & gravel
ACM 065	174	33°59.94'	132°26.78'	C1 Bryozoan sand & gravel
ACM 066	123	33°47.07'	132°35.12'	C1 Bryozoan sand & gravel
ACM 067	94	33°24.09'	132°51.11'	R1 Relict sand
ACM 068	82	33°06.98'	133°05.82'	R1 Relict sand
ACM 069	57	32°41.02'	133°24.93'	R2 Mollusc-rich relict sand
ACM 070	28.5	32°21.82'	133°39.31'	M2 Relict-rich quartzose skeletal sand & gravel
ACM 071	18.5	32°15.27'	133°29.73'	R1 Relict sand
ACM 072	50.5	32°18.03'	133°06.99'	R1 Relict sand/R2 Mollusc-rich relict sand
ACM 073	360	33°35.81'	131°40.92'	C7 Delicate branching bryozoan muddy sand
ACM 074	195	33°30.76'	131°43.33'	C1 Bryozoan sand & gravel
ACM 075	120	33°21.34'	131°48.92'	R3 Bryozoan-rich relict sand
ACM 076	91	32°55.31'	132°02.71'	R1 Relict sand
ACM 077	74.5	32°36.63'	132°13.81'	R1 Relict sand
ACM 078	63	32°12.24'	132°27.70'	M1 Relict-rich skeletal sand & gravel
ACM 079	45	32°00.11'	132°34.90'	C2 Skeletal sand and gravel
ACM 080	60	32°03.91'	131°43.07'	M1 Relict-rich skeletal sand & gravel
ACM 081	70	32°22.05'	131°31.00'	R1 Relict sand
ACM 082	89	32°52.09'	131°15.68'	R1 Relict sand
ACM 083	550–590	33°29.97'	130°51.94'	C8 Scaphopod, pteropod sand & mud

Sample #	Depth (m)	Latitude	Longitude	Facies
ACM 084	351	33°25.22'	130°51.97'	C8 Scaphopod, pteropod sand & mud
ACM 085	196	33°22.45'	130°51.80'	C7 Delicate branching bryozoan muddy sand
ACM 086	125	33°15.13'	130°52.10'	C5 Articulated coralline and intraclast sand
ACM 087	106	33°08.04'	130°52.18'	C5 Articulated coralline and intraclast sand
ACM 088	87	32°49.02'	130°52.00'	R2 Mollusc-rich relict sand
ACM 089	68	32°16.01'	130°51.92'	C2 Skeletal sand and gravel
ACM 090	50	31°38.52'	130°52.14'	M1 Relict-rich skeletal sand & gravel
ACM 091	38	31°30.76'	131°06.33'	M1 Relict-rich skeletal sand & gravel
ACM 092	50	31°38.94'	130°27.98'	M1 Relict-rich skeletal sand & gravel
ACM 093	56	31°57.89'	130°27.98'	M1 Relict-rich skeletal sand & gravel
ACM 094	66.5	32°30.03'	130°28.24'	C5 Articulated coralline and intraclast sand
ACM 095	94	32°56.14'	130°28.23'	C5 Articulated coralline and intraclast sand
ACM 096	150	33°15.98'	130°28.22'	C7 Delicate branching bryozoan muddy sand
ACM 097	260	33°22.35'	130°28.06'	C7 Delicate branching bryozoan muddy sand
ACM 098	504	33°26.18'	130°28.12'	C8 Scaphopod, pteropod sand & mud
ACM 100	495	33°29.88'	139°30.08'	C8 Scaphopod, pteropod sand & mud
ACM 101	342.5	33°25.03'	129°30.26'	C8 Scaphopod, pteropod sand & mud
ACM 102	200.5	33°21.47'	129°30.33'	C1 Bryozoan sand & gravel
ACM 103	127	33°11.48'	129°29.95'	C5 Articulated coralline and intraclast sand
ACM 104	80	32°55.98'	129°29.96'	C5 Articulated coralline and intraclast sand
ACM 105	65	32°38.06'	129°29.99'	C5 Articulated coralline and intraclast sand
ACM 106	51	32°10.05'	129°30.05'	C2 Skeletal sand & gravel
ACM 107	41	31°39.44'	129°30.02'	C2 Skeletal sand & gravel
PL 94-01	51	33°43.25'	134°54.98'	R2 Mollusc-rich relict sand
PL 94-02	72	33°56.19'	134°41.15'	R2 Mollusc-rich relict sand
PL 94-03	80	34°07.93'	134°30.08'	R2 Mollusc-rich relict sand
PL 94-04	90	34°20.83'	134°17.32'	R2 Mollusc-rich relict sand
PL 94-05	90	37°27.48'	134°10.14'	C5 Articulated coralline and intraclast sand
PL 94-06	110	34°40.34'	133°56.53'	C1 Bryozoan sand & gravel
PL 94-07	120	364°46.96'	133°49.60'	C5 Articulated coralline and intraclast sand
PL 94-08	160–220	34°.5'	133°42.7'	C1 Bryozoan sand & gravel
PL 94-8/9	250–160	34°55.10'	133°43.05'	C12 Coral, arborescent bryozoan gravel & mud
PL 94-10	450–500	34°55.63'	133°40.72'	C11 Spiculitic skeletal sandy mud
PL 94-12	45	34°44.92'	135°19.42'	R2 Mollusc-rich relict sand
PL 94-13	50	34°48.32'	135°16.32'	M1 Relict-rich skeletal sand & gravel
PL 94-14	95	34°56.93'	135°10.73'	C5 Articulated coralline and intraclast sand
PL 94-15	114	35°07.07'	135°03.15'	C5 Articulated coralline/C6 Fine skeletal sand
PL 94-16	125	35°14.24'	134°57.85'	C6 Fine skeletal sand
PL 94-17	129	35°16.54'	134°54.80'	C5 Articulated coralline and intraclast sand
PL 94-18	118	35°21.86'	134°51.26'	C6 Fine skeletal sand
PL 94-19	148	35°23.92'	134°49.94'	C6 Fine skeletal sand
PL 94-20	160–180	35°25.44'	134°49.12'	C7 Delicate branching bryozoan muddy sand
PL 94-21	220–250	35°26.10'	134°48.85'	C12 Coral, arborescent bryozoan gravel & mud
PL 94-22	320–350	35°26.84'	134°48.73'	C12 Coral, arborescent bryozoan gravel & mud

Appendix C   Seafloor Sediment Sample Sites–South Australian Sea Sector

Sample #	Depth (m)	Latitude	Longitude	Facies
ACM 009	30	38°07.93′S	140°45.74′E	C4 Encrusted rocky substrate
ACM 010	191	38°13.13′	140°24.13′	C7 Delicate branching bryozoan muddy sand
ACM 011	35	37°46.47′	140°12.92′	M1 Relict-rich skeletal sand & gravel
ACM 012	130	37°51.25′	139°59.80′	C6 Fine skeletal sand
ACM 013	179	37°52.49′	139°56.06′	C6 Fine skeletal sand
ACM 014	278	37°53.18′	139°54.06′	C7 Delicate branching bryozoan muddy sand
ACM 015	38	37°26.81′	139°49.18′	C1 Bryozoan sand & gravel
ACM 016	280–200	37°30.60′	139°24.94′	C12 Coral, arborescent bryozoan gravel & mud
ACM 018	130	37°27.51′	139°27.03′	M4 Relict-rich bryozoan sand
ACM 019	35	37°05.06′	139°37.25′	C1 Bryozoan sand & gravel
ACM 020	185–200	37°12.06′	139°00.19′	M4 Relict-rich bryozoan
ACM 021	255	37°11.25′	138°58.51′	C1 Bryozoan sand & gravel
ACM 023	50	36°50.23′	139°04.62′	M4 Relict-rich bryozoan sand
ACM 024	44	36°05.95′	139°18.05′	Q1 Calcareous quartz sand
ACM 025	42	35°46.98′	138°58.00′	Q1 Calcareous quartz sand
ACM 026	48	35°55.06′	137°54.07′	C1 Bryozoan sand & gravel
ACM 027	137	36°49.24′	137°26.82′	C1 Bryozoan sand & gravel
ACM 028	267	36°50.45′	137°27.17′	C1 Bryozoan sand & gravel
ACM 029	57	36°04.00′	137°03.14′	C1 Bryozoan sand & gravel
ACM 030	126	36°33.75′	136°51.48′	C1 Bryozoan sand & gravel
ACM 031	200	36°36.50′	136°50.58′	C1 Bryozoan sand & gravel
ACM 032	275	36°37.27′	136°51.43′	C1 Bryozoan sand & gravel
ACM 034	177	36°07.52′	135°53.81′	C1 Bryozoan sand & gravel
ACM 035	285	35°59.93′	135°42.42′	C7 Delicate branching bryozoan muddy sand
ACM 036	182	35°42.38′	135°27.12′	C7 Delicate branching bryozoan muddy sand
ACM 039	124	35°09.11′	134°46.26′	C5 Articulated coralline and intraclast sand
ACM 041	83	34°40.00′	134°54.08′	R3 Bryozoan-rich relict sand
ACM 041A	75	34°12.13′	134°52.22′	R2 Mollusc-rich relict sand
ACM 042	55–45	33°47.09′	134°50.15′	R2 Mollusc-rich relict sand
ACM 043	48	33°40.01′	134°49.97′	R2 Mollusc-rich relict sand
ACM 111	46	136°54.73′	34°56.69′	R3 Bryozoan-rich relict sand
PL 94-01	51	33°43.25′	134°54.98′	R2 Mollusc-rich relict sand
PL 94-02	72	33°56.19′	134°41.15′	R2 Mollusc-rich relict sand
PL 94-03	80	34°07.93′	134°30.08′	R2 Mollusc-rich relict sand
PL 94-04	90	34°20.83′	134°17.32′	R2 Mollusc-rich relict sand
PL 94-05	90	37°27.48′	134°10.14′	C5 Articulated coralline and intraclast sand
PL 94-06	110	34°40.34′	133°56.53′	C1 Bryozoan sand & gravel
PL 94-12	45	34°44.92′	135°19.42′	R2 Mollusc-rich relict sand
PL 94-13	50	34°48.32′	135°16.32′	M1 Relict-rich skeletal sand & gravel
PL 94-14	95	34°56.93′	135°10.73′	C5 Articulated coralline and intraclast sand
PL 94-15	114	35°07.07′	135°03.15′	C5 Articulated coralline/C6 Fine skeletal sand
PL 94-16	125	35°14.24′	134°57.85′	C6 Fine skeletal sand
PL 94-17	129	35°16.54′	134°54.80′	C5 Articulated coralline and intraclast sand
PL 94-18	118	35°21.86′	134°51.26′	C6 Fine skeletal sand
PL 94-19	148	35°23.92′	134°49.94′	C6 Fine skeletal sand
PL 94-20	160–180	35°25.44′	134°49.12′	C7 Delicate branching bryozoan muddy sand
PL 94-21	220–250	35°26.10′	134°48.85′	C12 Coral, arborescent bryozoan gravel & mud

Sample #	Depth (m)	Latitude	Longitude	Facies
PL 94-22	320–350	35°26.84'	134°48.73'	C12 Coral, arborescent bryozoan gravel & mud
PL 94-23	57	34°54.87'	135°45.94'	M1 Relict-rich skeletal sand & gravel
PL 94-24	81	35°05.91'	135°42.15'	M1 Relict-rich skeletal sand & gravel/C5 articulated coralline
PL 94-25	96	35°14.73'	135°39.67'	C5 Articulated coralline and intraclast sand
PL 94-26	112	35°24.91'	135°33.52'	M4 Relict-rich bryozoan sand
PL 94-27	126	35°31.50'	135°29.10'	C6 Fine skeletal sand
PL 94-28	165	35°38.56'	135°24.10'	C6 Fine skeletal sand
PL 94-29	166	35°39.44'	135°23.55'	C7 Delicate branching bryozoan muddy sand
PL 94-30	220	35°41.43'	135°22.72'	C7 Delicate branching bryozoan muddy sand
PL 94-31	300–243	35°41.52'	135°22.33'	C11 Spiculitic skeletal sandy mud
PL 94-32	380–320	35°42.41'	135°22.34'	C11 Spiculitic skeletal sandy mud
PL 94-35	460	36°04.32'	135°44.30'	C11 Spiculitic skeletal sandy mud
PL 94-36	300	36°03.10'	135°45.69'	C1 Bryozoan sand & gravel
PL 94-37	200	36°01.58'	135°45.82'	C6 Fine skeletal sand
PL 94-38	145	35°54.08'	135°49.83'	C6 Fine skeletal sand
PL 94-39	130	35°45.62'	135°54.17'	C6 Fine skeletal sand
PL 94-40	126	35°39.99'	135°56.50'	C6 Fine skeletal sand
PL 94-41	110	35°35.43'	135°58.78'	M4 Relict-rich bryozoan sand
PL 94-42	104	35°28.67'	136°01.81	R3 Bryozoan-rich relict sand
PL 94-43	90	35°20.52'	136°18.97'	C5 Articulated coralline and intraclast sand
PL 94-44	47	35°08.98'	136°41.98'	R2 Mollusc-rich relict sand
PL 94-45	78.5	35°23.94'	136°34.21'	C6 Fine skeletal sand
PL 94-46	95	35°34.42'	136°27.17'	C6 Fine skeletal sand
PL 94-47	115	35°45.64'	136°22.02'	C6 Fine skeletal sand
PL 94-48	120	35°55.28'	136°18.05'	R3 Bryozoan-rich relict sand
PL 94-49	115	36°05.72'	136°18.01'	M4 Relict-rich bryozoan sand
PL 94-50	114	36°15.88'	136°17.97'	R3 Bryozoan-rich relict sand
PL 94-51	115	36°19.35'	136°18.26'	R3 Bryozoan-rich relict sand
PL 94-52	140	36°28.02'	136°18.23'	C6 Fine skeletal sand
PL 94-53	205	36°30.64'	136°18.12'	C12 Coral, arborescent bryozoan gravel & mud
PL 94-54	357–204	36°31.37'	136°18.31'	C12 Coral, arborescent bryozoan gravel & mud
PL 94-56	86	35°36.46'	136°42.59'	C6 Fine skeletal sand
PL 94-57	64	35°29.02'	136°49.75'	C5 Articulated coralline and intraclast sand
PL 94-58	38	35°35.48'	137°09.10'	R2 Mollusc-rich relict sand
PL 94-59	40	35°29.31'	137°06.30'	R2 Mollusc-rich relict sand
PL 94-60	43	35°23.51'	137°05.03'	R2 Mollusc-rich relict sand
PL 94-61	29	35°17.11'	137°02.89'	C2 Skeletal sand & gravel
PL 94-62	26	35°17.56'	137°20.25'	M3 Relict-rich molluscan sand
PL 94-63	35	35°25.94'	137°21.87'	R2 Mollusc-rich relict sand
PL 94-64	27	35°32.47'	137°19.23'	M3 Relict-rich molluscan sand
PL 94-65	30	35°31.48'	137°41.11'	C2 Skeletal sand and gravel
PL 94-66	36	35°22.03'	137°39.52'	C2 Skeletal sand and gravel
PL 94-67	26	35°10.70'	137°30.80'	C2 Skeletal sand and gravel
PL 94-68	32	35°13.45'	137°46.41'	M3 Relict-rich molluscan sand
PL 94-69	38	35°21.42'	137°52.38'	M3 Relict-rich molluscan sand
PL 94-70	36	35°28.95'	137°59.00'	C10 Bivalve mud
PL 94-71	15	35°42.60'	137°49.64'	C9 Mollusc, coralline, benthic foraminifer gravel, sand & mud
VH89-01	171	36°56.57'	137°39.18'	C1 Bryozoan sand & gravel
VH89-02	127	36°55.59'	137°39.47'	C1 Bryozoan sand & gravel
VH89-03	123	36°55.46'	137°38.84'	C1 Bryozoan sand & gravel

Sample #	Depth (m)	Latitude	Longitude	Facies
VH89-04	67	36°45.75′	137°47.55′	M4 Relict-rich bryozoan sand
VH89-05	68	36°37.28′	137°57.5′	Q1 Calcareous quartz sand
VH-89-06	65	36°28.25′	138°07.20′	R4 Limestone gravel (paleostrandline)
VH89-07	62	36°19.50′	138°17.5′	Q1 Calcareous quartz sand
VH89-08	57	36°10.58′	138°27.18′	M2 Relict-rich, quartzose skeletal sand & gravel
VH89-09	52	36°01.88′	138°37.13′	R3 Bryozoan-rich relict sand
VH89-10	47	35°53′	138°47′	Q1 Calcareous quartz sand
VH89-11	42	35°44′	138°57.3′	Q1 Calcareous quartz sand
VH89-12	36	35°39′	138°51.2′	R4 Limestone gravel (paleostrandline)
VH89-13	42–45	35°42.5′	138°47.3′	R2 Mollusc-rich relict sand
VH89-14	43	35°43′	138°40.2′	R2 Mollusc-rich relict sand
VH89-15	37	36°02.31′	138°22.62′	C3 Coralline gravel
VH-89-16	38	36°02.44′	138°20.29′	C3 Coralline gravel
VH89-17	38	35°54.58′	138°16.67′	M4 Relict-rich bryozoan sand
VH89-18	24	35°55.36′	138°14.85′	C4 Encrusted rocky substrate
VH89-19	38	35°54.45′	138°11.92′	M2 Relict-rich, quartzose skeletal sand & gravel
VH89-20	42	35°54.77′	138°11.66′	M2 Relict-rich, quartzose skeletal sand & gravel
VH89-21	61	36°10.03′	138°16.44′	M2 Relict-rich, quartzose skeletal sand & gravel
VH89-22	45	36°09.48′	138°10.95′	C3 Coralline gravel
VH89-23	58	36°07.61′	138°00.21′	M4 Relict-rich bryozoan sand
VH89-24	65	36°14.75′	137°53.47′	R1 Relict sand
VH89-25	74	36°22.47′	137°46.03′	R2 Mollusc-rich relict sand
VH-89-26	81	36°30.43′	137°34.62′	R2 Mollusc-rich relict sand
VH89-27	90	36°38.15′	137°25.62′	M4 Relict-rich bryozoan sand
VH89-28	63	36°41.46′	137°52.72′	R4 Limestone gravel (paleostrandline)
VH89-29	144	37°02.60′	137°58.13′	C6 Fine skeletal sand
VH89-30	108	37°02.14′	138°01.47′	C1 Bryozoan sand & gravel
VH-89-31	85	37°00.51′	138°02.82′	C1 Bryozoan sand & gravel
VH89-32	73	36°54.40′	138°07.85′	M2 Relict-rich, quartzose skeletal sand & gravel
VH89-33	62	36°50.30′	138°12.73′	M2 Relict-rich, quartzose skeletal sand & gravel
VH-89-34	61	36°41.74′	138°22.45′	R4 Limestone gravel (paleostrandline)
VH89-35	62	36°32.75′	138°32.23′	Q1 Calcareous quartz sand
VH89-36	59	36°23.96′	138°42.21′	Q1 Calcareous quartz sand
VH89-37	53	36°15.26′	138°51.54′	Q1 Calcareous quartz sand
VH89-38	48	36°06.05′	137°01.46′	Q1 Calcareous quartz sand
VH89-39	46	35°57.97′	139°10.54′	Q1 Calcareous quartz sand
VH89-40	41	36°09.94′	139°25.86′	M2 Relict-rich, quartzose skeletal sand
VH89-41	50	36°18.59′	139°17.66′	Q1 Calcareous quartz sand
VH89-42	53	36°27.32′	139°07.47′	Q1 Calcareous quartz sand
VH89-43	55	36°35.91′	138°56.68′	Q1 Calcareous quartz sand
VH89-44	52	36°46.32′	138°46.57′	M2 Relict-rich, quartzose skeletal sand
VH89-45	57	36°50.19′	138°42.57′	M2 Relict-rich, quartzose skeletal sand
VH-89-46B	67	36°54.50′	138°36.94′	M2 Relict-rich, quartzose skeletal sand
VH89-47	77	36°58.46′	138°32.58′	M1 Relict-rich skeletal sand & gravel
VH89-48	98	37°02.79′	138°27.46′	C1 Bryozoan sand & gravel
VH89-49	164	37°08.56′	138°52.07′	C6 Fine skeletal sand
VH89-50	91	37°04.12′	138°56.31′	M1 Relict-rich skeletal sand & gravel
VH89-51	76	37°00.17′	139°01.13′	M2 Relict-rich, quartzose skeletal sand & gravel
VH-89-52	50	36°42.18′	139°10.83′	M4 Relict-rich bryozoan sand
VH89-53	47	36°42.18′	139°20.67′	M2 Relict-rich, quartzose skeletal sand & gravel
VH89-54	39	36°34.35′	139°30.21′	M2 Relict-rich, quartzose skeletal sand & gravel
VH89-55	29	36°57.84′	139°31.24′	M4 Relict-rich bryozoan sand

Sample #	Depth (m)	Latitude	Longitude	Facies
VH89-56	62	37°07.92'	139°22.37'	M2 Relict-rich, quartzose skeletal sand & gravel
VH89-57	74	37°11.48'	139°17.27'	M1 Relict-rich skeletal sand & gravel
VH89-58	141	37°14.86'	139°12.69'	C6 Fine skeletal sand
VH89-LB2	28	36°45.94'	139°38.42'	M3 Relict-rich molluscan sand
VH89-59	167	37°33.10'	139°25.39'	M1 Relict-rich skeletal sand & gravel
VH89-60	82	37°23.29'	139°30.81'	C1 Bryozoan sand & gravel
VH89-61	51	37°21.88'	139°36.00'	R4 Limestone gravel (paleostrandline)
VH89-62	62	37°20.39'	139°37.42'	R4 Limestone gravel (paleostrandline)
VH89-63	40	37°17.40'	139°41.07'	M2 Relict-rich, quartzose skeletal sand
VH-89-64	220	37°31.28'	139°25.84'	M1 Relict-rich skeletal sand & gravel
VH-89-65A	186	37°17.04'	139°11.07'	C6 Fine skeletal sand
VH-89-66	180	37°09.36'	138°49.36'	C1 Bryozoan sand & gravel
VH89-67	164	37°05.20'	138°24.72'	C12 Coral, arborescent bryozoan gravel & mud
VH89-68	113	36°43.55'	137°18.29'	C1 Bryozoan sand & gravel
VH-89-69A	176	36°46.07'	137°14.24'	C6 Fine skeletal sand
VH89-70	97	36°27.28'	137°07.67'	C6 Fine skeletal sand
VH89-71	130	36°35.68'	136°57.83'	C1 Bryozoan sand & gravel
VH89-72	129	36°37.37'	136°55.75'	C1 Bryozoan sand & gravel
VH89-73	84	36°14.61'	136°55.43'	R1 Relict sand
VH-89-74A	115	36°24.68'	136°43.37'	M1 Relict-rich skeletal sand & gravel
VH89-75	160	36°32.82'	136°33.32'	C6 Fine skeletal sand
VH89-76	196	36°34.12'	136°31.50'	C6 Fine skeletal sand
VH89-77	41	36°08.35'	138°10.99'	M4 Relict-rich bryozoan sand
VH89-84	49	35°54.40'	138°44.80'	Q1 Calcareous quartz sand
VH89-85	56	36°06.8'	138°28.6'	R4 Limestone gravel (paleostrandline)
VH89-88	38	35°54.26'	138°10.52'	M2 Relict-rich, quartzose skeletal sand & gravel
VH-89-89	38	35°50.4'	138°26.9'	R4 Limestone gravel (paleostrandline)
VH91-100	31	35°42.13'	138°22.01'	M1 Relict-rich skeletal sand & gravel
VH91-102D	128	36°54.83'	137°37.27'	C1 Bryozoan sand & gravel
VH91-103D	135	36°55.70'	137°37.26'	C1 Bryozoan sand & gravel
VH91-104	190	36°56.38'	137°36.95'	C7 Delicate branching bryozoan muddy sand
VH91-107D	425	36°57.19'	137°35.63'	C8 Scaphopod, pteropod sand & mud
VH91-108D	281	36°56.35'	137°36.06'	C7 Delicate branching bryozoan muddy sand
VH91-109D	106	36°53.42'	137°39.47'	C1 Bryozoan sand & gravel/C6 Fine skeletal sand
VH91-110D	99	36°53.10'	137°39.90'	C1 Bryozoan sand & gravel
VH91-111D	105	36°53.02'	137°39.95'	C6 Fine skeletal sand
VH91-112D	97	36°53.08'	137°39.61'	C1 Bryozoan sand & gravel
VH91-113D	90	36°49.58'	137°41.73'	C1 Bryozoan sand & gravel
VH91-114D	124	37°01.38'	137°56.38'	C1 Bryozoan sand & gravel
VH91-115G	582	37°12.24'	138°34.42'	C11 Spiculitic skeletal sandy mud
VH91-116G	434	37°09.90'	138°34.43'	C11 Spiculitic skeletal sandy mud
VH91-117G	305	37°07.74'	138°34.71'	C11 Spiculitic skeletal sandy mud
VH91-118D	200	37°07.11'	138°34.74'	C1 Bryozoan sand & gravel
VH91-119D	243	37°07.18'	138°34.77'	C1 Bryozoan sand & gravel
VH91-119G	189	37°07.12'	138°34.38'	C1 Bryozoan sand & gravel
VH91-120D	173	37°05.93'	138°33.82'	C1 Bryozoan sand & gravel
VH91-121D	148	37°04.90'	138°33.33'	C1 Bryozoan sand & gravel
VH91-122D	136	37°04.15'	138°33.56'	C6 Fine skeletal sand
VH91-123	174	37°06.94'	138°35.24'	C1 Bryozoan sand & gravel/C12 Coral, bryozoan gravel & mud
VH91-124D	256	37°07.61'	138°33.48'	C1 Bryozoan sand & gravel
VH91-127D	189	37°06.23'	138°04.34'	C1 Bryozoan sand & gravel

Sample #	Depth (m)	Latitude	Longitude	Facies
VH91-128D	179	37°06.07'	138°04.81'	C1 Bryozoan sand & gravel
VH91-129D	160	37°05.41'	138°04.64'	C1 Bryozoan sand & gravel
VH91-130D	113	37°03.71'	138°05.26'	C1 Bryozoan sand & gravel
VH91-131G	1048	37°16.47'	138°11.47'	C11 Spiculitic skeletal sandy mud
VH91-132D	200	37°04.32'	138°13.82'	C1 Bryozoan sand & gravel
VH91-133D	153	37°03.90'	138°13.93'	C6 Fine skeletal sand
VH91-134D	113	37°03.04'	138°14.16'	C1 Bryozoan sand & gravel
VH91-135D	99	37°01.97'	138°14.27'	C1 Bryozoan sand & gravel
VH91-136D	77	36°53.53'	137°57.84'	C1 Bryozoan sand & gravel
VH91-137D	86	36°57.63'	137°57'	C1 Bryozoan sand & gravel
VH91-138D	97	36°59.72'	137°56.62'	C6 Fine skeletal sand
VH91-139D	141	37°02.30'	137°56.54'	C6 Fine skeletal sand
VH91-140D	170	37°03.29'	137°55.79'	C6 Fine skeletal sand
VH91-141D	335	37°04.40'	137°55.52'	C7 Delicate branching bryozoan muddy sand
VH91-142D	429	37°05.86'	137°58.80'	C11 Spiculitic skeletal sandy mud
VH91-143	95	36°59.81'	137°56.81'	C6 Fine skeletal sand
VH91-144	154	37°03.92'	138°13.86'	C6 Fine skeletal sand
VH91-146D	98	36°27.18'	136°56.59'	M4 Relict-rich bryozoan sand
VH91-147D	108	36°29.55'	136°56.10'	M4 Relict-rich bryozoan sand
VH91-148D	123	36°33.88'	136°54.38'	M4 Relict-rich bryozoan sand
VH91-149D	129	36°36.37'	136°54.08'	C1 Bryozoan sand & gravel
VH91-150D	137	36°37.61'	136°54.36'	C1 Bryozoan sand & gravel
VH91-151D	28	37°21.57'	139°48.12'	M1 Relict-rich skeletal sand & gravel
VH91-152D	62	37°25.57'	139°38.48'	R4 Limestone gravel (paleostrandline)
VH91-153	95	37°30.75'	139°36.24'	C1 Bryozoan sand & gravel
VH91-154D	151	37°33.39'	139°31.99'	M1 Relict-rich skeletal sand & gravel
VH91-155D	175	37°34.14'	139°31.20'	M2 Relict-rich, quartzose skeletal sand & gravel
VH91-156D	29	37°36.21'	140°02.61'	C4 Encrusted rocky substrate
VH91-157D	60	37°38.66'	139°58.60'	C1 Bryozoan sand & gravel
VH91-158D	83	37°43.23'	139°53.75'	M2 Relict-rich, quartzose skeletal sand & gravel
VH91-159D	220–280	37°47.13'	139°47.67'	R2 Mollusc-rich relict sand
VH91-160D	285	37°47.79'	139°46.72'	M2 Relict-rich, quartzose skeletal sand & gravel
1884	22	34°25.70'S	136°11.80'E	C9 Mollusc, coralline benthic foraminifera gravel, sand & mud
1881	29	34°32.31'S	136°26.48'E	C3 Coralline gravel
1878	40	34°40.49'S	136°41.43'E	C3 Coralline gravel
1876	44	34°45.64'S	136°51.66'E	R2 Mollusc-rich relict sand
1875	36	34°47.49'S	136°57.37'E	C1 Bryozoan sand & gravel
1874	34	34°49.22'S	137°04.23'E	M1 Relict-rich skeletal sand & gravel
1885	17	34°34.04'S	135°59.84'E	C9 Mollusc, coralline benthic foraminifera gravel, sand & mud
1887	24	34°43.33'S	136°12.48'E	C3 Coralline gravel
1890	24	34°43.38'S	136°18.14'E	C3 Coralline gravel
1896	41	34°43.61'S	136°28.56'E	C2 Skeletal sand & gravel
1899	45	34°46.18'S	136°42.19'E	R2 Mollusc-rich relict sand
1903	8	34°53.34'S	137°04.74'E	C9 Mollusc, coralline benthic foraminifera gravel, sand & mud
1908	19	34°48.80'S	136°06.00'E	C3 Coralline gravel
1864	40	34°50.14'S	136°18.79'E	C2 Skeletal sand & gravel
1867	56	34°54.77'S	136°32.54'E	R2 Mollusc-rich relict sand
1870	48	34°59.24'S	136°46.17'E	R2 Mollusc-rich relict sand
1872	18	35°02'S	136°55'E	C4 Encrusted rocky substrate

Sample #	Depth (m)	Latitude	Longitude	Facies
1911	35	35°00′S	136°12′E	C1 Bryozoan sand & gravel
1915	47	35°05′S	136°24′E	R2 Mollusc-rich relict sand
1917	45	35°07′S	136°28′E	C3 Coralline gravel
1920	48	35°08′S	136°37′E	M3 Relict-rich molluscan sand
1923	20	35°12′S	136°50′E	R2 Mollusc-rich relict sand
1926	55	35°20′S	136°45′E	R2 Mollusc-rich relict sand
1927	55	35°26′S	136°51′E	R2 Mollusc-rich relict sand
1928	35	35°20′S	136°51′E	R2 Mollusc-rich relict sand

Appendix D   Seafloor Sample Sites– Bass Strait & Tasmania

Sample #	Depth (m)	Latitude	Longitude	Facies
1900	22	37°57′	147°50′	Q1 Calcareous quartz sand
1901	48	38°07′	147°50′	C6 Fine skeletal sand
1902	52	38°17′	147°50′	C2 Skeletal sand & gravel
1903	60	28°27′	147°50′	C2 Skeletal sand & gravel
1904	64	38°38.6′	147°50′	C2 Skeletal sand & gravel
1905	72	38°47′	147°50′	M1 Relict-rich skeletal sand & gravel
1906	62	38°59′	147°45′	C1 Bryozoan sand & gravel
1907	56	39°10′	147°42′	M1 Relict-rich skeletal sand & gravel
1908	60	39°14′	147°35′	C1 Bryozoan sand & gravel
1909	58	39°04′	147°35′	M1 Relict-rich skeletal sand & gravel
1910	61	38°54′	147°35′	C1 Bryozoan sand & gravel
1911	54	38°44′	147°35.8′	C1 Bryozoan sand & gravel
1912	49	38°33′	147°36.4′	M3 Relict-rich molluscan sand
1913	45	38°28.5′	147°35′	M3 Relict-rich molluscan sand
1914	35	38°13′	147°35′	M2 Relict-rich quartzose/M3 Relict-rich mollusc sand
1915	16	38°18.6′	147°20.1′	R4 Limestone gravel (paleostrandline)
1916	32	38°28.6′	147°20.3′	M2 Relict-rich quartzose/M1 relict-rich skeletal sand
1917	45	38°39′	147°19.5′	M1 Relict-rich skeletal sand & gravel
1918	57	38°50′	147°19′	C1 Bryozoan sand & gravel
1919	57	39°04′	147°18.5′	C1 Bryozoan sand & gravel
1920	58	39°14′	147°18′	C1 Bryozoan sand & gravel
1921	53	39°20′	147°50′	M1 Relict-rich skeletal sand & gravel
1922	42	39°30′	147°50′	C1 Bryozoan sand & gravel
1923				M1 Relict-rich skeletal sand & gravel
1924	38	39°48.5′	147°46.5′	Q1 Calcareous quartz sand
1925	40	39°59.5′	147°46′	C2 Skeletal sand & gravel
1926	34	40°08.5′	147°51′	C5 Articulated coralline & intraclast sand
1927	42	40°16′	147°45.5′	C5 Articulated coralline & intraclast sand
1928	36	40°25.5′	147°50′	C2 Skeletal sand & gravel
1929	32	40°35.5′	147°50.5′	C5 Articulated coralline & intraclast sand
1930	38	40°39.5′	148°07′	R4 Limestone gravel (paleostrandline)
1931	39	40°38.5′	148°30.5′	C6 Fine skeletal sand
1932	56	40°39′	148°43′	M2 Relict-rich quartzose skeletal sand & gravel

Sample #	Depth (m)	Latitude	Longitude	Facies
1933	17	40°07′	148°23′	Q1 Calcareous quartz sand
1934	46	40°10′	148°55′	M2 Relict-rich quartzose/C6 Fine skeletal sand
1935	118	40°10′	148°48′	C11 Spiculitic skeletal sandy mud
1936	97	40°23′	148°50′	C7 Delicate branching bryozoan muddy sand
1937	20	40°20′	148°37′	M2 Relict-rich quartzose skeletal sand & gravel
1951	46	40°10.6′	147°35′	M1 Relict-rich skeletal sand & gravel
1952	50	40°02′	147°34.8′	R3 Bryozoan-rich relict sand
1953	48	39°51.5′	147°34.5′	C1 Bryozoan sand & gravel
1954	58	39°51.9′	147°21.5′	C1 Bryozoan sand & gravel
1955	63	40°02′	147°21.5′	C1 Bryozoan sand & gravel
1956	64	40°11.5′	147°21.5′	C1 Bryozoan sand & gravel
1957	52	39°31.6′	147°33′	C1 Bryozoan sand & gravel/C2 Skeletal sand & gravel
1958	52	39°22.5′	147°33′	C1 Bryozoan sand & gravel/C2 Skeletal sand & gravel
1959	56	39°23′	147°20.3′	C1 Bryozoan sand & gravel
1960	55	39°33′	147°20′	C2 Skeletal sand & gravel
1961	59	39°43′	147°20′	C1 Bryozoan sand & gravel/C2 Skeletal sand & gravel
1962	70	39°53′	147°07.5′	C7 Delicate branching bryozoan muddy sand
1963	71	40°03′	147°07′	C7 Delicate branching bryozoan muddy sand
1964	72	40°13′	147°07′	C7 Delicate branching bryozoan muddy sand
1965	43	40°40.7′	147°35.5′	M2 Relict-rich quartzose skeletal sand & gravel
1966	46	40°31′	147°35′	M2 Relict-rich quartzose skeletal sand & gravel
1967	46	40°21′	147°35′	M5 Relict-rich fine skeletal sand
1968	55	40°31′	147°21′	M5 Relict-rich fine skeletal sand
1969	52	40°31′	147°22′	M4 Relict-rich bryozoan sand
1970	50	40°40.6′	147°21′	M2 Relict-rich quartzose skeletal sand & gravel
1971	40	40°50.5′	147°21.4′	M2 Relict-rich quartzose skeletal sand & gravel
1972	52	40°50.6′	147°08.4′	M5 Relict-rich fine skeletal sand
1973	66	40°40.7′	147°08.4′	C7 Delicate branching bryozoan muddy sand
1974	70	40°26′	147°08′	C7 Delicate branching bryozoan muddy sand
1975	71	39°42′	147°05′	C7 Delicate branching bryozoan muddy sand
1976	63	39°31.7′	147°05′	C7 Delicate branching bryozoan muddy sand/C1 Bryozoan sand & gravel
1977	62	39°21′	147°05′	C1 Bryozoan sand & gravel
1978	54	39°11′	147°05′	C1 Bryozoan sand & gravel
1979	55	39°01′	147°05′	C1 Bryozoan sand & gravel
1980	52	38°51′	147°05′	M1 Relict-rich skeletal sand & gravel
1981	52	38°58′	146°50′	M1 Relict-rich skeletal sand & gravel
1982	130	43°17.2′	148°00.5′	R3 Bryozoan-rich relict sand
1983	39	43°17.4′	148°07.2′	M4 Relict-rich bryozoan sand
1984	172	43°10′	148°12′	M4 Relict-rich bryozoan sand
1985	113	43°10′	148°06.7′	R3 Bryozoan-rich relict sand
1986	95	43°10′	148°01.4′	M4 Relict-rich bryozoan sand
1987	80	43°00.6′	148°00′	C7 Delicate branching bryozoan muddy sand
1988	97	43°00′	148°06.8′	C6 Fine skeletal sand
1989	122	43°00′	148°13.6′	C1 Bryozoan sand & gravel
1990	157	42°51′	148°20.4′	C1 Bryozoan sand & gravel
1991	99	42°50′	148°14′	C7 Delicate branching bryozoan muddy sand
1992	84	42°50′	148°07.3′	C11 Spiculitic skeletal sandy mud
1993	58	42°50′	147°59.8′	Q1 Calcareous quartz sand
1994	84	42°39.7′	148°11.6′	C1 Bryozoan sand & gravel
1995	106	42°39.6′	148°17.2′	C7 Delicate branching bryozoan muddy sand
1996	130	42°39.5′	148°24′	C7 Delicate branching bryozoan muddy sand
1997	44	42°30.5′	148°03.8′	M2 Relict-rich quartzose skeletal sand & gravel

Sample #	Depth (m)	Latitude	Longitude	Facies
1998	66	42°30′	148°10.4′	C6 Fine skeletal sand/C11 Spiculitic skeletal mud
2000	106	42°29.8′	148°23.2′	C7 Delicate branching bryozoan muddy sand
2001	184	42°30.2′	148°29.3′	C1 Bryozoan sand & gravel
2002	115	42°21.2′	148°31′	C7 Delicate branching bryozoan muddy sand
2003	45	42°20′	148°13′	Q1 Calcareous quartz sand
2004	42	42°20′	148°07.7′	Q1 Calcareous quartz sand
2005	33	42°20′	148°03′	Q1 Calcareous quartz sand
2006	18	42°14.8′	148°02.8′	Q1 Calcareous quartz sand
2007	24	42°14.5′	148°08.1′	Q1 Calcareous quartz sand
2008	16	42°14.6′	148°13.6′	Q1 Calcareous quartz sand
2009	15	42°10.2′	148°14.2′	Q1 Calcareous quartz sand
2010	16	42°10.2′	148°10.4′	Q1 Calcareous quartz sand
2011	14	42°10.2′	148°06.2′	Q1 Calcareous quartz sand
2012	148	42°00′	148°35.5′	C1 Bryozoan sand & gravel
2013	88	41°59.5′	148°29.5′	C7 Delicate branching bryozoan muddy sand
2014	70	42°00′	148°23.0′	M5 Relict-rich fine skeletal sand
2015	28	42°00′	148°18′	C2 Skeletal sand & gravel
2016	64	42°10.3′	148°21.7′	M5 Relict-rich fine skeletal sand
2017	205	42°10′	148°34.7′	C7 Delicate branching bryozoan muddy sand
2018	104	42°10.2′	148°29.2′	C7 Delicate branching bryozoan muddy sand
2020	100	42°20′	148°26.3′	M4 Relict-rich bryozoan sand
2021	73	42°20′	148°21.4′	C1 Bryozoan sand & gravel M5 Relict-rich fine sand
2024	53	43°14.5′	147°27.8′	Q1 Calcareous quartz sand
2025	33	41°50′	148°17.3′	Q1 Calcareous quartz sand
2026	60	41°50′	148°23.3′	M2 Relict-rich quartzose skeletal sand & gravel
2027	84	41°50′	148°28.9′	C7 Delicate branching bryozoan muddy sand
2028	128	41°49.9′	148°35.3′	C1 Bryozoan sand & gravel
2030	113	41°39.8′	148°32.1′	C1 Bryozoan sand & gravel
2031	69	41°40′	148°25.1′	M4 Relict-rich bryozoan sand
2032	27	41°40′	148°18.4′	Q1 Calcareous quartz sand
2033	31	41°30′	148°17.5′	Q1 Calcareous quartz sand
2034	71	41°30.2′	148°23.6′	M5 Relict-rich fine skeletal sand
2035	113	41°30′	148°30′	C11 Spiculitic skeletal sandy mud
2037	73	41°20.1′	148°23.4′	C11 Spiculitic skeletal sandy mud
2038	110	41°20.6′	148°30′	C11 Spiculitic skeletal sandy mud
2039	121	41°20′	148°37′	C7 Delicate branching bryozoan muddy sand
2040	161	41°10′	148°38.6′	C7 Delicate branching bryozoan muddy sand
2041	110	41°10′	148°32.2′	C11 Spiculitic skeletal sandy mud
2042	95	41°10.1′	148°25.7′	C11 Spiculitic skeletal sandy mud
2043	60	41°09.8′	148°19.2′	Q1 Calcareous quartz sand
2044	60	41°00′	148°24.3′	Q1 Calcareous quartz sand
2045	97	41°02′	148°31.5′	C8 Scaphopod, pteropod sand & mud
2046	119	41°00′	148°38.3′	C6 Fine skeletal sand
2047	33	40°48.7′	148°20.1′	Q1 Calcareous quartz sand
2048	51	40°48.7′	148°27′	M2 Relict-rich quartzose skeletal sand & gravel
2049	62	40°49.5′	148°32.1′	M2 Relict-rich quartzose skeletal sand & gravel
2050	82	40°49.6′	148°39.9′	M2 Relict-rich quartzose skeletal sand & gravel
2051	399	40°50.6′	148°46.5′	C12 Coral, arborescent bryozoan gravel & mud
2052	113	41°45.5′	148°31.0′	C7 Delicate branching bryozoan muddy sand
2054	161	43°28.8′	147°58.0′	C1 Bryozoan sand & gravel
2055	62	43°28.6′	147°22.6′	M5 Relict-rich fine skeletal sand
2056	121	43°35.5′	147°32.3′	M4 Relict-rich bryozoan sand

Sample #	Depth (m)	Latitude	Longitude	Facies
2057	146	43°40.5'	147°40.3'	C1 Bryozoan sand & gravel/M4 Relict-rich bryozoan sand
2058	212	43°47.0'	147°48.5'	C1 Bryozoan sand & gravel
2059	175	43°58'	147°30'	C1 Bryozoan sand & gravel/C7 Delicate BR bryozoan muddy sand
2060	84	43°33.6'	147°06'	Q1 Calcareous quartz sand
2061	128	43°43.5'	147°07.1'	M5 Relict-rich fine skeletal sand
2062	148	43°53.2'	147°08.3'	M5 Relict-rich fine skeletal sand
2063	168	44°02.9'	147°10'	C7 Delicate branching bryozoan muddy sand
2064	154	43°47.9'	147°25.6'	C1 Bryozoan sand & gravel/M4 Relict-rich bryozoan sand
2065	95	43°39.5'	147°20.5'	C1 Bryozoan sand & gravel
2066	104	43°40.4'	146°50.4'	M2 Relict-rich quartzose skeletal sand & gravel
2067	124	43°46.5'	146°50.5'	M4 Relict-rich bryozoan sand
2068	168	43°55.0'	146°51'	M5 Relict-rich fine skeletal sand
2069	176	44°02.2'	146°50.5'	C7 Delicate branching bryozoan muddy sand
2070	58	43°35.5'	146°33.5'	Q1 Calcareous quartz sand
2071	115	43°42'	146°33'	M5 Relict-rich fine skeletal sand
2072	159	43°49.5'	146°33.5'	M5 Relict-rich fine skeletal sand
2073	159	43°57'	146°33.7'	M5 Relict-rich fine skeletal sand
2074	168	43°58.5'	146°19.1'	C1 Bryozoan sand & gravel
2075	165	43°50.6'	146°18.5'	M5 Relict-rich fine skeletal sand
2076	108	43°42.2'	146°18.6'	M4 Relict-rich bryozoan sand
2077	97	43°20.3'	147°37.7'	C2 Skeletal sand & gravel
2078	133	43°24.6'	147°48.8'	M5 Relict-rich fine skeletal sand
2079	53	43°33.5'	146°14.2'	M2 Relict-rich quartzose/M5 Relict-rich fine sand
2080	119	43°38.5'	146°07.8'	M5 Relict-rich fine skeletal sand
2081	159	43°44.0'	146°00.5'	C1 Bryozoan sand & gravel
2082	161	43°33.5'	145°52.1'	C1 Bryozoan sand & gravel
2083	104	43°31.5'	145°55.8'	M5 Relict-rich fine skeletal sand
2085	82	43°20.3'	145°48.2'	M5 Relict-rich fine skeletal sand
2086	144	43°22.5'	145°44.5'	M5 Relict-rich fine skeletal sand
2087	159	43°24.2'	145°41.2'	C1 Bryozoan sand & gravel
2088	62	43°12.2'	145°43.3'	R4 Limestone gravel (paleostrandline)
2089	132	43°13.8'	145°36.9'	M5 Relict-rich fine skeletal sand
2090	155	43°15.0'	145°30.6'	C1 Bryozoan sand & gravel
2091	190	43°16.2'	145°23.7'	C1 Bryozoan sand & gravel
2092	154	43°06.4'	145°16.1'	C1 Bryozoan sand & gravel
2093	135	43°05'	145°26'	M5 Relict-rich fine skeletal sand
2094	73	43°04.1'	145°35.7'	M2 Relict-rich quartzose skeletal sand & gravel
2095	84	42°58.2'	145°26.6'	Q1 Calcareous quartz sand
2096	132	42°58.1'	145°15.5'	C1 Bryozoan sand & gravel
2097	188	42°58.2'	145°05'	C1 Bryozoan sand & gravel
2098	91	42°51.1'	145°19.5'	M4 Relict-rich bryozoan sand
2099	124	42°51.1'	145°09.9'	M5 Relict-rich fine skeletal sand
2100	146	42°51.2'	145°00.6'	C1 Bryozoan sand & gravel
2102	90	42°39.5'	145°09.6'	M4 Relict-rich bryozoan sand
2104	88	42°30'	145°09.1'	R3 Bryozoan-rich relict sand
2105	104	42°30'	145°01.0'	C1 Bryozoan sand & gravel
2106	154	42°30.2'	144°52.5'	C1 Bryozoan sand & gravel
2107	44	40°30.5'	144°37.5'	C6 Fine skeletal sand/M2 Relict-rich quartzose
2108	50	40°30'	144°23.4'	R4 Limestone gravel (paleostrandline)
2109	71	40°29.8'	144°09.7'	R2 Mollusc-rich relict sand

Sample #	Depth (m)	Latitude	Longitude	Facies
2110	58	40°20′	144°10′	R2 Mollusc-rich relict sand
2111	55	40°20′	144°22.9′	C1 Bryozoan sand & gravel
2112	55	40°20′	144°36.4′	C1 Bryozoan sand & gravel
2113	49	40°19.8′	144°49.5′	R2 Mollusc-rich relict sand
2114	49	40°20.2′	145°02.5′	M5 Relict-rich fine skeletal sand
2115	53	40°10.2′	145°00.7′	C6 Fine skeletal sand
2116	55	41°00′	144°33.7′	C1 Bryozoan sand & gravel
2117	80	41°01.2′	144°21.5′	C1 Bryozoan sand & gravel
2118	104	41°00′	144°07.5′	M4 Relict-rich bryozoan sand
2119	170	41°00′	143°55′	C1 Bryozoan sand & gravel
2120	132	41°09.4′	144°10.6′	C1 Bryozoan sand & gravel
2121	88	41°09.2′	144°24.2′	C1 Bryozoan sand & gravel
2122	119	41°29.5′	144°24.4′	C1 Bryozoan sand & gravel
2123	91	41°29.5′	144°36.2′	M4 Relict-rich bryozoan sand
2124	49	41°30.3′	144°45.8′	R3 Bryozoan-rich relict sand
2125	60	41°39.8′	144°47.3′	M4 Relict-rich bryozoan sand
2126	130	41°39.5′	144°37.1′	M4 Relict-rich bryozoan sand
2128	170	41°50′	144°34.6′	C1 Bryozoan sand & gravel
2129	86	41°49.5′	144°46′	C1 Bryozoan sand & gravel
2130	69	41°50′	144°57.1′	Q1 Calcareous quartz sand
2131	155	41°58.3′	144°37.3′	C6 Fine skeletal sand
2132	132	42°00.2′	144°51.8′	M4 Relict-rich bryozoan sand
2133	88	42°00.5′	145°00.6′	Q1 Calcareous quartz sand
2135	37	42°10.2′	145°10.4′	Q1 Calcareous quartz sand
2136	28	44°10.2′	144°57.2′	C1 Bryozoan sand & gravel
2137	161	44°09.6′	144°43.8′	C1 Bryozoan sand & gravel
2138	170	44°19.8′	144°51.0′	C1 Bryozoan sand & gravel
2139	122	44°20.2′	145°00.3′	M4 Relict-rich bryozoan sand
2140	90	42°20′	145°08.3′	Q1 Calcareous quartz sand
2141	80	41°11.2′	144°35.6′	M1 Relict-rich skeletal sand & gravel
2142	30	41°20.3′	144°39.8′	R3 Bryozoan-rich relict sand
2143	128	41°19.6′	144°26.6′	M4 Relict-rich bryozoan sand
2144	49	40°10.2′	144°06.5′	M2 Relict-rich quartzose skeletal sand & gravel
2146	59	40°10′	144°32.6′	C6 Fine skeletal sand
2147	59	40°10′	144°45.8′	C1 Bryozoan sand & gravel
2148	51	40°09.2′	145°11.6′	M5 Relict-rich fine skeletal sand
2149	64	40°09.5′	145°25.5′	C6 Fine skeletal sand
2150	128	40°22.5′	143°39′	M1 Relict-rich skeletal sand & gravel
2152	104	40°09.5′	143°30.6′	C6 Fine skeletal sand
2153	106	40°20′	143°27.5′	M4 Relict-rich bryozoan sand
2154	86	40°20.5′	143°41′	M2 Relict-rich quartzose skeletal sand & gravel
2155	59	40°21.2′	143°53.6′	Q1 Calcareous quartz sand
2157	95	40°13.8′	143°35.7′	C1 Bryozoan sand & gravel
2158	110	40°24.8′	143°34.3′	M1 Relict-rich skeletal sand & gravel
2159	108	40°36.5′	143°37′	C1 Bryozoan sand & gravel
2160	90	40°36.4′	143°47′	C1 Bryozoan sand & gravel
2161	33	40°00.1′	144°13.7′	Q1 Calcareous quartz sand
2162	46	40°00′	144°26.5′	M2 Relict-rich quartzose skeletal sand & gravel
2163	46	40°00′	144°38.5′	M3 Relict-rich molluscan sand
2164	53	40°00′	144°52.6′	C6 Fine skeletal sand
2165	55	40°00′	145°06.0′	C6 Fine skeletal sand
2166	64	40°00′	145°19.0′	M5 Relict-rich fine skeletal sand

Sample #	Depth (m)	Latitude	Longitude	Facies
2167	73	40°00′	145°32.5′	C8 Scaphopod, pteropod muddy sand
2168	27	39°49.6′	144°11.9′	M2 Relict-rich quartzose skeletal sand & gravel
2169	37	39°49.9′	144°25.3′	M2 Relict-rich quartzose skeletal sand & gravel
2170	46	39°50′	144°38.7′	M2 Relict-rich quartzose/C6 Fine skeletal sand
2171	49	39°50′	144°51.3′	C6 Fine skeletal sand
2172	51	39°50′	145°04.6′	C2 Skeletal sand & gravel
2173	59	39°50′	145°18.0′	M5 Relict-rich fine skeletal sand
2174	68	39°50′	145°31.0′	C1 Bryozoan sand & gravel
2175	71	38°40′	144°01.1′	M5 Relict-rich fine skeletal sand
2176	73	38°40′	144°14.0′	R2 Mollusc-rich relict sand
2177	77	38°40′	144°27.0′	R2 Mollusc-rich relict sand
2178	77	38°40.5′	144°39.4′	R2 Mollusc-rich relict sand
2179	75	38°40.5′	144°47.4′	M2 Relict-rich quartzose skeletal sand & gravel
2180	73	38°40′	144°53.8′	R2 Mollusc-rich relict sand
2181	73	38°40′	145°06.7′	R2 Mollusc-rich relict sand
2182	73	38°40′	145°19.2′	C10 Bivalve mud
2183	44	38°40.3′	145°30.2′	Q1 Calcareous quartz sand
2184	77	38°50′	143°58.5′	R2 Mollusc-rich relict sand
2185	75	38°49.6′	144°11.2′	R3 Bryozoan-rich relict sand
2186	79	38°50′	144°24′	M5 Relict-rich fine skeletal sand
2187	73	38°50′	144°36.6′	R1 Relict sand
2188	73	38°50′	144°49.3′	M5 Relict-rich fine skeletal sand
2189	72	38°49.2′	145°02.4′	M1 Relict-rich skeletal sand & gravel
2190	73	38°49.6′	145°14.0′	M1 Relict-rich skeletal sand & gravel
2191	71	38°52.5′	145°30′	M1 Relict-rich skeletal sand & gravel
2192	84	39°10.0′	144°05.3′	R2 Mollusc-rich relict sand
2193	75	39°09.5′	144°19.0′	M5 Relict-rich fine skeletal sand
2194	66	39°09.5′	144°33.6′	C6 Fine skeletal sand/R1 Relict sand
2195	62	39°09.0′	144°47.6′	M3 Relict-rich molluscan sand
2196	68	39°09.0′	145°02.2′	C6 Fine skeletal sand
2197	71	39°08.5′	145°15.9′	M1 Relict-rich skeletal sand & gravel
2198	73	39°08′	145°31.0′	M5 Relict-rich fine skeletal sand
2199	71	39°08′	145°39.5′	M1 Relict-rich skeletal sand & gravel/R1 Relict sand

References

Alderman AR, Skinner HCW (1957) Dolomite sedimentation in the south-east of South Australia. Am J Sci 255:561–567

Alexandersson ET (1978) Destructive diagenesis of carbonate sediments in the eastern Skagerrak, North Sea. Geology 6:324–327

Alexandersson ET (1979) Marine maceration of skeletal carbonates in the Skagerrak, North Sea. Sedimentology 26:845–852

Alsharhan AS, Kendall CGSC (2003) Holocene coastal carbonates and evaporites of the southern Arabian Gulf and their ancient analogues. Earth-Sci Rev 61:191–243

Amorosi A (1995) Glaucony and sequence stratigraphy; a conceptual framework of distribution in siliciclastic sequences. J Sediment Res 65:419–425

Amorosi A (2003) Glaucony and verdine. In: Middleton GV (ed) Encyclopedia of sediments and sedimentary rocks. Kluwer Academy Publishers, Dordrecht, pp 331–333

Andres MS, Bernasconi SM, McKenzie JA, Roehl U (2003) Southern ocean deglacial record supports global Younger Dryas. Earth Planet Sci Lett 216:515–524

Andres MS, McKenzie JA (2004) Late Pleistocene oxygen and carbon isotope stratigraphy in bulk- and fine-fraction carbonate from the Great Australian Bight, ODP Leg 182, Site 1127. Proc Ocean Drill Prog Sci Results (CD ROM), p 182

Baines PG, Edwards RJ, Fandry CB (1983) Observations of a new baroclinic current along the western continental slope of Bass Strait. Aust J Mar Freshw Res 34:155–157

Barker PM (2004) The circulation and formation of water masses south of Australia and the interannual wind variability along the southern Australia coast. Unpublished PhD thesis, University of Melbourne, p 351

Barnett EJ, Harvey N, Belperio AP, Bourman RP (1997) Sea-level indicators from a Holocene, tide-dominated coastal succession, Port Pirie, South Australia. Trans R Soc S Aust 121:125–135

Bates RL, Jackson JA (1984) Dictionary of geological terms. Anchor Press, New York, p 571

Bathurst RGC (1975) Carbonate sediments and their diagenesis. Elsevier Science Publ. Co., New York, p 658

Bein J, Taylor ML (1981) The Eyre Sub-basin: recent exploration results. Aust Petrogr Explor Assoc J 21:91–98

Belperio AP (1993) Land subsidence and sea level rise in the Port Adelaide estuary; implications for monitoring the greenhouse effect. Aust J Earth Sci 40:359–368

Belperio AP (1995) Quaternary. In: Drexel JF, Preiss VP (eds) The geology of South Australia. The Phanerozoic, Geological Survey of South Australia, Mines and Energy South Australia, vol 2. Bulletin 54, pp 219–280

Belperio AP, Gostin VA, Cann JH, Murray-Wallace CV (1988) Sediment organism zonation and the evolution of Holocene tidal sequences in southern Australia. In: de Boer PL et al. (eds) Tide-Influenced sedimentary environments and facies. D. Riedel, Dordrecht, pp 475–497

Belperio AP, Hails JR, Gostin VA, Polach HA (1984) The stratigraphy of coastal carbonate banks and Holocene sea-levels of northern Spencer Gulf, South Australia. Mar Geol 61:297–313

Bernecker T, Partridge AD, Webb JA (1997) Mid-late Tertiary deep-water temperate carbonate deposition, offshore Gippsland Basin, southeastern Australia. Soc Sediment Geol 56:221–236 (Spec Publ)

Betzler C, Brachert TC, Nebelsick J (1997) The warm temperate carbonate province: a review of facies, zonations and delineations. Courier Forschungsinstitut Senckenberg 201:83–99

Beu AG, Henderson RA, Nelson CS (1972) Notes on the taphonomy and paleoecology of New Zealand Tertiary Spatangoida. NZ J Geol Geophys 15:275–286

Birch WD (2003) Geology of Victoria. Geological Society of Australia, Melbourne, Special Publication no. 23, p 842

Black A (1853) Black's general atlas of the world. Adam and Charles Black, Edinburgh, 60 plates, pp 12–63

Blom WM, Alsop DB (1988) Carbonate mud sedimentation on a temperate shelf: Bass Basin, southeastern Australia. Sediment Geol 60:269–280

Bock PE, Cook P (2004) A review of Australian Conescharellinidae (Bryozoa, Cheilostomata). Mem Mus Vic 61:135–182

Bock PE, Cook PL (1998) A new species of multiphased Corbulipora MacGillivray, 1895 (Bryozoa, Cribriomorpha) from southwestern Australia. Rec S Aust Mus 30:63–68

Boggs SJ (1995) Principles of sedimentology and stratigraphy, 2nd edn. Prentice Hall, Englewood Cliffs, p 774

Bone Y (1978) The geology of Wardang Island, Yorke Peninsula, South Australia. Unpublished Bachelor's thesis, University of Adelaide, Adelaide, South Australia, p 138

Bone Y (1984) Wardang Island–A refuge for Marginopora vertebralis? Trans R Soc S Aust 108:127–128

Bone Y (1991) Population explosion of the bryozoan (Membranipora aciculata) in the Coorong Lagoon in late 1989. Aust J Earth Sci 38:121–123

Bone Y (2009) Geology and geomorphology. In: Jennings JT (ed) Natural history of the Riverland and Murraylands.

N. P. James, Y. Bone, *Neritic Carbonate Sediments in a Temperate Realm,*
DOI 10.1007/978-90-481-9289-2, © Springer Science+Business Media B.V. 2011

Occasional Publications of the Royal Society of South Australia Inc., Adelaide, pp 1–49

Bone Y, Deer L, Edwards SA, Campbell EM (2006) Adelaide Coastal Waters Study–Sediment Budget, ACWS Technical Report no 16, CSIRO, p 65

Bone Y, James NP (1993) Bryozoans as carbonate sediment producers on the cool-water Lacepede Shelf, southern Australia. Sediment Geol 86:247–271

Bone Y, James NP (1997) Bryozoan stable isotope survey from the cool-water Lacepede shelf, southern Australia. In: James NP, Clarke JAD (eds) Cool water carbonates. Spec Publ–Soc Sediment Geol, SEPM (Society for Sedimentary Geology, Tulsa, OK) 56:93–105

Bone Y, James NP (2002) Bryozoans from deep-water reef mounds: Great Australian Bight, Australia. In: Wyse J, Jackson JBC, Buttler R, Spencer Jones M (eds) Bryozoans 2001. Balkner Press, Lisse, pp 9–14

Bone Y, James NP, Kyser TK (1992) Synsedimentary detrital dolomite in Quaternary cool-water carbonate sediments, Lacepede shelf. S Aust Geol 20:109–112

Bone Y, Wass RE (1990) Sub-recent bryozoan-serpulid buildups in the Coorong Lagoon, South Australia. Aust J Earth Sci 37:207–214

Boreen TD, James NP (1993) Holocene sediment dynamics on a cool-water carbonate shelf: Otway, southeastern Australia. J Sediment Petrol 63:574–588

Boreen T, James NP, Heggie D, Wilson C (1993) Surficial cool-water carbonate sediments on the Otway continental margin, southeastern Australia. Mar Geol 112:35–56

Bosellini A, Ginsburg RN (1971) Form and internal structure of Recent algal nodules (rhodolites) from Bermuda. J Geol 79:669–682

Boutakoff N (1963) The geology and geomorphology of the Portland area. Memoir—Geological Survey of Victoria, Melbourne, p 171

Bowers DG, Lennon G (1987) Observations of stratified flow over a bottom gradient in a coastal sea. Cont Shelf Res 7:1105–1121

Bowler JM (1976) Aridity in Australia; age, origins and expression in aeolian landforms and sediments. Earth-Sci Rev 12:279–310

Bowler JM, Teller JT (1986) Quaternary evaporites and hydrological changes, Lake Tyrrell, north-west Victoria. Aust J Earth Sci 33:43–63

Boyd R, Ruming K, Roberts JJ (2004) Geomorphology and surficial sediments of the southeast Australian continental margin. Aust J Earth Sci 51:743–764

Brooks GR, Hine AC, Mallinson D, Drexler TM (2004) Texture and composition of Quaternary upper-slope sediments in the Great Australian Bight; Sites 1130 and 1132. Proc Ocean Drill Prog Sci Results (CD ROM) 182:1–21

Brown KM (2005) Calcareous epiphytes on modern seagrasses as carbonate sediment producers in shallow cool-water marine environments, South Australia. Unpublished PhD thesis, University of Adelaide, Adelaide, p 247

Brown CR, Stephenson AE (1991) Geology of the Murray Basin, Southeastern Australia. Australian Government Publishing Service, Canberra, p 321

Burchette TP, Wright VP (1992) Carbonate ramp depositional systems. Sediment Geol 79(1–4):3–57

Burdige DJ (1993) The biogeochemistry of manganese and iron reduction in marine sediments. Earth-Sci Rev 35:249–284

Burne RV (1982) Relative fall of Holocene sea level and coastal progradation, northeastern Spencer Gulf, South Australia. Aust BMR J Aust Geol Geophys 7:35–45

Burne RV, Colwell JB (1982) Temperate carbonate sediments of northern Spencer Gulf, South Australia: a high salinity 'foramol' province. Sedimentology 29:223–238

Bye JAT (1972) Oceanic circulation south of Australia. In: Hayes DE (ed) Antarctic Oceanology II: The Australian–New Zealand Sector; Antarctic Research Series no 19. American Geophysical Union, Washington DC, pp 95–100

Bye JAT (1976) Physical oceanography of Gulf St-Vincent and Investigator Strat. In: Twidale CR, Tyler MJ, Webb BP (eds) Natural history of the Adelaide region, vol 1. Occasional Publications of Royal Society of South Australia, Adelaide, pp 143–160

Bye JAT (1981) Exchange processes for upper Spencer Gulf, South Australia. Trans R Soc S Aust 105:59–66

Bye JAT (1983) The general circulation in a dissipative ocean basin with longshore wind stresses. J Phys Oceanogr 13:1553–1563

Callahan E (1972) The structure and circulation of deep water in the Antarctic. Deep Sea Res 19:563–576

Cann JH, Clarke JDA (1993) The significance of (Marginopora vertebralis) (foraminifera) in surficial sediments at Esperance, Western Australia, and in last interglacial sediments in northern Spencer Gulf, South Australia. Mar Geol 111:171–187

Cann JH, Gostin VA (1985) Coastal sedimentary facies and foraminiferal biofacies of the St. Kilda Formation at Port Gawler, South Australia. Trans R Soc S Aust 109:121–142

Cann JH, Belperio AP, Gostin VA, Murray-Wallace CV (1988) Sea-level history, 45,000 to 30,000 yr B.P., inferred from benthic foraminifera, Gulf St. Vincent, South Australia. Quat Res 29:153–175

Cann JH, Belperio AP, Gostin VA, Rice RL (1993) Contemporary benthic foraminifera in Gulf St Vincent, South Australia, and a refined late Pleistocene sea-level history. Aust J Earth Sci 40:197–211

Cann JH, Murray-Wallace CV, Belperio AP, Brenchley AJ (1999) Evolution of Holocene coastal environments near Robe, southeastern South Australia. Quat Int 56:81–97

Cann JH, Bourman RP, Barnett EJ (2000) Holocene Foraminifera as indicators of relative estuarine-lagoonal and oceanic influences in estuarine sediments of the River Murray, South Australia. Quat Res 53:378–391

Cann JH, Harvey N, Barnett EJ, Belperio AP, Bourman RP (2002) Foraminiferal biofacies eco-succession and Holocene sealevels, Port Pirie, South Australia. Mar Micropaleontol 44(1–2):31–55

Cann JH, Murray-Wallace CV, Riggs NJ, Belperio AP (2006) Successive foraminiferal faunas and inferred palaeoenvironments associated with the postglacial (Holocene) marine transgression, Gulf St Vincent, South Australia. Holocene 16:224–234

Cann JH, Scardigno MF, Jago JB (2009) Mangroves as an agent of rapid coastal change in a tidal-dominated environment, Gulf St Vincent, South Australia: implications for coastal management. Aust J Earth Sci 56:927–938

Carannante G, Esteban M, Milliman JD, Simone L (1988) Carbonate lithofacies as paleolatitude indicators: problems and limitations. Sediment Geol 60:333–346

Carannante G, Cherchi A, Simone L (1995) Chlorozoan versus foramol lithofacies in Upper Cretaceous rudist limestones. Palaeogeogr Palaeoclimatol Palaeoecol 119:137–154

Carter L, McCave IN, Williams MJM (2009) Circulation and water masses of the Southern Ocean: a review. In: Florindo F, Siegert M (eds) Antarctic climate evolution, developments in earth and environmental sciences 8. Elsevier, Amsterdam, pp 85–114

Chappell J, Shackleton NJ (1986) Oxygen isotopes and sea level. Nature 324:137–140

Chave KE (1952) A solid solution between calcite and aragonite. J Geol 60:190–192

Cirano M, Middleton JF (2004) Aspects of the Mean Wintertime Circulation along Australia's Southern Shelves: numerical Studies. J Phys Oceanogr 34:668–684

Collins LB (1988) Sediments and history of the Rottnest Shelf, southwest Australia: a swell-dominated, non-tropical carbonate margin. Sediment Geol 60:15–50

Collins LB, Zhu ZR, Wyrwoll K-H, Hatcher BG, Playford PE, Chen JH, Eisenhauer A, Wasserburg GJ (1993) Late Quaternary evolution of coral reefs on a cool-water carbonate margin: the Abrolhos carbonate platforms, southwest Australia. Mar Geol 110:203–212

Condie SA, Dunn JR (2006) Seasonal characteristics of the surface mixed layer in the Australasian region: implications for primary production regimes and biogeography. Mar Freshw Res 57:569–590

Conolly JR, von der Borch CC (1967) Sedimentation and physiography of the sea floor south of Australia. Sediment Geol 1:181–220

Conroy T, Cook P, Bock PE (2001) New species of Otionellina and Selenaria (Bryozoa-Cheilostroma) from the South West Shelf, Western Australia. Trans R Soc S Aust 125(1):15–23

Cook PL, Chimonides PJ (1978) Observations on living colonies of Selenaria (Bryozoa, Cheilostomata) I. Cah Biol Mar 19:147–158

Cook PL, Chimonides PJ (1981) Morphology and systematics of some rooted cheilostome Bryozoa. J Nat Hist 15:97–134

Cook PJ, Colwell JB, Firman JB, Lindsay JM, Schwebel DA, von der Borch CC (1977) The late Cainozoic sequence of Southeast South Australia and Pleistocene sea-level changes. BMR J Aust Geol Geophys 2:81–88

Crawford CM, Edgar GJ, Cresswell G (2000) The Tasmanian region. In: Sheppard CRC (ed) Seas at the millenium: an environmental evaluation, vol II. Regional chapters: The Indian Ocean to the Pacific. Pergamon Press, Amsterdam, pp 647–660

Cresswell G (2000) Currents of the continental shelf and upper slope of Tasmania. Pap Proc R Soc Tasman 133(3):21–30

Cresswell GR (1991) The Leeuwin Current–observations and recent models. In: Pearce AF, Walker DI (eds) The Leeuwin Current. J R Soc Western Aust 74:1–14

Cresswell GR, Peterson JL (1993) The Leeuwin Current South of Western Australia. Aust J Mar Freshw Res 44:285–303

Currie DR, McClatchie S, Middleton JF, Nayar S (in press) Biophysical factors affecting the distribution of demersal fish around a submarine canyon off the Bonney Coast, South Australia. Deep Sea Res

Dalrymple RW (2010) Interpreting sedimentary successions: facies, facies analysis and facies models. In: James NP, Dalrymple RW (eds) Facies models 4. Geological Association of Canada, St. John's, Newfoundland, pp 3–18

Daly SJ, Fanning CM (1993) Archaean. In: Drexel JF, Preiss WV, Parker AJ (eds) The geology of South Australia. The Precambrian, vol 1. Bulletin–Geological Survey of South Australia, Adelaide, pp 32–49

Davies JL (1980) Geographical variation in coastal development. Longman, London, p 371

Davies HL, Clarke JDS, Stagg HMJ, Shafik S, McGowran B, Alley NF, Willcox JB (1989) Maastrichtian and younger sediments from the Great Australian Bight, BMR J Aust Geol Geophys, Report no 288, p 52

De Deckker P, Bauld P, Burne RV (1982) Pillie Lake, Eyre Peninsula, South Australia; modern environment and biota, dolomite sedimentation, and Holocene history. Trans R Soc S Aust 106(Pt 4):169–181

Dickinson JA, Wallace MW, Holdgate GR, Gallagher SJ, Thomas L (2002) Origin and timing of the Miocene-Pliocene unconformity in Southeast Australia. J Sediment Res 72:288–303

Drexel JF, Preiss WV, Parker AJ (1993) The geology of South Australia. The Precambrian, vol 1. Geological Survey of South Australia, Adelaide, vol 54, p 242

Duddy IR (2003) Mesozoic. In: Birch WD (ed) Geology of Victoria. Geological Society of Australia, Melbourne, Special Publication no 23, pp 239–288

Dutkiewicz A, von der Borch CC (1995) Lake Greenly, Eyre Peninsula, South Australia; sedimentology, palaeoclimatic and palaeohydrologic cycles. Palaeogeogr Palaeoclimatol Palaeoecol 113:43–56

Edgar GJ (2001) Australian marine habitats in temperate waters. New Holland Publishers (Australia) Pty Ltd, Sydney, p 224

Edyvane KS (1999) Conserving marine biodiversity in South Australia. Part 2: Identification of areas of high conservation value in South Australia. South Australian Research and Development Institute, South Australia, p 283

Evans SR, Middleton JF (1998) A regional model of shelf circulation near Bass Strait: a new upwelling mechanism. J Phys Oceanogr 28:1439–1457

Exon NF, Kennett JP, Malone M (2004) Proceedings of the Ocean Drilling Program. Ocean Drilling Program CD-ROM, College Station, Texas

Exon NF, Hill PJ, Mitchell C, Post A (2005) Nature and origin of the submarine Albany canyons off southwest Australia. Aust J Earth Sci 52:101–115

Fairbanks RG (1989) A 17,000-year glacio-eustatic sealevel record: influence of glacial melting rates on the Younger Dryas event and deep-ocean circulation. Nature 342:637–642

Fandry CB (1983) Model for the three-dimensional structure of wind-driven and tidal circulation in Bass Strait. Aust J Mar Freshw Res 34:121–141

Farrow GE, Fyfe JA (1988) Bioerosion and carbonate mud production on high-latitude shelves. Sediment Geol 60:281–297

Feary DA, James NP (1998) Seismic stratigraphy and geological evolution of the Cenozoic, cool-water Eucla platform, Great Australian Bight. AAPG Bulletin 82:792–816

Feary D, Boreen TD, James NP, Bone Y, Birch G, Lanyon R, Shafik S (1992) Preliminary post-cruise report, Rig Seismic Research Cruise 1991, Sediments of the Great Australian Bight, pp 163. Bureau of Mineral Resources, Southern Margin Project 121.27

Feary D, Hine AC, Malone M (2000a) Great Australian Bight: Cenozoic cool-water carbonates. Proceedings of the Ocean Drilling Program, Initial Reports, College Station, Texas, p 58

Feary D, Hine AC, Malone M, et al. (2000b) Great Australian Bight: Cenozoic cool-water carbonates. College Station, Texas, p 58

Feary DA, Hine AC, James NP, Malone MJ (2004) Leg 182 synthesis; exposed secrets of the Great Australian Bight. In: Hine AC, Feary D, Malone M (eds) Proceedings of the Ocean Drilling Program, Scientific Results (CD ROM), 182, Ocean Drilling Program. College Station, Texas, pp 1–30

Ferguson J, Burne RV (1981) Interactions between saline redbed groundwaters and peritidal carbonates, Spencer Gulf, South Australia: significance for models of stratiform copper ore genesis. J Aust Geolo Geophys 6:319–325

Ferguson J, Burne RV, Chambers LA (1983) Iron mineralization of peritidal carbonate sediments by continental groundwaters, Fisherman Bay, South Australia. Sediment Geol 34:41–57

Fielding CR, Frank TD, Birgenheier LP, Rygel MC, Jones AT, Roberts J (2008) Stratigraphic record and facies associations of the late Paleozoic ice age in eastern Australia (New South Wales and Queensland), vol 441. Geological Society of America Special Papers, pp 41–57

Flottmann T, James P (1997) Influence of basin architecture on the style of inversion and fold-thrust belt tectonics–the southern Adelaide Fold-Thrust Belt, South Australia. J Struct Geol 19:1093–1110

Flügel E (2004) Microfacies of carbonate rocks: analysis, interpretation and application. Springer, Berlin, p 976

Foden JD, Turner SP, Morrison RS (1990) Tectonic implications of Delamerian magmatism in South Australia and western Victoria. Geol Soc Aust 16:465–482 (Spec Publ)

Foden J, Elburg MA, Dougherty-Page J, Burtt A (2006) The timing and duration of the Delamerian Orogeny; correlation with the Ross Orogen and implications for Gondwana assembly. J Geol 114:189–210

Folk RL (1959) Practical petrographic classification of limestones. Am Assoc Pet Geol Bull 43:1–38

Fornos JJ, Ahr WM (1997) Temperate carbonates on a modern, low-energy, isolated ramp: the Balearic platform, Spain. J Sediment Res 67:364–373

Fox DR, Batley GE, Blackburn D, Bone Y, Bryars S, Cheshire A, Collings G, Ellis D, Fairweather P, Fallowfield H, Harris GJ, Henderson B, Kämpf J, Nayar S, Pattriaratchi C, Petrusevics P, Townsend M, Westphalen G, Wilkinson J (2007) The Adelaide Coastal waters study. South Australian Protection Authority, South Australia, p 43

Fraser AR, Tilbury LA (1979) Structure and stratigraphy of the Ceduna Terrace region, Great Australian Bight. APEA J 19(Pt 1):53–65

Freiwald A (1995) Bacteria-induced carbonate degradation; a taphonomic case study of Cibicides lobatulus from a high-boreal carbonate setting. Palaios 10:337–346

Froelich PN, Klinkhammer GP, Bender ML, Luedtke N, Heath GR, Cullen D, Dauphin P, Hammond D, Hartman B,

Maynard V (1979) Early oxidation of organic matter in pelagic sediments of the eastern equatorial Atlantic; suboxic diagenesis. Geochim Cosmochim Acta 43:1075–1090

Fuller MK, Bone Y, Gostin VA, von der Borch CC (1994) Holocene cool-water carbonate and terrigenous sediments of southern Spencer Gulf, South Australia. Aust J Earth Sci 41:353–363

Galloway WE (1975) Process framework for describing the morphologic and stratigraphic evolution of deltaic depositional systems. In: Broussard ML (ed) Deltas: models for exploration. Houston Geological Society, Houston, pp 87–96

Gammon PR, James NP (2003) Paleoenvironmental controls on upper Eocene biosiliceous neritic sediments, southern Australia. J Sediment Res 73:957–972

Gammon PR, James NP, Clarke JDA, Bone Y (2000) Sedimentology and lithostratigraphy of upper Eocene sponge-rich sediments, southern Western Australia. Aust J Earth Sci 47:1087–1103

Gentilli J (1971) Climates of Australia and New Zealand. Elsevier, Amsterdam, p 403

Gersbach GH, Pattiaratchi CB, Ivey GN, Cresswell GR (1999) Upwelling on the south-west coast of Australia–source of the Capes Current? Cont Shelf Res 19:363–400

Gibbs CF, Tomczak MJ, Longmore AR (1986) The nutrient regime of Bass Strait. Aust J Mar Freshw Res 37:451–466

Ginsburg RN, Bosellini A (1973) Form and internal structure of Recent algal nodules (rhodolites) from Bermuda; a reply. J Geol 81:239

Ginsburg RN, James NP (1974) Holocene carbonate sediments of continental shelves. In: Burke CA, Drake CL (eds) The geology of continental margins. Springer-Verlag, New York, pp 137–157

Godfrey JS, Ridgway KR (1985) The large-scale environment of the Poleward-flowing Leeuwin Current, western Australia: longshore steric height gradients, wind stresses and geostrophic flow. J Phys Oceanogr 15:481–495

Godfrey JS, Jones ISF, Maxwell JG, Scott BD (1980) On the winter cascade from Bass Strait into the Tasman Sea. Aust J Mar Freshw Res 31:275–286

Godfrey JS, Vaudrey DJ, Hahn SD (1986) Observations of the Shelf-Edge Current South of Australia, Winter 1982. J Phys Oceanogr 16:668–679

Golubic S, Perkins RD, Lukas KJ (1975) Boring microorganisms and microborings in carbonate substrates. In: Frey RW (ed) The study of trace fossils. Springer-Verlag, New York, pp 229–259

Golubic S, Seong-Joo L, Browne K (2000) Cyanobacteria: architects of sedimentary structures. In: Riding RE, Awramik SM (eds) Microbial sediments. Springer, Berlin, pp 233–249

Gostin VA, Hails JR, Belperio AP (1984) The sedimentary framework of northern Spencer Gulf, South Australia. Mar Geol 61:111–138

Gostin VA, Belperio AP, Cann JH (1988) The Holocene nontropical coastal and shelf carbonate province of southern Australia. Sediment Geol 60:51–70

Greenhalgh SA, Love D, Malpas K, McDougall R (1994) South Australian earthquakes, 1980–1992. Aust J Earth Sci 41:483–495

Griffin D, Thompson PA, Bax NJ, Bradford RW, Hallegraeff GM (1997) The 1995 mass mortality of pilchard: no role found for physical or biological oceanographic factors in Australia. Aust J Mar Freshw Res 48:27–42

Hageman SJ, Bone Y, McGowran B, James NP (1995) Modern bryozoan assemblages and distribution on the cool-water Lacepede Shelf, southern Australian margin. Aust J Earth Sci 42:571–580

Hageman SJ, Bone Y, McGowran B, James NP (1996) Bryozoan species distribution on the cool-water Lacepede Shelf, southern Australia. In: Gordon DP, Smith AJ (eds) Bryozoans in time and space. Proceedings of the 10th International Bryozoology Association. Wellington, New Zealand, pp 109–116

Hageman SJ, Bock PE, Bone Y, McGowran B (1998) Bryozoan growth habits: classification and analysis. J Paleontol 72:418–436

Hageman SJ, James NP, Bone Y (2000) Cool-water carbonate production from Epizoic Bryozoans on Ephemeral substrates. Palaios 15:33–48

Hageman SJ, Lukasik JJ, McGowran B and Bone Y (2003) Paleoenvironmental significance of Celleporaria (bryozoa) from modern and Tertiary cool-water carbonates of Southern Australia. Palaios 18:510–527

Hahn SD (1986) Physical structure of the waters of the South Australian continental shelf. PhD, Flinders University of South Australia, p 364

Hails JR, Gostin VA (1984) The Spencer Gulf region. Mar Geol 61:167–179

Hails JR, Belperio AP, Gostin VA (1984a) Quaternary sea levels, northern Spencer Gulf, Australia. Mar Geol 61:373–389

Hails JR, Belperio AP, Gostin VA, Sargent GE (1984b) The submarine Quaternary stratigraphy of northern Spencer Gulf, South Australia. Mar Geol 61:345–372

Halfar J, Godinez-Orta L, Mutti M, Valdez-Holguin JE, Borges JM (2004) Nutrient and temperature controls on modern carbonate production; an example from the Gulf of California, Mexico. Geology 32:213–216

Halfar J, Godinez Orta L, Mutti M, Valdez-Holguin JE, Borges JM (2006a) Carbonates calibrated against oceanographic parameters along a latitudinal transect in the Gulf of California, Mexico. Sedimentology 53:297–320

Halfar J, Strasser M, Riegl B, Godinez-Orta L (2006b) Oceanography, sedimentology and acoustic mapping of a bryomol carbonate factory in the northern Gulf of California, Mexico. In: Pedley HM, Carannante G (eds) Cool-water carbonates: depositional systems and paleoenvironmental controls. Geological Society of London, London, Special Publications no 255, pp 197–215

Hallock P (1984) Distribution of selected species of living algal symbiont-bearing foraminifera on two Pacific coral reefs. J Foraminiferal Res 14:250–261

Hardie LA (1977) Sedimentation on the modern carbonate tidal flats of northwest Andros Island, Bahamas. The Johns Hopkins University Press, Baltimore, p 202

Harris G, Nilsson C, Clementson L, Thomas D (1987) The water masses of the east coast of Tasmania: seasonal and interannual variability and the influence on phytoplankton biomass and productivity. Aust J Mar Freshw Res 38:569–590

Heggie DT, O'Brien GW (1988) Hydrocarbon gas geochemistry in sediments of the Fuels Project 9131.20, Preliminary Post-Cruise Report, Record 19883/2, p 149

Hemer MA (2006) The magnitude and frequency of combined flow bed shear stress as a measure of exposure on the Australian continental shelf. Cont Shelf Res 26:1258–1280

Hemer MA, Bye JAT (1999) The swell climate of the South Australian Sea. Trans R Soc S Aust 123:107–113

Henderson T (1997) Mineralogy and stable isotopic composition of Recent carbonate sediments and waters from five small lakes, south-eastern, South Australia. Unpublished B.Sc Honours Thesis, University of South Australia, Adelaide, p 82

Henrich R, Wefer G (1986) Dissolution of biogenic carbonates: effects of skeletal structure. Mar Geol 71:341–362

Herzfeld M (1997) The annual cycle of sea surface temperature in the Great Australian Bight. Prog Oceanogr 39:1–27

Herzfeld M, Tomczak M (1997) Numerical modelling of sea surface temperature and circulation in the Great Australian Bight. Prog Oceanogr 39:29–78

Hill KA, Durrand C (1993) The western Otway Basin: an overview of the rift and drift history using serial composite seismic profiles. PESA J 21:67–78

Hill KC, Hill KA, Cooper GT, O'Sullivan AJ, Richardson MJ (1995) Inversion around the Bass Basin, SE Australia. In: Buchanan JG, Buchanan PG (eds) Basin inversion. Geological Society, London, Special Publication no 88, pp 525–547

Hill PJ, De Deckker P, Exon NF (2005) Geomorphology and evolution of the gigantic Murray canyons on the Australian southern margin. Aust J Earth Sci 52:117–136

Hill PJ, De Deckker P, von der Borch CC, Murray-Wallace CV (2009) Ancestral Murray river on the Lacepede Shelf, southern Australia: late Quaternary migrations of a major river outlet and strandline development. Aust J Earth Sci 56:135–157

Hillman K, Walker DI, Larkum AWD, McComb AJ (1989) Productivity and nutrient limitation. In: Larkum AWD, McComb AJ, Shepherd SA (eds) Biology of seagrasses. A treatise on the biology of seagrasses with special reference to the Australian region, Aquatic plant studies 2. Elsevier, Amsterdam, pp 635–685

Hine AC, Brooks GR, Mallinson D, Brunner CA, James NP, Feary DA, Holbourn AE, Drexler TM, Howd P (2004) Late Pleistocene-Holocene sedimentation along the upper slope of the Great Australian Bight. Proc Ocean Drill Prog Sci Results (CD ROM), p 182

Hocking RM (1990) Eucla Basin. Memoir–Geological Survey of Western Australia 3:548–561

Holbourn A, Kuhnt W, James N (2002) Late Pleistocene bryozoan reef mounds of the Great Australian Bight; isotope stratigraphy and benthic foraminiferal record. Paleoceanography 17:1042–1057

Holdgate GR, Gallagher SJ (2003) Tertiary. In: Birch WD (ed) Geology of Victoria. Geological Society of Australia, Melbourne, Special Publication no 23, pp 289–336

Holdgate GR, Geurin B, Wallace MW, Gallagher SJ (2001) Marine geology of Port Phillip, Victoria. Aust J Earth Sci 48:439–455

Hufford GE, McCartney KA, Donohue KA (1997) Northern boundary currents and adjacent recirculations off southwestern Australia. Geophys Res Lett 24:2797–2800

Huntley DJ, Hutton JT, Prescott JR (1993) The stranded beach-dune sequence of South-east Australia. Quat Sci Rev 12:1–20

Imbrie J, Hays JD, Martinson DG, McIntyre A, Morley JJ, Pisias NG, Prell WL, Shackleton NJ (1984) The orbital theory of Pleistocene climate: support from a revised chronology of the marine d18O record. In: Berger A, Imbrie J, Hays J, Kukla G, Saltzman B (eds) Milankovitch and Climate, Part I (NATO ASI Series C, vol. 126). D. Reidel Publishing Co., Dordrecht, pp 269–305

James NP (1997) The cool-water carbonate depositional realm. In: James NP, Clarke MJ (eds) Cool-water carbonates, 56, SEPM Special Publication, pp 1–20

James NP, Bone Y (1989) Petrogenesis of Cenozoic, temperate water calcarenites, South Australia: a model for meteoric/shallow burial diagenesis of shallow water calcite sediments. J Sediment Petrol 59:191–203

James NP, Bone Y (1991) Origin of a cool-water, Oligo-Miocene deep shelf limestone, Eucla Platform, southern Australia. Sedimentology 38:323–341

James NP, Bone Y (1994) Paleoecology of cool-water, subtidal cycles in mid-Cenozoic limestones, Eucla Platform, southern Australia. Palaios 9:457–476

James NP, Bone Y (2007) A late Pliocene-early Pleistocene, inner-shelf, subtropical, seagrass-dominated carbonate: Roe Calcarenite, Great Australian Bight, Western Australia. Palaios 22:343–359

James NP, Choquette PW (1990) Limestone diagenesis, the meteoric environment. In: McIlreath I, Morrow D (eds) Sediment Diagenesis. Geological Association of Canada Reprint Series, pp 36–74

James NP, Clarke JDA (eds) (1997) Cool-water carbonates. Society for Sedimentary Geology, Special Publication no 56, Tulsa, OK, p 440

James NP, Kendall AC (1992) Introduction to carbonate and evaporite facies models. In: Walker RG, James NP (eds) Facies models; response to sea level change Geological Association of Canada, St. Johns, pp 265–275

James NP, Lukasik JJ (2010) Cool- and cold-water carbonates. In: James NP, Dalrymple RW (eds) Facies models 4. Geological Association of Canada, St. John's, Newfoundland, pp 369–398

James NP, von der Borch CC (1991) Carbonate shelf edge off southern Australia: a prograding open-platform margin. Geology 19:1005–1008

James NP, Bone Y, Kyser TK (1991) Shallow burial dolomitization of mid-Cenozoic, cool-water, calcitic, deep-shelf limestones, southern Australia. AAPG Bulletin 75:602

James NP, Bone Y, von der Borch CC, Gostin VA (1992) Modern carbonate and terrigenous clastic sediments on a cool water, high energy, mid-latitude shelf: Lacepede, southern Australia. Sedimentology 39:877–903

James NP, Boreen TD, Bone Y, Feary DA (1994) Holocene carbonate sedimentation on the West Eucla Shelf, Great Australian Bight; a shaved shelf. Sediment Geol 90:161–177

James NP, Bone Y, Hageman DA, Feary DA, Gostin VA (1997) Cool-water carbonate sedimentation during the terminal Quaternary sea-level cycle: Lincoln shelf, southern Australia. SEPM 56:51–75 (Spec Publ)

James NP, Collins LB, Bone Y, Hallock P (1999) Subtropical carbonates in a temperate realm; modern sediments on the Southwest Australian shelf. J Sediment Res 69:1297–1321

James NP, Feary DA, Surlyk F, Simo JAT, Betzler C, Holbourn AE, Li Q, Matsuda H, Machiyama H, Brooks GR, Andres

MS, Hine AC, Malone MJ, Ocean Drilling Program Leg 182 Scientific Party (2000) Quaternary bryozoan reef mounds in cool-water, upper slope environments: Great Australian Bight. Geology 28:647–650

James NP, Bone Y, Collins LB, Kyser TK (2001) Surficial sediments of the Great Australian Bight; facies dynamics and oceanography on a vast cool-water carbonate shelf. J Sediment Res 71:549–567

James NP, Feary DA, Betzler C, Bone Y, Holbourn AE, Li Q, Machiyama H, Simo JAT, Surlyk F (2004) Origin of Late Pleistocene Bryozoan Reef Mounds; Great Australian Bight. J Sediment Res 74:20–48

James NP, Bone Y, Kyser TK (2005) Where has all the aragonite gone? Mineralogy of Holocene neritic cool-water carbonates, Southern Australia. J Sediment Res 75:454–463

James NP, Bone Y, Carter RM, Murray-Wallace CV (2006) Origin of the late Neogene Roe Plains and their calcarenite veneer; implications for sedimentology and tectonics in the Great Australian Bight. Aust J Earth Sci 53:407–419

James NP, Martindale RC, Malcolm I, Bone Y, Marshall J (2008) Surficial sediments on the continental shelf of Tasmania, Australia. Sediment Geol 211:33–52

James NP, Bone Y, Brown KM, Cheshire N (2009) Calcareous epiphyte production in cool-water carbonate depositional environments; Southern Australia. In: Swart P, Eberli G (eds) Advances in carbonate sedimentology. International Association of Sedimentologists, Special Publication no 41, pp 123–148

Jensen-Schmidt B, Cockshell CD, Boult PJ (eds) (2002) Structural and tectonic setting. Minerals and Energy Resources, Government of South Australia, Primary Industries and Resources SA, Adelaide, p 53

Johnson D (2004) The geology of Australia. Cambridge University Press, Cambridge, p 276

Jones HA, Davies PJ (1983a) Superficial sediments of the Tasmanian continental margin and part of Bass Strait. BMR Geol Geophys Bull 218:25

Jones HA, Davies PJ (1983b) Superficial sediments of the Tasmanian continental shelf and part of Bass Strait. Bull Aust Bur Miner Resour Geol Geophys 218:25

Kämpf J, Doubell M, Griffin D, Matthews RL, Ward TM (2004) Evidence of a large seasonal coastal upwelling system along the southern shelf of Australia. Geophys Res Lett 31(L09310):1–4

Kirkman H, Kuo J (1990) Pattern and process in southern Western Australian seagrasses. Aquat Bot 37:367–382

Ku TCW, Walter LM, Coleman ML, Blake RE, Martini AM (1999) Coupling between sulfur recycling and syndepositional carbonate dissolution; evidence from oxygen and sulfur isotope composition of pore water sulfate, South Florida Platform, U.S.A. Geochim Cosmochim Acta 63:2529–2546

Kyser TK, James NP, Bone Y (2002) Shallow burial dolomitization and dedolomitization of Cenozoic cool-water limestones, southern Australia; geochemistry and origin. J Sediment Res 72:146–157

Lambeck K, Nakada M (1990) Late Pleistocene and Holocene sea-level change along the Australian coast. Glob Planet Change 3:143–176

Last WM (1992) Petrology of modern carbonate hardgrounds from East Basin Lake, a saline maar lake, southern Australia. Sediment Geol 81:215–229

Leach AS, Wallace MW (2001) Cenozoic submarine canyon systems in cool water carbonates from the Otway Basin, Victoria, Australia. In: Hill KC, Bernecker T (eds) Eastern Australasian Basin symposium, a refocused energy perspective for the future. Petroleum Exploration Society of Australia, Melbourne, Special Publication no 1, Melbourne, pp 465–473

Lennon GP, Du BL, Wisniewski GM (1987) Minimizing the impact of salinity intrusion due to sea level rise. American Society of Civil Engineers, New York, pp 437–442

Levings AH, Gill PC (2009) Water temperature and its effect on the life history of the southern Australian giant crab Pseudocarcinus gigas. Proceedings of Conference on the effects of climate change on cool water crabs, Anchorage, Alaska, March (in press)

Lewis R (1981) Seasonal upwelling along the south-eastern coastline of South Australia. Mar Freshw Res 32:843–854

Lewis R, Aust T, Clarke S, Edyvane KS, Fisher R, Fotheringham D, McLennan B, McShane P, Newland N, Noye J, Neverauskas V, Steffensen D, White M, Zacharin W, Zeidler W (1997) Description, use and management of South Australia's marine and estuarine environment. South Australian Marine and Estuarine Strategy Committee, Adelaide, p 124

Li Q, McGowran B, James NP, Bone Y (1996a) Foraminiferal biofacies on the mid-latitude Lincoln Shelf, South Australia; oceanographic and sedimentological implications. Mar Geol 129:285–312

Li Q, McGowran B, James NP, Bone Y, Cann JH (1996b) Mixed foraminiferal biofacies on the mesotrophic, mid-latitude Lacepede Shelf, South Australia. Palaios 11:176–191

Li Q, James NP, McGowran B, Bone Y, Cann J (1998) Synergenetic influence of water masses and Kangaroo Island barrier on foraminiferal distribution, Lincoln and Lacepede Shelves, South Australia: a synthesis. Alcheringa 22:153–176

Li Q, James NP, Bone Y, McGowran B (1999) Palaeoceanographic significance of recent foraminiferal biofacies on the southern shelf of Western Australia; a preliminary study. Palaeogeogr Palaeoclimatol Palaeoecol 147:101–120

Lindsay JM, Alley NF (1995) St Vincent Basin. In: Drexel JF, Preiss VP (eds) The geology of South Australia. The Phanerozoic, vol 2. SA Geol Surv Bull 54:163–172

Lindsay JM, Harris WK (1975) Fossiliferous marine and non-marine Cainozoic rocks from the eastern Eucla Basin, South Australia. Miner Res Revue, Southern Australia 138:29–42

Logan BW, Cebulski DE (1970) Sedimentary environments of Shark Bay, Western Australia. In: Logan BW, Davies G, Read JF, Cebulski DE (eds) Carbonate sedimentation and environments, Shark Bay, Western Australia. Memoir–American Association of Petroleum Geologists, Tulsa 13:1–37

Longhurst A (1998) Ecological geography of the sea. Academic Press, San Diego, p 398

Lowry DC (1970) Geology of the Western Australian part of the Eucla Basin. Geolo Surv Western Aust Bull 122:201

Ludbrook NH (1976) The Glanville Formation at Port Adelaide. Quarterly Geological Notes–South Australia, Geological Survey, pp 4–7

Lukasik JJ, James NP (2003) Deepening-upward subtidal cycles, Murray Basin, South Australia. J Sediment Res 73:653–671

Lynch-Stieglitz J, Fairbanks RG, Charles CD (1994) Glacial-interglacial history of Antarctic intermediate water: relative strengths of Antarctic versus Indian ocean sources. Paleoceanography 9:7–29

Lyne VD, Thresher RE (1994) Dispersal and advection of Macruronus novaezealandiae (Gadiformes: Merlucciidae) larvae off Tasmania: simulation of the effects of physical forcing on larval distribution. In: Sammarco PW, Heron ML (eds) Bio–physics of marine larval dispersal, coastal and estuarine studies 45. American Geophysics Union, Washington DC, pp 109–136

Malikides M, Harris PT, Jenkins CJ, Keene JB (1988) Carbonate sandwaves in Bass Strait. Aust J Earth Sci 35:303–311

Malikides M, Harris PT, Tate PM (1989) Sediment transport and flow over sandwaves in a non-rectilinear tidal environment; Bass Strait, Australia. Cont Shelf Res 9:203–221

Marshall JF (1983) Geochemistry of iron-rich sediments on the outer continental shelf off northern New South Wales. Mar Geol 51:163–175

Marshall JF, Davies PJ (1978) Skeletal carbonate variation on the continental shelf of eastern Australia. Aust BMR J Geolo Geophys 3:85–92

Marshall JF, Tsuji Y, Matsuda H, Davies PJ, Iryu Y, Honda N, Satoh Y (1998) Quaternary and Tertiary subtropical carbonate platform development on the continental margin of southern Queensland, Australia. In: Camoin GF, Davies PJ (eds) Reefs and Carbonate Platforms in the Pacific and Indian Oceans, Special Publications of the International Association of Sedimentologists, London, pp 163–195

Marshallsay PG, Radok R (1972) Drift cards in the southern and adjacent oceans, Horace Lamb Centre for Oceanic Research, Research Paper 52, Flinders University of South Australia, p 86

Martin RE (1998) One long experiment. Columbia University Press, New York, p 262

Mazzoleni AG, Bone Y, Gostin VA (1995) Cathodoluminescence of aragonitic gastropods and cement in Old Man Lake thrombolites, southeastern South Australia. Aust J Earth Sci 42:497–500

McCartney MS (1977) Subantarctic mode water. In: Angel VM (ed) A voyage of discovery: George Deacon 70th anniversary volume, Supplement to Deep-Sea Research. Pergamon Press, Oxford, pp 103–119

McCartney MS (1982) The subtropical recirculation of Mode Waters. J Mar Res 40(suppl):427–463

McCartney MS, Donohue KA (2007) A cyclonic gyre in the Atlantic Australian Basin. Prog Oceanogr 75:675–750

McClatchie S, Middleton JF, Ward TM (2006) Water mass analysis and alongshore variation in upwelling intensity in the eastern Great Australian Bight. J Geophys Res 111

McGowran B, Holdgate GR, Li Q, Gallagher SJ (2004) Cenozoic stratigraphic succession in southeastern Australia. Aust J Earth Sci 51:459–496

McGowran B, Li Q, Moss G (1997) The Cenozoic neritic record in southern Australia: the biogeohistorical framework. In: James NP, Clarke JAD (eds) Cool-Water Carbonates, SEPM Spec. Publ. 56, pp 185–204

McKinney FK (2007) The northern Adriatic ecosystem; deep time in a shallow sea. Columbia University Press, New York, p 299

McKinney F, Jackson J (1989) Bryozoan evolution. Unwin Hyman, London, p 238

Middleton JF, Black KP (1994) The low frequency circulation in and around Bass Strait: a numerical study. Cont Shelf Res 14:1495–1521

Middleton JF, Bye JAT (2007) A review of the shelf-slope circulation along Australia's Southern Shelves: Cape Leeuwin to Portland. Prog Oceanogr 75:1–41

Middleton JF, Cirano M (1999) Wind-forced downwelling slope currents: a numerical study. J Phys Oceanogr 29:1723–1743

Middleton JF, Cirano M (2002) A northern boundary current along Australia's southern shelves: the Flinders current. J Geophys Res 107(12):1–11

Middleton JF, Platov G (2003) The mean summertime circulation along Australia's southern shelves: a numerical study. J Phys Oceanogr 33:2270–2287

Millikan M (1994) Quaternary geology of the American River area, Kangaroo Island, South Australia. Unpublished BSc Honours thesis, University of Adelaide, Adelaide, South Australia, p 90

Milnes AR, Ludbrook NH (1986) Provenance of microfossils in aeolian calcarenites and calcretes in southern South Australia. Aust J Earth Sci 33:145–159

Morse JW, Arvidson RS (2002) The dissolution kinetics of major sedimentary carbonate minerals. Earth-Sci Rev 58:51–84

Motoda S, Kawamura T, Taniguchi A (1978) Differences in productivities between the Great Australian Bight and the Gulf of Carpentaria, Australia, in summer. Mar Biol 46:93–99

Muir M, Lock DE, von der Borch CC (1980) The Coorong model for penecontemporaneous dolomite formation in the Middle Proterozoic McArthur Group, Northern Territory, Australia. In: Zenger DH, Dunham JB, Ethington RL (eds) Concepts and models of dolomitization, SEPM Special Publication no 28, Tulsa, OK, pp 51–68

Murray JW (1991) Ecology and paleoecology of benthic foraminifera. Longman Scientific, Harlow, p 397

Murray-Wallace CV (2002) Pleistocene coastal stratigraphy, sea-level highstands and neotectonism of the southern Australian passive continental margin; a review; Sea-level changes and neotectonics. J Quat Sci 17:469–489

Murray-Wallace CV, Belperio AP (1991) The last interglacial shoreline in Australia; a review. Quat Sci Rev 10:441–461

Murray-Wallace CV, Belperio AP, Picker K, Kimber RWL (1991) Coastal aminostratigraphy of the last interglaciation in southern Australia. Quat Res 35:63–71

Murray-Wallace CV, Belperio AP, Bourman RP, Cann JH, Price DM (1999) Facies architecture of a last interglacial barrier; a model for Quaternary barrier development from the Coorong to Mount Gambier coastal plain, southeastern Australia. Mar Geol 158:177–195

Murray-Wallace CV, Brooke BP, Cann JH, Belperio AP, Bourman RP (2001) Whole-rock aminostratigraphy of the Coorong coastal plain, South Australia; towards a 1 million year record of sea-level highstands. J Geol Soc London, 158(Pt 1):111–124

Nelson CS (1978) Temperate shelf carbonate sediments in the Cenozoic of New Zealand. Sedimentology 25:737–771

Nelson CS (1988a) An introductory perspective on non-tropical shelf carbonates. In: Nelson CS (ed) Non-tropical shelf carbonates: modern and ancient. Sediment Geol 60:3–12

Nelson CS (ed) (1988b) Non-tropical shelf carbonates-modern and ancient. Sediment Geol, p 367

Nelson CS, James NP (2000) Marine cements in mid-Tertiary cool-water shelf limestones of New Zealand and southern Australia. Sedimentology 47:609–629

Nelson CS, Hyden FM, Keane SL, Leask WL, Gordon DP (1988a) Application of bryozoan zoarial growth-form studies in facies analysis of non-tropical carbonate deposits in New Zealand. In: Nelson CS (ed) Non-tropical shelf carbonates: modern and ancient. Sediment Geol 60:301–322

Nelson CS, Keane SL, Head PS (1988b) Non-tropical carbonate deposits on the modern New Zealand shelf. In: Nelson CS (ed) Non-tropical shelf carbonates: modern and ancient. Sediment Geol 60:71–94

Newell BS (1961) Hydrology of S-E Australian waters: Bass Strait and New South Wales tuna fishing area, CSIRO Division of Fisheries and Oceanography Technical Paper no 10, p 21

Newell BS (1974) Hydrology of south-east Australian waters, CSIRO Division of Fisheries and Oceanography Technical Paper no 10, p 21

Norvick MS, Smith MA (2001) Mapping the plate tectonic reconstruction of southern and southeastern Australia and implications for petroleum systems. Aust Petroleum Prod Explor Assoc J 41:15–35

Nunes R, Lennon G (1986) Physical property distributions and seasonal trends in Spencer Gulf, South Australia: an inverse estuary. Mar Freshw Res 37:39–53

Nunes Vaz RA, Lennon GW, Bowers DG (1990) Physical behaviour of a large, negative or inverse estuary. Cont Shelf Res 10:227–304

Odin GS, Fullagar PD (1988) Geological significance of the glaucony facies. Dev Sedimentol 45:295–332

Passlow V (1997) Slope sedimentation and shelf to basin sediment transfer; a cool-water carbonate example from the Otway margin, southeastern Australia. Soc Sediment Geol 56:107–125 (Spec Publ)

Patterson WP, Walter LM (1994) Depletion of ^{13}C in seawater Ω CO_2 on modern carbonate platforms: significance for the carbon isotopic record of carbonates. Geology 22:885–888

Pearce AF (1991) Eastern boundary currents of the southern hemisphere. In: Pearce AF, Walker DI (eds) The Leeuwin Current. R Soc Western Aust J 74:34–45

Pearce A, Pattiaratchi C (1999) The Capes Current: a summer countercurrent flowing past Cape Leeuwin and Cape Naturaliste, Western Australia. Cont Shelf Res 19:401–420

Pedley HM, Carannante G (eds) (2006) Cool-water carbonates; depositional systems and palaeoenvironmental controls. Geological Society of London, London, Special Publication no 255, London, p 373

Petrusevics P (1993) SST fronts in inverse estuaries, South Australia-indicators of reduced gulf-shelf exchange. Mar Freshw Res 44:305–323

Petrusevics P, Bye JAT, Fahlbusch V, Hammat J, Tippins DR, van Wijk E (2009) High salinity winter outflow from a mega inverse-estuary–the Great Australian Bight. Cont Shelf Res 29:371–380

Pratt BR (2010) Peritidal. In: James NP, Dalrymple RW (eds) Facies models 4. Geological Association of Canada, St. John's, Newfoundland, pp 399–420

Price RC, Nicholls IA, Gray CM (2003) Cainozoic igneous activity. In: Birch WD (ed) Geology of Victoria. Geological Society of Australia, Melbourne, Special Publication no 23, pp 362–375

Pufahl PK, James NP, Bone Y, Lukasik JJ (2004) Pliocene sedimentation in a shallow, cool-water, estuarine gulf, Murray Basin, South Australia. Sedimentology 51:997–1027

Purser BH (ed) (1973) The Persian Gulf. Springer-Verlag, New York, p 471

Quilty PG (1977) Cenozoic sedimentation cycles in Western Australia. Geology 5:336–340

Raghukumar C, Rao VPC, Iyer SD (1989) Precipitation of iron in windowpane oyster shells by marine shell-boring cyanobacteria. Geomicrobiol J 7:235–244

Rahimpour-Bonab H, Bone Y, Moussavi-Harami R, Turnbull K (1997) Geochemical comparisons of modern cool-water calcareous biota, Lacepede shelf, South Australia, with their tropical counterparts. Soc Sediment Geol 56:77–91 (Spec Publ)

Reeckmann SA, Gill ED (1981) Rates of vadose diagenesis in Quaternary dune and shallow marine calcarenites, Warnambool, Victoria, Australia. Sediment Geol 30:157–172

Reid CM, James NP, Kyser TK, Barrett N, Hirst AJ (2008) Modern estuarine siliceous spiculites, Tasmania, Australia: a non-polar link to Phanerozoic spiculitic cherts. Geology 36:107–110

Richardson JR (1987) Brachiopods from carbonate sands of the Australian shelf. Proc R Soc Vic 99:37–50

Richardson LE, Kyser TK, James NP, Bone Y (2009) Analysis of hydrographic and stable isotope data to determine water masses, circulation, and mixing in the eastern Great Australian Bight. J Geophys Res 114:14

Ridgway KR, Condie SA (2004) The 5500-km-long boundary flow off western and southern Australia. J Geophys Res 109:1–18

Rivers JM, James NP, Kyser TK (2005) Geochemical attributes and early diagenesis of neritic cool-water carbonates; central southern Australian margin. Abstracts: Annual Meeting. Am Assoc Pet Geol 14:A118

Rivers JM, James NP, Kyser TK, Bone Y (2007) Genesis of palimpsest cool-water carbonate sediment on the continental margin of southern Australia. J Sediment Res 77:480–494

Rivers JM, James NP, Kyser TK (2008) Early Diagenesis of carbonates on a cool-water carbonate shelf, Southern Australia. J Sediment Res 78:784–802

Rivers JM, Kyser TK, James NP (2009) Isotopic composition of a large photosymbiotic foraminifer: evidence for hypersaline environments across the Great Australian Bight during the late Pleistocene. Sediment Geol 213:113–120

Rochford DJ (1977) Monitoring of coastal sea temperatures around Australia. Search (Syd) 8:167–169

Rosen MR, Miser DE, Warren JK (1988) Sedimentology, mineralogy and isotopic analysis of Pellet lake, Coorong region, South Australia. Sedimentology 35:105–122

Rosen MR, Miser DE, Starcher MA, Warren JK (1989) Formation of dolomite in the Coorong region, South Australia. Geochim Cosmochim Acta 53:661–669

Ruddiman WF (2001) Earth's climate; past and future. W. H. Freeman and Company, New York, p 465

Ryan DA, Brooke BP, Collins LB, Kendrick GA, Baxter KJ, Bickers AN, Siwabessy PJW, Pattiaratchi CB (2007) The influence of geomorphology and sedimentary processes on shallow-water benthic habitat distribution: Esperance Bay, Western Australia. Estuar Coast Shelf Sci 72:379–386

Ryan DA, Brooke BP, Collins LB, Spooner MI, Siwabessy PJW (2008) Formation, morphology and preservation of high-energy carbonate lithofacies: evolution of the cool-water Recherche Archipelago inner shelf, south-western Australia. Sediment Geol 207:41–55

Sanderson PG, Eliot I, Hegge B, Maxwell S (2000) Regional variation of coastal morphology in southwestern Australia: a synthesis. Geomorphology 34:73–88

Sandery PA, Kämpf J (2005) Winter-spring flushing of Bass Strait, South-Eastern Australia: a numerical modelling study. Estuar Coast Shelf Sci 63:23–31

Sandiford M (2003a) Geomorphic constraints on the late Neogene tectonics of the Otway Range, Victoria. Aust J Earth Sci 50:69–80

Sandiford M (2003b) Neotectonics of southeastern Australia; linking Quaternary faulting record with seismicity and in situ stress. In: Hillis RR, Muller RD (eds) Evolution and dynamics of the Australian Plate. Geol Soc Aust, Special Publication no 22, pp 101–113

Saxena S, Betzler C (2003) Genetic sequence stratigraphy of cool water slope carbonates (Pleistocene Eucla shelf, southern Australia). Int J Earth Sci 92:482–493

Schahinger RB (1987) Structure of coastal upwelling events observed off the south-east coast of South Australia during February 1983-April 1984. Aust J Mar Freshw Res 38:439–459

Schlager W (1981) The paradox of drowned reefs and carbonate platforms. Geol Soc Am Bull 92:197–211

Schodlok MP, Tomczak MJ, White NJ (1997) Deep sections through the South Australian Basin and across the Australian Discordance, Flinders Institute for Atmospheric and Marine Sciences, Research Report no 55, p 45

Schwebel DAH (1983) Quaternary dune systems. In: Michael JT (ed) Natural history of the South East, vol 3. Occasional Publications of the Royal Society of South Australia, Adelaide, pp 15–24

Shackleton NJ, van Andel TH, Boyle EA, Jansen E, Labeyrie L, Leinen M, McKenzie J, Mayer L, Sundquist E (1990) Contributions from the oceanic record to the study of global change on three time scales; Report of Working Group I, Interlaken workshop for past global changes. Glob Planet Change 2:5–37

Sheard MJ (1986) Some volcanological observations at Mount Schank, southeast South Australia. Geological Survey of South Australia–Quarterly Geological Notes 100:14–20

Shepherd SA (1983) Benthic communities of upper Spencer Gulf, South Australia. Trans R Soc S Aust 107:69–86

Shepherd SA, Hails JR (1984) The dynamics of a megaripple field in northern Spencer Gulf, South Australia. Mar Geol 61:249–263

Shepherd SA, Robertson EL (1989) Regional studies–Seagrasses of South Australia, western Victoria and Bass Strait. In: Larkum AWD, McComb AJ, Shepherd SA (eds) Biology of seagrasses: a Treatise on the biology of seagrasses with special reference to the Australian region. Elsevier, Amsterdam, pp 211–229

Shepherd SA, Sprigg RC (1976) Substrate, sediments and subtidal ecology of Gulf St. Vincent and Investigator Strait. In: Twidale CR, Tyler MJ, Webb BP (eds) Natural history of the Adelaide region. Royal Society of South Australia, Adelaide, pp 161–174

Shepherd SA, Womersley HBS (1976) The subtidal algal and seagrass ecology of St. Francis Island, South Australia. Trans R Soc S Aust 100(Pt 4):177–191

Shepherd SA, McComb AJ, Bulthuis DA, Neverauskas V, Steffensen D, West RR (1989) Decline of seagrasses. In: Larkum AWD, McComb AJ, Shepherd SA (eds) Biology of seagrasses. Elsevier, Amsterdam, pp 346–393

Short AD (1988) Holocene coastal dune formation in southern Australia; a case study. Sediment Geol 55:121–142

Short AD, Hesp PA (1982) Wave, beach and dune interactions in southeastern Australia. Mar Geol 48:259–284

Short AD, Wright LD (1984) Morphodynamics of high energy beaches: an Australian perspective. In: Thom BG (ed) Coastal geomorphology in Australia. Academic Press, New York, pp 43–68

Simo JA, Slatter NM (2004) Sedimentology of a Pleistocene middle slope cool-water carbonate platform, Great Australian Bight, ODP Leg 182. Proc Ocean Drill Prog Sci Results (CD ROM) 182:1–15

Smith AM, Nelson CS (2003) Effects of early sea-floor processes on the taphonomy of temperate shelf skeletal carbonate deposits. Earth-Sci Rev 63:1–31

Smith SV, Veeh HH (1989) Mass balance of biogeochemically active materials (C, N, P) in a hypersaline gulf. Estuar Coast Shelf Sci 29:195–215

Smith AM, Key MM, Gordon DP (2006) Skeletal mineralogy of bryozoans: taxonomic and temporal patterns. Earth Sci Rev 78:287

Smith AM, Nelson CS, Danaher PJ (1992) Dissolution behaviour of bryozoan sediments: taphonomic implications for nontropical shelf carbonates. Palaeogeogr Palaeoclimatol Palaeoecol 93:213–226

Smith AM, Nelson CS, Spencer HG (1998) Skeletal carbonate mineralogy of New Zealand bryozoans. Mar Geol 151:27–46

Sprigg RC (1952) The geology of the South-East Province, South Australia, with special reference to Quaternary coastline migrations and modern beach developments. Geological Survey of South Australia, Adelaide, p 120

Sprigg RC (1979) Stranded and submerged sea-beach systems of southeast South Australia and the aeolian desert cycle. Sediment Geol 22:53–96

Sprigg M, Bone Y (1993) Bryozoans in Coorong-type lagoons in southern Australia. Trans R Soc S Aust 117:87–95

Stagg HMJ, Cockshell CD, Willcox JB, Hill AC, Needham DJL, Thomas B, O'Brien GW, Hough LP (1990) Basins of the Great Australian Bight region: geology and petroleum potential. Australian Govt Pub Service, Canberra, p 391

Sullivan LA, Koppi TJ (1998) Iron staining of quartz beach sand in southeastern Australia. J Coast Res 14:992–999

Sverdrup HU, Johnson MW, Fleming J (1942) The oceans; their physics, chemistry and general biology. Prentice-Hall, Englewood Cliffs, p 532

Swart PK, James NP, Mallinson D, Malone MJ, Matsuda H, Simo T, Hine AC, Feary DA, Malone MJ, Andres M, Betzler C, Brooks GR, Brunner CA, Fuller M, Molina Garza RS, Holbourn AE, Huuse M, Isern AR, James NP, Ladner BC, Li Q, Machiyama H, Mallinson DJ, Matsuda H, Mitterer RM, Robin C, Russell JL, Shafik S, Simo JA, Smart PL, Spence GH, Surlyk FC, Swart PK, Wortmann UG (2004) Carbonate mineralogy of sites drilled during Leg 182. Proc Ocean Drill Prog Sci Results (CD ROM), p 182

Taylor PD, James NP, Bone Y, Kuklinski P, Kyser TK (2009) Evolving Mineralogy of Cheilostome Bryozoans. Palaios 24:440–452

Thom BG (1984a) Coastal geomorphology in Australia. Academic, Sydney, p 342

Thom BG (1984b) Geomorphic research on the coast of Australia; a preview. Academic, Sydney, pp 1–21

Tomczak M, Godfrey TS (1994) Regional oceanography: an introduction. Pergamon Press, Oxford, p 422

Totterdell JM, Bradshaw BE (2004) The structural framework and tectonic evolution of the Bight Basin. Petroleum Exploration Society of Australia Special Publication, vol 2, pp 41–61

Tucker ME, Wright VP (1990) Carbonate Sedimentology. Blackwell Scientific Publications, Oxford, p 482

Van Der Zee C, Roberts DR, Rancourt DG, Slomp CP (2003) Nanogoethite is the dominant reactive oxyhydroxide phase in lake and marine sediments. Geology 31:993–996

Veevers JJ (1986) Breakup of Australia and Antarctica estimated as Mid-Cretaceous (95+ or −5 Ma) from magnetic and seismic data at the continental margin. Earth Planet Sci Lett 77:91–99

Veevers JJ (2000) Billion-year earth history of Australia and neighbours in Gondwanaland. Gemoc Press, Sydney, p 388

Veevers JJ, Eittreim SL (1988) Reconstruction of Antarctica and Australia at breakup (95+/−5 Ma) and before rifting (160 Ma). Aust J Earth Sci 35:355–362

von der Borch CC (1965) The distribution and preliminary geochemistry of modern carbonate sediments of the Coorong area, South Australia. Geochim Cosmochim Acta 29:781–799

von der Borch CC (1968) Southern Australian submarine canyons; their distribution and ages. Mar Geol 6:267–279

von der Borch CC (1976) Stratigraphy and formation of Holocene dolomitic carbonate deposits of the Coorong area, South Australia. J Sediment Petrol 46:952–966

von der Borch CC, Lock DE (1979) Geological significance of Coorong dolomites. Sedimentology 26:813–824

von der Borch CC, Conolly JR, Dietz RS (1970) Sedimentation and structure of the continental margin in the vicinity of the Otway Basin, southern Australia. Mar Geol 8:59–83

von der Borch CC, Hughes-Clarke JE (1993) Slope morphology adjacent to the cool-water carbonate shelf of South Australia; GLORIA and Seabeam imaging. Aust J Earth Sci 40:57–64

von der Borch CC, Lock DE, Schwebel D (1975) Ground-water formation of dolomite in the Coorong region of South Australia. Geology 3:283–285

von der Borch CC, Bolton B, Warren JK (1977) Environmental setting and microstructure of subfossil lithified stromatolites associated with evaporites, Marion Lake, South Australia. Sedimentology 24:693–708

Walter LM, Burton EA (1990) Dissolution of Recent platform carbonate sediments in marine pore fluids. Am J Sci 290:601–643

Walter LM, Morse JW (1984) Reactive surface area of skeletal carbonates during dissolution: effect of grain size. J Sediment Petrol 54:1081–1090

Walter LM, Morse JW (1985) The dissolution kinetics of shallow marine carbonates in seawater: a laboratory study. Geochim Cosmochim Acta 49:1503–1513

Walter LM, Bischof SA, Patterson WP, Lyons TW (1993) Dissolution and recrystallization in modern shelf carbonates; evidence from pore water and solid phase chemistry. Philos Trans–R Soc London, Phys Sci Eng 344:27–36

Ward TM, McLeay LJ, Dimmlich WF, Rogers PJ, McClatchy S, Matthews R, Kampf J, van Ruth PD (2006) Pelagic ecology of a northern boundary current system: effects of upwelling on the production and distribution of sardine (Sardinops sagax), anchovy (Engraulis australis) and southern bluefin tuna (Thunnus maccoyii) in the Great Australian Bight. Fish Oceanogr 15:191–207

Warren JK (1982a) The hydrologic setting, occurrence and significance of gypsum in late Quaternary salt lakes in South Australia. Sedimentology 29:609–637

Warren JK (1982b) The hydrological significance of Holocene tepees, stromatolites and boxwork limestones in coastal salinas in South Australia. J Sediment Petrol 52:1171–1201

Warren JK (1983) Pedogenic calcrete as it occurs in Quaternary calcareous dunes in coastal South Australia. J Sediment Petrol 53:787–796

Warren JK (1988) Sedimentology of Coorong dolomite in the Salt Creek region, South Australia. Carbonates and Evaporites 3:175–199

Warren JK (1991) Sulfate dominated sea-marginal and platform evaporative settings. In: Melvin JL (ed) Evaporites: petroleum and mineral resources, developments in sedimentology 50. Elsevier Science Publishers, New York, pp 69–187

Wass RE, Conolly JR, MacIntyre RJ (1970) Bryozoan carbonate sand continuous along southern Australia. Mar Geol 9:63–73

Webster I, Golding TJ, Dyson N (1979) Hydrological features of the near-shelf waters of Fremantle, Western Australia, during 1974, CSIRO Australian Division of Fisheries and Oceanography, Report no 106, Melbourne, Victoria, Australia, p 30

Willcox JB, Stagg HMJ (1990) Australia's southern margin: a product of oblique extension. Tectonophysics 173:269–281

Williams MAJ (2000) Quaternary Australia: extremes in the last Glacial-Interglacial Cycle. In: Veevers JJ (ed) Billion-year earth history of Australia and neighbours in Gondwanaland. GEMOC Press, Sydney, pp 55–59

Williams MAJ (2001) Quaternary climatic changes in Australia and their environmental effects. In: Gostin VA (ed) Gondwana to greenhouse: Australian environmental geoscience. Geological Society of Australia Inc., Melbourne, Special Publication no 21, pp 3–11

Williams M, Dunkerley D, De Deckker P, Kershaw S, Chappell J (1998) Quaternary environments, 2nd edn. Arnold, London, p 352

Willman CE, VandenBerg AHM, Morand V (2002) Evolution of the southeastern Lachlan Fold Belt in Victoria. Aust J Earth Sci 49:271–289

Wilson C (1991) Geology of the Quaternary bridgewater formation of southwest and central South Australia. Unpublished PhD thesis, Flinders University, Adelaide, Australia, p 341

Wilson JL (1975) Carbonate facies in geologic history. Springer Verlag, New York, p 471

Winston JE (1988) Life histories of free-living bryozoans. Nat Geogr Res Explor 4:528–539

Winston JE (2009) Cold comfort: systematics and biology of Antarctic bryozoans. In: Krupnik M, Lang A, Miller SE (eds) Smithsonian at the Poles. Contribution to International Polar Year Science. Smithsonian Institution Scholarly Press, Washington DC, pp 205–221

Wirtz P (1994) Quaternary geology of the Pelican Lagoon area, Kangaroo Island, South Australia. Unpublished BSc Honours Thesis, University of Adelaide, Adelaide, p 90

Womersley HBS (1981a) Biogeography of Australian marine macroalgae. In: Clayton MN, King RJ (eds) Marine botany: an Australasian perspective. Longman Cheshire, Melbourne, pp 292–307

Womersley HBS (1981b) Marine ecology and the zonation of temperature coasts. In: Clayton MN, King RJ (eds) Marine botany: an Australasian perspective. Longman Cheshire, Melbourne, pp 211–240

Womersley HBS (1984) The marine benthic flora of Southern Australia, Part I. In: Handbook of the flora and fauna of South Australia. Government Printer, Adelaide, p 329

Womersley HBS (1987) The marine benthic flora of Southern Australia, Part II. In: Handbook of the Flora and Fauna of South Australia, Government Printer, Adelaide, p 484

Womersley HBS (1996) The marine benthic flora of southern Australia, Part IIIB. In: Handbook of the Flora and Fauna of South Australia. Government Printer, Adelaide, p 432

Woo M, Pattiaratchi C (2008) Hydrography and water masses off the western Australian coast. Deep Sea Res (Part I Oceanogr Res Pap) 55:1090–1104

Wood JD, Terray E (2005) Ocean current, temperature, and wave observations at the Amrit-1 location offshore of southern Victoria, Australia, Final Report April to October 2004. Ocean Data Technologies, Inc. Technical report prepared for Santos, Ltd. and Unocal Corporation, p 88

Wray JL (1977) Calcareous Algae. Elsevier Publishing Co., New York, p 185

Wright DT (2000) Benthic microbial communities and dolomite formation in marine and lacustrine environments; a new dolomite model. Spec Publ–Soc Sediment Geol 66:7–20

Wright DT, Camoin GF (1999) The role of sulphate-reducing bacteria and cyanobacteria in dolomite formation in distal ephemeral lakes of the Coorong region, South Australia. Sediment Geol 126:147–157

Wyrtki K, Bennett EB, Rochford DJ (1971) Oceanographic atlas of the International Indian Ocean expedition. National Science Foundation, Washington DC, p 531

Yagunova EB, Ostrovsky AN (2008) Encrusting bryozoan colonies on stones and algae: variability of zooidal size and its possible causes. J Mar Biolog Assoc U K 88:901–908

Young HR, Nelson CS (1988) Endolithic biodegradation of cool-water skeletal carbonates on Scott shelf, northwestern Vancouver Island, Canada. In: Nelson CS (ed) Non-tropical shelf carbonates; modern and ancient. Sediment Geol 60:251–268

Young J, Nishida T, Stanley C (1999) A preliminary survey of the summer hydrography and plankton biomass of the eastern Great Australian Bight, Australia. Southern Bluefin Tuna Recruitment Monitoring and Tagging Program. In: Report of the Eleventh Workshop, CSIRO, Hobart, p 15

Zann LP (2000) The Australian region: an overview. In: Sheppard CRC (ed) Seas at the Millennium: an environmental evaluation, vol II: Regional Chapters: The Indian Ocean to the Pacific. pp 579–592

Zheng H, Wyrwoll K-H, Li Z, Powell CM (1998) Onset of aridity in southern Western Australia–a preliminary palaeomagnetic appraisal. Glob Planet Change 18:175–187

Index